Physics and Chemistry of
Crystalline Lithium Niobate

The Adam Hilger Series on Optics and Optoelectronics

Series Editors: **E R Pike** FRS and **W T Welford** FRS

The Adam Hilger Series on Optics and Optoelectronics

Physics and Chemistry of Crystalline Lithium Niobate

A M Prokhorov and Yu S Kuz'minov

General Physics Institute,
USSR Academy of Sciences, Moscow

Translated from the Russian by
Dr T M Pyankova and O A Zilbert

Adam Hilger, Bristol and New York

British Library Cataloguing in Publication Data

Prokhorov, A. M.
    Physics and chemistry of crystalline lithium niobate.
    1. Crystals
    I. Kuz'minov, Yu S.
    548

    ISBN 0-85274-002-6

Library of Congress Cataloging-in-Publication Data are available

Published under the Adam Hilger imprint by IOP Publishing Ltd
Techno House, Redcliffe Way, Bristol BS1 6NX, England
335 East 45th Street, New York, NY 10017-3483, USA

Typeset by KEYTEC, Bridport, Dorset
Printed in Great Britain by J W Arrowsmith Ltd, Bristol

Contents

Series Editors' Preface

Optics has been a major field of pure and applied physics since the mid 1960s. Lasers have transformed the work of, for example, spectroscopists, metrologists, communication engineers and instrument designers in addition to leading to many detailed developments in the quantum theory of light. Computers have revolutionized the subject of optical design and at the same time new requirements such as laser scanners, very large telescopes and diffractive optical systems have stimulated developments in aberration theory. The increasing use of what were previously not very familiar regions of the spectrum, e.g. the thermal infrared band, has led to the development of new optical materials as well as new optical designs. New detectors have led to better methods of extracting the information from the available signals. These are only some of the reasons for having an *Adam Hilger Series on Optics and Optoelectronics*.

The name Adam Hilger, in fact, is that of one of the most famous precision optical instrument companies in the UK; the company existed as a separate entity until the mid 1940s. As an optical instrument firm Adam Hilger had always published books on optics, perhaps the most notable being Frank Twyman's *Prism and Lens Making*.

Since the purchase of the book publishing company by The Institute of Physics in 1976 their list has been expanded into all areas of physics and related subjects. Books on optics and quantum optics have continued to comprise a significant part of Adam Hilger's output, however, and the present series has some twenty titles in print or to be published shortly. These constitute an essential library for all who work in the optical field.

Preface

The ferroelectric lithium niobate crystal is unique concerning its set of properties and its wide applicability in science and engineering. This book offers a sufficiently detailed treatment of the crystal, its physico-chemical, ferroelectric, electrophysical, nonlinear optical, and other properties, as well as the technology of its manufacture. Due account of the specificities of this material is a contributing factor in its appropriate utilization in devices and scientific experiments. The emphasis of the book is lithium niobate's applicability to laser radiation control.

The authors do not only confine themselves to practical recommendations regarding the production and use of lithium niobate. They also furnish explanations of the physical and physico-chemical processes induced in the crystal by high temperatures, environmental conditions, radiation, or electric fields. Thought is given to the impact of impurities and non-stoichiometry.

The book aims at highlighting the achievements of Soviet scientists who have actually contributed very much to the development of this material.

The book will serve, we think, to introduce the exciting prospect of lithium niobate to a wide variety of interests—research workers, instrument developers, and students majoring in the relevant fields.

In conclusion, the authors wish to thank the translators of the Russian book—Oleg A Zilbert for chapters 1–4 and Dr Tamara M Pyankova for chapters 5–9.

A M Prokhorov
Yu S Kuz'minov
Moscow 1989

List of Symbols

B	birefringence
c	speed of electromagnetic radiation in free space
d	thickness along the applied field
d_{ij}	piezoelectric strain coefficient
$\underline{d}_{ij}$	nonlinear coefficient
$\bar{D}$	electric displacement vector
D	diffusion coefficient
e	electron charge
$\bar{E}$	electric field vector
E_g	width of the forbidden band
f	Lorentz factor
g	latent crystallization heat
$\bar{g}$	growth vector
g_{ij}	quadratic electro-optic coefficient
$\bar{G}$	gradient of temperature
h	Planck's constant
h	height
H	activation energy
$\bar{H}$	magnetic field vector
I_0	intensity of light incident onto a crystal
j	current density
k	Boltzmann constant
$\bar{k}$	wavevector
K_G	Glass constant
$\bar{K} = \pi n/d$	vector of spatial periodicity
l	length
l_D	Debye screening length
l_c	coherence length
m	mass of electron
m^*	effective electron mass
n	density of the charge
n_e	extraordinary refractive index
n_o	ordinary refractive index
N_a	concentration of the compensating acceptor impurity

N_c	density of states in the conduction band
N_0	electron density
$\bar{p}$	vector in the direction of drawing
p_{O_2}	partial oxygen pressure
p_{ij}	photoelastic (strain) coefficient
$\bar{P}_s$	spontaneous polarization vector
P_i	polarization component
Q	mechanical loss
r	beam radius
r_{ij}	linear electro-optic coefficient
R	resistance
RDS	regular domain structure
s_{kl}	second-rank strain tensor
S_0	average oscillator strength
t	time
t_s	repolarization time
T_C	Curie temperature
T_m	phase matching temperature
T	temperature
v	bulk acoustic wave velocity
V	electrical potential
$V_{\lambda/2}$	half-wave voltage
V_s	growth rate
x	strain
X_{kl}	stress
X_{ijk}	nonlinear susceptibility coefficient
α	photoionization cross section
α	absorption coefficient
α_1	thermoelectric coefficient
α_v	crystallization EMF
α_{ij}	cosines of the angle
α_{ijk}	differential optical polarizability
α_{ijk}	photovoltaic tensor
β	probability for thermal ionization
β	quantum yield
γ	capillary tension
γ	pyrocoefficient
Γ	phase difference of two waves in the crystal
δ	thickness of the boundary layer
δ_{ij}	Miller's constant (nonlinear susceptibility)
Δ	birefringence
ε_{ij}	permittivity tensor
ε_0	permittivity of vacuum
ε^t	unclamped permittivity
ε^s	clamped permittivity
ε_C	permittivity at Curie point
ζ_{ijk}	change in polarizability

η	packing density
η	conversion efficiency
η	degree of sample unipolarity
θ_m	phase matching angle
λ	wavelength
λ_s, λ_L	heat conduction coefficient for crystal and melt
λ_0	average oscillator position
λ_0	fundamental wavelength
Λ	separation of diffraction grating
μ	electron mobility
μ_H	Hall mobility electrons
ν_{res}	resonance oscillation frequency
π_{ij}	piezo-optical constant
ρ	sample density
ρ	space charge density
ρ_m	melt density
σ	effective surface charge density
σ	density of dislocations
σ	stress tensor
σ_d	dark conductivity
σ_{ph}	photoconductivity
τ	relaxation time
φ	potential difference
$\varphi = 2\pi l\beta/\lambda$	phase
χ	affinity for the electron
χ_e	dielectric susceptibility
ω	rotation speed
ω_0	angular resonant frequency
Ω	true angle between the optic axes of crystal

Introduction

The use of ferroelectric crystals to control laser emission is based on the nonlinear effects which occur in these crystals as a laser beam passes through them. The research work in this field did not in fact start until 1961, when Franken *et al* (1961) succeeded in generating a second harmonic of radiation from a ruby laser on a quartz crystal. These investigations have subsequently become rather extensive, and at the present time nonlinear optical crystals are being widely used for developing new sources of coherent visible radiation, frequency conversion and the detection and various transformations of signals and images.

One of the advantages that can be expected from using lasers in communication systems is their capability of passing tremendous amounts of information in the visible and infrared regions of the spectrum. To exploit this advantage fully, one should develop means for sufficiently wide-band and efficient modulation, deflection, commutation and frequency conversion of the signals thus transmitted. These tasks have, in turn, stimulated intensive studies aimed at producing crystals characterized by small optical losses: crystals whose properties can be changed by the application of an electric field, and which may thus definitely affect the characteristic of optical beams.

Crystals which may find application in nonlinear optics should meet the following requirements: they should be transparent in the entire range of wave interaction; and for them, the phase-matching condition must be fulfilled, i.e. such crystals should have a rather high birefringence and good optical quality.

One should mention yet another factor which significantly affects the optical quality of a nonlinear crystal, namely the laser radiation itself, whose passage through such a crystal is manifested by a change in the crystal's refractive index.

From what has been said above it is clear that progress in quantum electronics strongly depends on our ability, both to produce already known, and find new, kinds of nonlinear crystals with high optical quality and with excellent electro-optic and nonlinear optic coefficients.

Crystals of potassium hydrophosphate (PHP) and their PHP-based analogues, in which potassium is replaced by the ammonium group and hydrogen by its heavy isotope deuterium, were the first to be employed for laser emission control. Crystals of the PHP type are grown from aqueous solutions, the growth technology having reached a high level of perfection. These crystals, whose optical quality is quite good, have so far been used rather widely. They have some inherent disadvantages however, of which the main ones are their relatively low electro-optic and nonlinear optic coefficients. And it is exactly these characteristics that are responsible for the electro-optic effect and the radiation conversion efficiency.

Quite recently there have been synthesized, and rather thoroughly explored, some ferroelectric single crystals of the niobates and tantalates of alkali and alkaline earth metals, which have excellent electro-optic, piezoelectric, pyroelectric and nonlinear characteristics. The physical properties of these crystals allow their wide application in devices used for modulation, deflection and frequency conversion of laser radiation, as well as for the holographic storage of information.

A very typical representative of this class of crystals is lithium niobate (LN), first synthesized in 1965 by S A Fedulov in the USSR and by A A Ballman in the USA. Although lithium niobate has now been studied for over 20 years, it never ceases to surprise us by revealing again and again its new properties, which at once find application in science and technology. The necessity of producing more and more of this material has increased every year. At the same time, the sensitivity of lithium niobate to laser radiation, as well as its inherent optical in-homogeneities, impose limitations on its application in optical devices. An understanding of the specific character of this unique crystal, and of its advantages and disadvantages, will thus enable us to make the right decisions on its possible application in research work and in the manufacture of scientific instruments. It is exactly for the purpose of giving a detailed and thorough analysis of the physical and chemical properties of crystalline lithium niobate, as well as the technological aspects of the growth of its single crystals, that this book has been written.

The volume contains nine chapters. The first chapter deals with the physical and chemical properties, as well as the crystal structure, of lithium metaniobate. The second chapter describes the synthesis of its crystals, taking account of the peculiarity of the material under study. The third chapter is concerned with crystal structure defects, peculiar to this compound, which may arise both in the process of crystal growth and under the influence of various physico-chemical factors. All these chapters are closely related to each other.

Lithium niobate is a ferroelectric, which means that only those crystals representing a single domain may be used for practical purposes. Accordingly, bringing LN crystals to a single-domain state (e.g. by poling) is an essential step in the technology of this material. The fourth chapter is therefore devoted to a discussion of the domain structure and the processes that accompany the attainment of the single-domain state.

The fifth chapter discusses the electrophysical properties of LN, which are intimately related to the domain structure and defects of the crystal.

The sixth and seventh chapters describe the electro-optic and non-linear optic characteristics of LN, establishing a close relation between them and the ferroelectric properties of the material, which is due to the crystal structure of LN being made up of oxygen octahedra, NbO_6.

Chapter 8 concerns itself with the effect laser radiation has on LN crystals, i.e. the optically induced refractive index damage. The phenomenon is explained by adopting various models. The relations between laser radiation intensity, photon energy, the photo-conduction of the crystal and its sensitivity to various optical radiations are given.

The ninth chapter discusses some practical methods of quantitative assessment of the optical inhomogeneity of LN crystals and the criteria of their applicability in electro-optics and nonlinear optics.

Mention should be made of the fact that the book not only offers practical recommendations to be used in the technology of these crystals, but also provides an insight into the physical processes that take place in the material under the influence of such factors as temperature, radiation, electric fields, impurities and stoichiometry violation.

1 Physico-chemical Properties of Lithium Metaniobate

1.1 Phase diagram for the $Li_2O-Nb_2O_5$ system

Relatively few studies of this system have been made. Lapitsky (1952) and independently Sue (1937) were the first to obtain anhydrous lithium metaniobate by heating a mixture of lithium carbonate, niobium pentoxide and lithium flouride in a silver crucible at a temperature of 700 °C, and also by heating an equimolar mixture of lithium carbonate with niobium pentoxide. The small crystals of lithium metaniobate thus obtained had the shape of a prism and were yellowish in colour. The density of the crystals varied from 4.283 to 4.308 g cm^{-3}, depending on the method by which they had been obtained. According to Lapitsky, LN crystals can be melted at a temperature of 1164 ± 2 °C without being decomposed.

Lithium metaniobate was also obtained by Wainer and Wentworth (1952), who aimed to study its dielectric properties. Reisman and Holtzberg (1958a,b), who studied the $Li_2O-Nb_2O_5$ system by using the methods of DTA, X-ray phase analysis and density measurements, gave its first phase diagram.

As a result of a critical analysis of all available data on the $Li_2O-Nb_2O_5$ system, Shapiro *et al* (1978) compiled its complete phase diagram, which is shown in figure 1.1.

Of greatest interest to us is the range from 37 to 60 mol.% Li_2O, where lithium metaniobate is the first phase to crystallize from the melt, and also the compounds $LiNb_3O_8$ and Li_3NbO_4, because they can be formed in the course of the decomposition of the originally formed solid solutions.

Lithium orthoniobate Li_3NbO_4 was first described by Reisman and

1

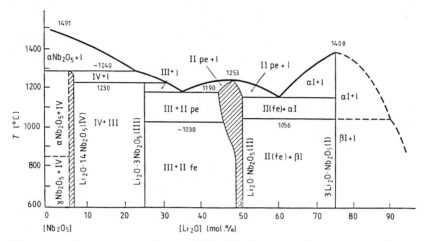

Figure 1.1. Phase diagram for the $Li_2O-Nb_2O_5$ system (Reisman and Holtzberg 1958a,b).

Holtzberg (1958a,b) as an individual phase with a congruent melting temperature of 1408 °C. Subsequently Wiston and Smith (1965) and also Svaasend *et al* (1974) have shown that the system has a body-centred cubic unit cell, containing eight formula units, with the parameter $a = 8.4300$ Å. Wiston and Smith (1965) note that above 1300 °C the compound decomposes to $LiNbO_3$ and possibly to Li_2O.

The compound $LiNb_3O_8$ is congruently melted at 1230 °C (Svaasend *et al* 1974), and has a monoclinic unit cell containing 44 formula units with the following parameters, given by Lundberg (1971) and Svaasend *et al* (1973): $a = 15.262$ Å; $b = 5.033$ Å; $c = 7.457$ Å; $\beta = 107.34$ °C. Svaasend *et al* (1973) also reported that they had obtained single crystals of this compound. The crystal density quoted by them coincides with that calculated from X-ray data, the crystal purity having been controlled by a mass spectrographic analysis. This has led these authors to conclude that the formula of this compound is $LiNb_3O_8$, rather than $Li_2Nb_8O_{12}$ as reported previously (Reisman and Holtzberg 1958a,b, Wiston and Smith 1965).

Lithium metaniobate $LiNbO_3$ is the best studied phase in the $Li_2O-Nb_2O_5$ system. This is due to the fact that its single crystals display a unique combination of the electrophysical, optical and nonlinear optical properties, and are being widely used currently. It is of interest to note that it is just the problem of raising the quality of such single crystals and improving the reproducibility of their properties that gave rise to the investigations (Svaasend *et al* 1974, Lerner *et al* 1968, Carruthers *et al* 1971, Holman *et al* 1978, Bridenbaugh 1973, Peterson and Carruthers 1969) whose results lie at the root of the modern concept that lithium niobate is a variable-composition phase.

As was found by Reisman and Holtzberg (1958a,b), in the composition range of 36 mol.% to 60 mol.% Li_2O, lithium metaniobate ($LiNbO_3$) crystallizes; this is congruently melted at 1253 °C. Accordingly, lithium metaniobate single crystals had for a long time been grown from the stoichiometric melt of the composition of $Li_2O \cdot Nb_2O_5$ and all studies had been carried out using single crystals whose composition was thought to be identical to that of the initial melt (Fedulov *et al* 1965, Ballman 1965, Nassau and Levinstein 1965). The inhomogeneity of the optical properties of the crystals and a large spread of their Curie temperature T_c that was soon revealed (Bergman *et al* 1968) have drawn the investigators' attention to the inconstancy of the crystal composition. It was then assumed that the crystals may contain an excess of one of the components or impurities. A chemical analysis of the crystals and melts, made by Bergman *et al* (1968), has enabled them to establish that, as the molar ratio $Li_2O/Nb_2O_5 = R$ varies in a melt from 1.2 to 0.8, the corresponding variation in the crystal is from 1.04 to 0.96. The variation in properties that accompanies the variation in composition is probably due to the incorporation of an excess of lithium, up to $R \approx 1.2$, in the crystal lattice. At the International Conference in Birmingham in 1968, the French scientists Lerner *et al* (1968) reported the results they obtained when studying the $Li_2O{-}Nb_2O_5$ system in the vicinity of the $LiNbO_3$ composition. Their precise measurements of the lattice parameters and the differential thermal analysis they had carried out enabled them to conclude that near the stoichiometric composition there exists a range of solid solutions. Figure 1.2 illustrates an equilibrium state diagram near the stoichiometric composition point, which has been compiled using X-ray and DTA data. From this diagram it follows that the congruently melted composition corresponds to the molar percentage of Li_2O of 48 or 49. As a single crystal is grown, the liquid-phase composition undergoes a change. In that part of the diagram where the solidus curve is characterized by a gentle slope ($R < 1$), small changes in the composition of the liquid phase produce rather significant changes in the composition of the solid phase being formed. Similar, though less well pronounced, changes are observed when single crystals are grown from melts with $R > 1$. The departure of the congruent composition from the stoichiometric one has been confirmed by Byer *et al* (1970), who discovered a high optic homogeneity in crystals grown from melts deficient in lithium. Having analysed the variation in crystal composition that takes place in the process of crystal growth from a stoichiometric melt, they have come to the conclusion that high-quality crystals with a constant value of birefingence may be obtained from the congruent melt, because only then will crystal composition be practically independent of temperature variations at the growth front and of the reduction of the mass of the melt in the course of crystallization.

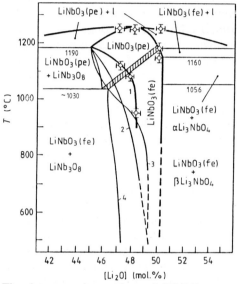

Figure 1.2. The homogeneity region of $LiNbO_3$, as refined by: 1, Holman (1978); 2, Scott and Burns (1972); 3, Svaasend *et al* (1974); 4, Lerner *et al* (1968).

The deviation of congruent composition from stoichiometric has been found in studying lithium metaniobate crystals by the NMR technique involving [93]Nb nuclei (Peterson and Carnevale 1972). Those authors have explained this deviation by the ease of formation of lithium vacancies, which may be expected if we consider the difference between the bond strength for Nb–O and for Li–O, as suggested by Peterson and Bridenbaugh (1968). According to Peterson and Carnevale (1972), the congruent melt contains 48.6 mol.% Li_2O. It should be noted that stoichiometric composition crystals may be grown from melts with 58 mol.% Li_2O.

Congruent melt compositions for lithium niobate vary little from author to author, corresponding, according to the most recent data available (Byer *et al* 1974, Ivanova *et al* 1980), to 48.5 ± 0.1 mol.% Li_2O in the crystal (see figure 1.2). The parameters of a unit cell in a congruently melted crystal are as follows: $a = 5.1508$ Å; $c = 13.864$ Å; density $= 4.659$ g cm^{-3} (Lerner *et al* 1968).

The homogeneity region for the system at 1190 °C has a maximum width of ~ 6 mol.% in the range between 44.5 and 50.5 mol.% Li_2O. As the temperature is lowered to room temperature, the homogeneity region narrows to 2 mol.% (see figure 1.2).

With the decreasing content of lithium oxide in the solid solution, the density slightly increases (by ~ 0.2%); simultaneously, the unit-cell volume enlarges. One may therefore conclude that the most probable mechanism for the formation of solid solutions is the partial substitution of Nb ions for Li ones, with a simultaneous formation of the corresponding number of lithium vacancies. The chemical formula for a solid solution may be written, in Kröger's notation (1964), as follows: $Li_{1-2x}^{+}Nb_{1+\varepsilon}^{5+}(V_{Li}^{-})_{x+\varepsilon}O_3^{2-}$.

To obtain the homogeneity region for $LiNbO_3$, Carruthers *et al* (1971) used, along with the data of a conventional differential thermal analysis, the data on ^{93}Nb NMR peak broadening with increasing Li_2O content in crystals and also those on the Curie temperature inside the homogeneity region. Their results support those of Lerner *et al* (1968).

Figure 1.3 illustrates the dependence of crystal composition upon melt composition according to Carruthers *et al* (1971). The studies of the departure of $LiNbO_3$ and $LiTaO_3$ from their stoichiometric composition, carried out by Scott and Burns (1972) on ceramic specimens using a Raman powder spectroscopic technique, have shown that under long annealing (of several days) the composition of the solid solution differs from that given by Lerner *et al* (1968) or Carruthers *et al* (1971). Scott

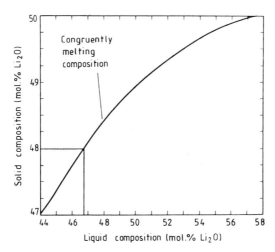

Figure 1.3. LN crystal composition–melt composition diagram (Carruthers *et al* 1971). The rectangle at bottom left is the two-phase region at room temperature under equilibrium conditions.

and Burns (1972) therefore suggest that the congruent composition becomes unstable below 800 °C, at which point a second phase of the $LiNb_3O_8$ composition arises, so that at room temperature the stoichiometric composition phase with a deviation of about 0.5 mol.% is stable.

Svaasend *et al* (1974) have studied the properties of crystals subjected to long annealing (for 100 to 1000 h) at various temperatures (below 1000 °C) at normal pressure. Upon cooling down to room temperature, the optical quality of the crystals significantly deteriorated due to the appearance of milky white opalescent regions. In crystals grown from melts with a molar content of Li_2O less than 48 mol.%, the opalescence arose as early as after a 10 h period of annealing at 800 °C, whereas for molar contents of 48.6 mol.% and above a much longer annealing period was required (more than 500 h), the other factors being the same. X-ray studies have disclosed that, in this case, the $LiNb_3O_8$ phase arises. It is noteworthy that under a second annealing of the same specimens at temperatures above 1000 °C the scattering centres have completely disappeared, the X-ray patterns displaying the line due to metaniobate alone.

The above-mentioned experiments have thus revealed that LN crystals grown from the congruent composition melt (48.6 mol.% Li_2O) are unstable if kept for a long time at a temperature of about 910 °C.

One should however mention that, with the currently accepted cooling rate ($\sim 100 °C h^{-1}$) for lithium metaniobate crystals, no precipitation of a second phase is observed, and the crystal composition corresponds to that of the solid phase, as determined from the diagram of figure 1.3.

The position of the boundary of the $LiNbO_3$ homogeneity region was found from the change in the weight of a ceramic or crystalline sample of a known composition under isothermal annealing in an atmosphere with a given partial pressure of lithium oxide vapours to a constant weight (Holman *et al* 1978). The annealing was carried out in sealed crucibles pressed from a mixture of $LiNbO_3$ with $LiNb_3O_8$ to determine the boundaries of the homogeneity region on the side of excessive niobium, or with Li_3NbO_4 to do this on the side of the lithium excess. The error of determination of the phase boundary position in composition does not exceed ± 0.1 mol.%. The method seems therefore to be the most exact of those known at present. Its applicability is, however, limited by the temperature range between 1050 and 1140 °C, in which the equilibrium with the gaseous phase sets in rather rapidly, yet no melting of the solid solution takes place.

By presenting all the available data on the position of the homogeneity region boundaries for LN in a single figure (see figure 1.3) one can readily notice how controversial they are.

For example, according to Scott and Burns (1972), the solid solution of congruent composition LN is decomposed below 800 °C, whereas Svaasend *et al* (1974) give 950 °C as the temperature at which a second phase, $LiNb_3O_8$, separates out. To resolve this discrepancy is of great practical importance, because the right annealing regime means good optical quality, as well as good mechanical properties, of the single crystals.

The phase diagram near the $LiNbO_3$ composition was refined by D'yakov (1982), who made use of a DTA technique and a composition dependence of crystal lattice parameters.

An X-ray determination of the LN unit cell parameters has shown that the hexagonal cell parameters vary as a function of the melt composition utilized to grow crystals (figure 1.4). The absolute value of the parameters and their dependence upon the composition of the initial melt is in good agreement, within the error of measurement, with the data of Lerner *et al* (1968).

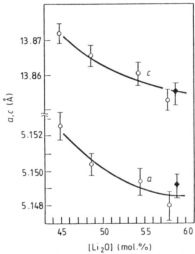

Figure 1.4. Variation in the parameters *a* and *c* of the LN hexagonal unit cell as a function of the melt composition (D'yakov 1982): ○ Czochralski-grown crystals; ◆ crystals grown by a modified Kiropoulous technique.

In temperature diagrams for LN crystals grown from melts whose composition strongly deviates from the congruent one, the decomposition of LN and the formation of solid solutions are observed.

A colourless and inclusion-free LN crystal, grown from a melt with 45 mol.% Li_2O, was slightly opalescent. During the first heating, the DTA curve for the specimen exhibits a weak yet rather distinct peak at

1125 °C, which corresponds to dissolving micro-inclusions of $LiNb_3O_8$ and the formation of a homogeneous solid solution (figure 1.5, curve A). The melting temperature corresponds to 1246 °C.

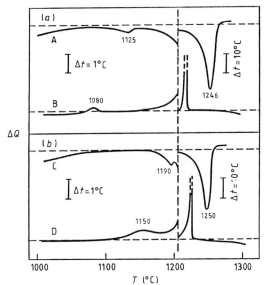

Figure 1.5. Thermograms of LN single crystals (D'yakov 1982): (*a*) crystals grown from a 45 mol.% Li_2O melt (A, heating; B, cooling); (*b*) crystals grown from a 58 mol.% Li_2O melt (C, heating; D, cooling).

The cooling curve (figure 1.5, curve B) shows a weak peak very gently sloping on the side of low temperatures, which corresponds to the solid solution decay at 1080 °C.

A crystal grown from a melt with 58 mol.% Li_2O was transparent and colourless, without any inclusion of the melt. As the crystal was heated for the first time (figure 1.5, curve C), one could observe, simultaneously with the onset of melting, a weak peak at 1190 °C, corresponding to the decomposition of the solid solution. The process of melting ended at 1250 °C.

The curves for cooling (figure 1.5, curve D) and reheating display a weak peak at 1150 °C ± 5 °C, corresponding to the $LiNbO_3$ + Li_3NbO_4 eutectic. During a second heating the solid solutions were observed to decay between 1170 and 1190 °C.

To determine the $LiNbO_3$ homogeneity region from a phase diagram for the Li_2O–Nb_2O_5 systems, D'yakov (1982) made use of the measurements of the lattice parameters for LN single crystals and also of the results of a thermal analysis. He thus obtained a number of points on curves that contain the solid solution region (figure 1.2). The position of these points corresponds, within the experimental error, to the liquidus and solidus curves constructed by Lerner *et al* (1968) and to the boundary of the decomposition region of solid solutions according to Holman *et al* (1978).

When analysing the results of lattice parameter determination (figure 1.4), attention is drawn to the fact that the lattice parameters of a crystal grown from a melt with 59 mol.% Li_2O using the Kiropoulous technique correspond to the lower content of Li_2O than those of a Czochralski-grown crystal from a melt with 58 mol.% Li_2O. This may be explained by the difference in the growth rate of the crystals.

As was pointed out by Carruthers *et al* (1971), the main factor which determines the rate of crystal growth from melts with a great excess of Li_2O is the rate of diffusion of Li in the melt. The melt at the growing crystal facet is enriched in lithium and the greater the rate of crystallization the greater the enrichment.

Accordingly, a crystal grown by the Czochralski technique at a rate of $2.5 \, mm \, h^{-1}$ will crystallize from the melt with a higher content of Li_2O than a Kiropoulous-grown crystal at a rate of $0.5 \, mm \, h^{-1}$. This accounts for the difference in the parameters for a unit crystal cell observed when LN is grown by different methods (see figure 1.4).

It is of interest to analyse the wide range of the variable-composition phase in the $Li_2O–Nb_2O_5$ system from a thermodynamic viewpoint. Considering the equilibrium conditions for non-stoichiometric chemical compounds, Anderson (1946) noted that the tendency to form variable-composition phases is shown by compounds containing well polarized cations and anions, in which bonding forces are essentially covalent. The degree of intrinsic disorder in such compounds is much larger than in purely ionic systems. The same feature of complex compounds has also been mentioned by Palatnik *et al* (1969), who point out that in ternary compounds with two types of cations of unequal ionicity, the tendency to form variable-composition phases increases with the increasing portion of the covalent-bond component. Various authors (Anderson 1946, Palatnik *et al* 1969, Ormont 1969) also noted that the tendency to form phases of variable composition is manifested by those compounds which contain an ion whose valence is readily changeable.

The stability of a variable phase is determined by the steepness of the free energy curve (relative to the composition). The wide range of homogeneity requires that there should be a long gentle slope about the extremum point of the curve (Ormont 1969). It has been found (Alberts and Haase 1969) that such a shape of the free energy curve is attained at a high degree of the inherent disorder, i.e. at a high concentration of structural defects of various kinds—vacancies, interstitial ions, interchangeable cations in cationic lattices etc. In more complex compounds, the part of structural units may be played by complex anionic or cationic groups in melts and crystals, owing to a large contribution provided by the covalent bonds in these groups to the total lattice energy. In these conditions, relatively small amounts of energy are expended to give rise to cation defects.

In lithium metaniobate, NbO_6 octahedra are distinguished by a high degree of covalence (Peterson and Bridenbaugh 1968). The existence of complex structural groups in lithium metaniobate is confirmed by the results obtained by Bol'shakov *et al* (1969), who studied the temperature and time dependences of the kinematic viscosity and density of LN melts. He indicates that at temperatures close to T_M there still remain, in the melt, the homopolar bonds characteristic of the solid phase, these bonds being completely broken only after a long isothermal heating of the melt at a temperature much higher than the melting point. Bearing all this in mind, one may assume that the formation of Li vacancies in a metaniobate lattice should occur with a small expenditure of energy, i.e. the free energy curve should have a rather gentle slope, its minimum being shifted towards the excess of Nb_2O_5. This conclusion has been reached by Carruthers *et al* (1971), who quoted the work of Alberts and Haase (1969). The former authors also noted that, since the position of a congruently melted composition is primarily determined by that of the minimum on the free energy curve, one may expect that, for solid phases containing complex anionic and cationic groups, the congruent composition will be somewhat displaced with respect to the stoichiometric one. In the case of lithium metaniobate, this shift constitutes 1.4 mol.% Nb_2O_5.

In conclusion, we once again stress that lithium metaniobate is a typical compound of variable composition, whose structure is normally characterized by a high concentration of inherent defects.

Metaniobates of alkali metals are slightly soluble in water. Table 1.1 presents the values of the solubility of $LiNbO_3$ and $LiTaO_3$ as a function of temperature. From the solubility data the following thermodynamic functions have been calculated: the activity product, the free energy

Table 1.1. Solubility (in $mol\,l^{-1}$) of $LiNbO_3$ and $LiTaO_3$ in water (according to Goroshchenko 1965).

T (°C)	0	25	50	75	100
$LiNbO_3$	2.3×10^{-4}	2.8×10^{-4}	4.3×10^{-4}	6.0×10^{-4}	7.4×10^{-4}
$LiTaO_3$	5.14×10^{-5}	1.05×10^{-4}	2.29×10^{-4}	3.81×10^{-4}	5.09×10^{-4}

variation and the crystal lattice energy variation (see table 1.2). The simultaneous solution of the Kapustinsky equation and Fiance equation has made it possible to determine the radius of NbO_3^- ions (2.04 Å) and TaO_3^- ions (1.83Å). The hydration heats for these ions are 95 and 110 kcal mol^{-1} respectively.

Table 1.2. Thermodynamic functions for lithium niobate and tantalate at 20°C (according to Goroshchenko 1965).

Compound	Solubility (mol l^{-1})	Activity product	Free energy variation F_0 (kcal mol^{-1})	Solubility heat (kcal mol^{-1})	Crystal lattice energy v_k (kcal mol^{-1})
LiNbO$_3$	2.566×10^{-4}	6.344×10^{-8}	9.7	6.2	228.5
LiTaO$_3$	8.974×10^{-5}	7.878×10^{-9}	10.9	10.5	249.1

The temperature dependences of the specific heat, heat conduction and linear expansion coefficient for lithium niobate have been studied by Zhdanova *et al* (1968). The averaged specific heat data are listed in table 1.3.

Table 1.3. Averaged values for the heat capacity of LiNbO$_3$ (cal mol^{-1} K^{-1}) (according to Zhdanova *et al* 1968).

T (K)	Averaged	T (K)	Averaged
80	4.80	240	20.40
90	6.30	250	20.90
100	7.70	260	21.35
110	8.90	270	21.85
120	10.10	280	22.20
130	11.25	290	22.55
140	12.35	300	22.90
150	13.40	310	23.25
160	14.40	320	23.55
170	15.40	330	23.85
180	16.35	340	24.10
190	17.30	350	24.40
200	18.15	360	24.65
210	18.75	370	24.85
220	19.35	380	25.05
230	19.90	390	25.35

The specific heat has been calculated according to the Einstein model

$$C = (5)(3Nk)\left(\frac{\theta_E}{T}\right)^2 \frac{\exp(\theta_E/T)}{[\exp(\theta_E/T) - 1]^2} \tag{1.1}$$

using $\theta_E = 514$ K. The values were converted to C_P and corrected for anharmonicity to give

$$C_P = C + (1.53 \times 10^{-3}T). \tag{1.2}$$

The result is shown in figure 1.6, where the computed values of C_P are shown as full circles. The curve represents the experimentally measured specific heat over the temperature ranges 4–10 K, 20–300 K and 300–1250 K. The good agreement of the computed and measured specific heats indicates that the simple vibrational model on which θ_E was computed from the NQR data is adequate over the temperature range 21–515 K (Schempp *et al* 1970).

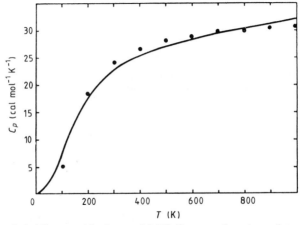

Figure 1.6. The specific heat of LiNbO$_3$ as a function of temperature. •, calculated from an Einstein model using $\theta_E = 514$ K; full curve, experimental values (Schempp *et al* 1970).

Figure 1.7 illustrates the temperature dependence of the heat conduction of lithium niobate, measured both parallel and normal to the trigonal axis; the figure also shows the temperature dependence of the heat resistance of the compound.

The temperature dependence of the linear expansion coefficient of lithium niobate, measured both along the trigonal axis and normal to it, which is of importance for the practical application of single crystals of the compound, is shown in figure 1.8. From the figure one can conclude that lithium niobate single crystals are characterized by a well pronounced anisotropy of its linear expansion coefficient.

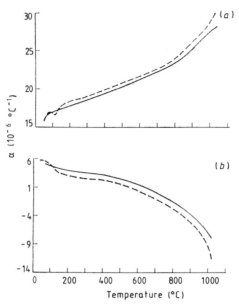

Figure 1.7. Coefficient of thermal expansion along (*a*) the *a* axis and (*b*) the *c* axis for stoichiometric (full curve) and congruent (broken curve) LiNbO$_3$ (4° C min^{-1}) (according to Gallagher and O'Bryan 1985).

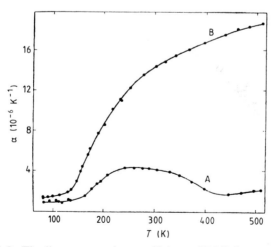

Figure 1.8. The linear expansion coefficient of LiNbO$_3$ as a function of temperature (Zhdanova *et al* 1968): A, along the trigonal axis; B, normal to the trigonal axis.

1.2 Crystal structure of lithium metaniobate

The crystal structure of LN was first studied by Zachariasen (1928), who used an X-ray technique. He found that at room temperature this crystal belongs to the space group $R\bar{3}(C_{3i}^2)$, suggesting that its structure is similar to that of ilmenite ($FeTiO_3$). The discovery of the ferroelectric properties of LN (see Matthias and Remeika 1949), however, indicated that its structure should not be centrally symmetric. In 1952, the results of Zachariasen (1928) were therefore revised by Bailey, who demonstrated that LN should belong to the space group $R3c(C_{3v}^6)$. He failed, however, to determine unambiguously the position of the niobium ion with respect to the ions of lithium and oxygen. Accordingly, he proposed two possible structural models. An analysis of Bailey's data made by Megaw (1954, 1956) led her to conclude that the structure of LN at room temperature is a greatly distorted perovskite structure ($CaTiO_3$), while at temperatures above the Curie point it should be the undistorted perovskite structure. However, ion displacements needed for the suggested rearrangement in a phase transition turned out to be rather large, leading Megaw to propose to place LN with 'frozen' ferroelectrics, in which ion rearrangements are only possible at high temperatures, when the probability of large ionic displacements is fairly high.

The neutron diffraction analysis by Shiosaki and Mitsui (1963) confirmed the R3c group for lithium metaniobate. Yet the coordinates of the atoms remained uncertain, as before. Those authors believed that the ferroelectric phase transition in the crystal should be associated with the displacement of the niobium ions. With the advent of high-quality single crystals and advances in X-ray techniques, a complete structure analysis of LN became possible (Abrahams *et al* (1966a,b,c). They conclusively determined its space group as R3c, and also established its unit cell parameters. A hexagonal cell contains six formula units, its parameters being as follows: $c_H = 13.8631 \pm 0.0004$ Å; $a_H = 5.14829 \pm 0.00002$ Å. A rhombohedral cell contains two formula units, the cell parameters being $a_R = 5.4944$ Å, $\alpha = 55°52'$.

The oxygen structure carcass is built on the motif of the closest hexagonal packing. There are six flat oxygen layers in a hexagonal unit cell. The tetrahedral voids in such a carcass remain vacant, while the octahedral ones are largely (two-thirds of them) occupied by cations. The oxygen octahedra, stretched in the direction of a three-fold axis, have common facets. The oxygen atoms do not lie above one another along the three-fold axis, but are arranged in a screw-like fashion (figure 1.9). In the neighbouring columns, the octahedra are joined by their edges. In this respect, the lithium metaniobate structure is greatly different from the perovskite one, in which the octahedra are only joined by their vertices.

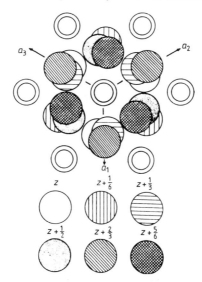

Figure 1.9. Projection of the LiNbO$_3$ crystal structure upon the plane of the (0001) basis (Abrahams *et al* 1966a,b,c). Differently shaded circles show oxygen ions at different levels relative to the plane of the drawing; light double circles stand for projections of positions of metal ions upon the drawing plane.

The succession of the cations in the columns of octahedra along a three-fold axis is as follows: Li, Nb, an empty octahedron (figure 1.10). In the ferroelectric phase, the cations are displaced from the centres of the octahedra. The flat oxygen layers are at a distance of 2.310 Å from each other (in the closest oxygen packing, the interlayer spacing is equal to $2^{1/2}R_{O^{-2}} = 1.87$ Å). The niobium ion is 0.897 Å from the nearest oxygen plane and 1.413 Å from the next-nearest oxygen plane, while for lithium these distances are 0.714 Å and 1.597 Å respectively.

The oxygen octahedra are distorted: the O–O spacing in the oxygen plane which is closest to the niobium ion is 2.879 Å. The lithium ion also distorts the oxygen octahedron. Thus the distance between the oxygen ions lying in the face closest to the lithium ion is greater (3.362 Å) than the corresponding distance in the more distant face. The octahedra occupied by niobium have two characteristic Nb–O spacings, as do the lithium-occupied and the vacant octahedra (table 1.4 and figure 1.11). If one compares these distances to the sum of the ionic radii for Nb^{5+} and O^{2-} (2.01 Å) and for Li$^+$ and O^{2-} (2.00 Å), as quoted by Ahrens (1952), then it will be obvious that the lithium ion is rather freely placed within the octahedron, at a distance of 2.068 Å from an oxygen ion in one oxygen triplet and 2.238 Å from an oxygen ion in another triplet. At the same time, the niobium ion is, within its octahedron, at a distance of about 1.89 Å from each oxygen ion in one

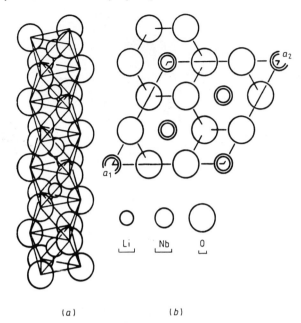

Figure 1.10. Crystal structure of $LiNbO_3$ (Abrahams *et al* 1966a,b,c): (*a*) a succession of the distorted octahedrons along the polar *c* axis; (*b*) an idealized arrangement of the atoms in a unit cell along the *c* axis.

of the ion triplets, which is considerably less than the sum of the corresponding ionic radii. This is indicative of the overlapping electronic shells of the ions and the formation of the covalent bonds.

The distinguishing feature of the lithium-occupied octahedron and its empty neighbour is the abnormally large distance between the oxygen ions in the layer common to both octahedra, viz. 3.362 Å. As a simple geometrical calculation shows, the radius of the sphere which may pass through the structural channel between three oxygen ions, placed at such a distance from each other, equals

$$3.362(\sqrt{3}/2)(2/3) - R_{O^{-2}} \approx 0.62 \text{ Å}$$

which is less than the ionic radius for Li^+ (R_{Li^+} being 0.68 Å). This turns out to be essential in an interpretation of the phase transition mechanism.

A transition from the ferroelectric to the paraelectric phase takes place at an unusually high temperature for the known ferroelectrics (1150 °C to 1180 °C for crystals grown out of stoichiometric melts: Carruthers *et al* (1971), Bergman *et al* (1968)).The space group for the

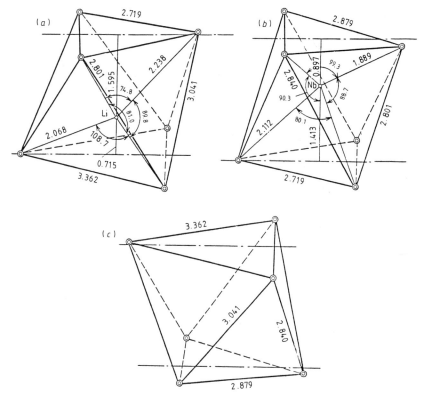

Figure 1.11. Schematic arrangement of the ions in the oxygen octahedrons (distance between the atoms in Å, angles in degrees). The schemes have been constructed using the data of Abrahams *et al* (1966a,b,c): (*a*) Li-occupied octahedron; (*b*) Nb-occupied octahedron; (*c*) vacant octahedron.

paraelectric phase is $R\bar{3}$, although Megaw (1956) has pointed out that the space group $R\bar{3}c$ is equally possible. This uncertainty could not be resolved using the X-ray data alone. Niizeki *et al* (1967) confirmed the latter possibility, proceeding from their observation of growth figures which appeared in growing single crystals, as well as from the mutual orientation of the domains arising in a phase transition.

A model of the transition of LN from the paraelectric to the ferroelectric phase was proposed by Abrahams *et al* (1966c). In the phase transition, the sublattices of positive ions of lithium and niobium are displaced relative to the sublattice of oxygen anions. The direction of the displacement of the cations determines the direction of the spontaneous polarization vector (P_s) in the ferroelectric phase, viz. [0001].

Table 1.4. Interatomic distances and the angles in the lattice of $LiNbO_3$ at $24°C$ (according to Nassau *et al* 1966).

Interatomic distances ($\mathring{A}$)

Nb–Nb	3.765 ± 0	O–O	2.719 ± 4
Li–Li	3.765 ± 0		2.801 ± 1
Nb–O	1.889 ± 3		2.840 ± 1
	2.212 ± 4		2.879 ± 4
			3.042 ± 2
			3.362 ± 4
Nb–Li	3.010 ± 31		
	3.054 ± 7		
	3.381 ± 15	Li–O	2.068 ± 11
	3.922 ± 31		2.238 ± 23

Angles between the atoms (deg)

O–Nb–O	80.1 ± 2	O–Li–O	74.8 ± 9
	88.7 ± 1		81.0 ± 3
	90.3 ± 1		89.8 ± 4
	99.3 ± 2		108.7 ± 9
Average	89.6		88.6

Abrahams *et al* (1968) have pointed out that it is the position of the metal ions in the structure of the ferroelectric phase that gives rise to a dipole moment. As crystals of lithium niobate are cooled down from the Curie point, there may occur two opposite directions of displacement for the metal ions, which correspond to 180° electric domains (Abrahams *et al* 1966c) (see figures 1.12(*a*), (*b*)). It has been suggested (Nassau *et al* 1966) that the end of a single-domain crystal which is positively charged in cooling, due to the pyro-effect, would be referred to as positive. It is possible to discriminate between the positive and the negative ends of such a crystal by means of etching or from the intensity of X-ray reflections. The negative end is etched more rapidly than positive (Nassau *et al* 1966, Evlanova and Rashkovich 1974), and its X-ray reflection is less distinct.

In order to change the polarization of single-domain crystals or to attain a single domain by changing the polarization of some of the domains making up a many-domain crystal, it is necessary to allow the ions of lithium and niobium to pass through the oxygen layers. For lithium niobate, the distance from a vertex of an oxygen triangle to the midpoint on the median has been calculated by Niizeki *et al* (1967) for the temperature range from 297 to 1473 K on the basis of the data of Abrahams *et al* (1966a,b,c) (see figure 1.13). This distance is larger than

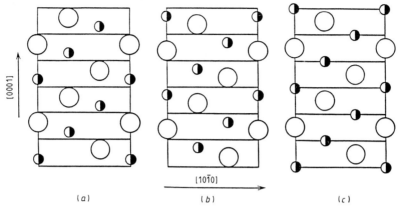

Figure 1.12. Projection of a LiNbO$_3$ unit cell upon the plane of the (1210) prism (Abrahams *et al* 1966a,b,c). The horizontal line shows the oxygen layers; ○ Nb^{+5}; ◑ Li$^+$. (*a*) a positively charged domain; (*b*) a negatively charged domain; (*c*) a para-phase.

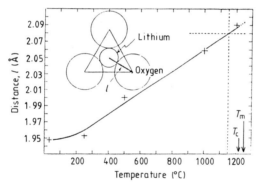

Figure 1.13. The temperature dependence of the distance from the top to the centre of the triangle made by the oxygen ions, which surround the lithium ion, in the structure of LN (Niizeki *et al* 1967).

the sum of the radii of the ions of lithium and oxygen at temperatures above 1423 K. For this reason, there can be no change in polarization of lithium niobate at these relatively low temperatures, which accounts for the name of a 'frozen' ferroelectric given to it by Megaw (1954). At the same time, at rather high temperatures, the direction of the spontaneous polarization of the crystal can be reversed (see Evlanova and Rashkovich 1974) without any external influence (e.g. without the application of an electric field).

Table 1.5 presents the values for the angles between the crystallographic planes in a hexagonal unit cell of lithium niobate, according to

Nassau *et al* (1966). The stereographic projection of the crystallographic planes and axes of lithium metaniobate in hexagonal coordinates is illustrated in figure 1.14.

Table 1.5. Angles φ between the crystallographic planes in the LN structure (data of Nassau *et al* 1966).

(0001)	φ (deg)	(0001)	φ (deg)	(11$\bar{2}$0)	φ (deg)
(10$\bar{1}$1)	72,17	(11$\bar{2}$1)	79,48	(21$\bar{3}$0)	10,89
(10$\bar{1}$2)	57,24	(11$\bar{2}$2)	69,62	(52$\bar{7}$0)	13,90
(10$\bar{1}$3)	46,02	(11$\bar{2}$3)	60,87	(31$\bar{4}$0)	16,10
(10$\bar{1}$4)	37,85	(11$\bar{2}$4)	53,39	(41$\bar{5}$0)	19,11
(10$\bar{1}$5)	31,87	(11$\bar{2}$5)	47,12	(61$\bar{7}$0)	22,41
(10$\bar{1}$6)	27,89	(11$\bar{2}$6)	41,90		
(10$\bar{1}$7)	23,95	(11$\bar{2}$8)	33,94		
(10$\bar{1}$8)	21,23	(11$\bar{2}$.12)	24,16		
(10$\bar{1}$.10)	17,27	(11$\bar{2}$.15)	19,74		
(10$\bar{1}$.16)	11,00	(11$\bar{2}$.18)	16,65		

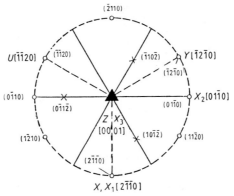

Figure 1.14. The stereographic projection of the point symmetry group 3m and the choice of the crystallographic (X, Y, U, Z) and crystal physical (X_1, X_2, X_3) axes.

The choice of the origin of the coordinates along the c axis in structures belonging to the space symmetry group R3c is arbitrary. In structures of the pseudo-ilmenite type, it is customary to place that origin on the niobium ion. The coordinates of the atoms in the LN structure at room temperature are listed in table 1.6 (the data of Abrahams *et al* 1966a,b,c).

At elevated temperatures up to 1473 K, the relative positions of the atoms in the LN structure remain unchanged, the shape of a unit cell at these temperatures being retained (table 1.7).

Table 1.6. Coordinates of the atoms in the LN structure at 297 K (data of Abrahams *et al* 1966a,b,c).

Atom	Coordinate		
	x	y	z
Nb	0	0	0
O	0.0492 ± 0.0004	0.3446 ± 0.0005	0.0647 ± 0.0004
Li	0	0	0.2829 ± 0.0023

Table 1.7. Coordinates of the atoms in the LN structure in the temperature range from 523 to 1473 K (data of Bergman *et al* 1968).

Temp. (K)	Coordinate			
	$x(O)$	$y(O)$	$z(O)$	$z(Li)$
523	0.0398 ± 0.0067	0.3238 ± 0.0089	0.0656 ± 0.0015	0.2752 ± 0.0068
773	0.0518 ± 0.0046	0.3318 ± 0.0065	0.0655 ± 0.0010	0.2758 ± 0.0037
1023	0.0415 ± 0.0113	0.3073 ± 0.0188	0.0683 ± 0.0024	0.2803 ± 0.0063
1273	0.0549 ± 0.0123	0.3253 ± 0.0198	0.0677 ± 0.0023	0.2798 ± 0.0038
1473	0.0634 ± 0.0201	0.3375 ± 0.0315	0.0702 ± 0.0057	—

Figure 1.15 shows X-ray diffraction patterns for lithium niobate at temperatures of 20 and 700 °C.

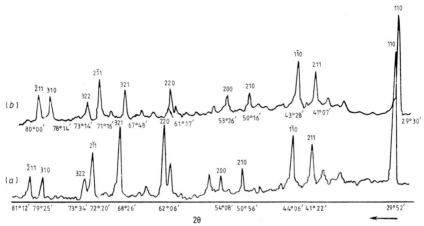

Figure 1.15. X-ray diffraction pattern of LiNbO₃: (*a*) at 20 °C; (*b*) at 700 °C (Ismailzade 1965).

1.3 Crystal-chemical features of metaniobates of alkali metals

The physico-chemical properties of these compounds are closely related
to their crystal-chemical features.

The crystal-chemical analysis made by Wood (1951), who used the
Goldschmidt ionic radii and stability criteria for various structures, has
revealed that this purely geometric approach leads to a large number of
exceptions and discrepancies with the experimental data.

A closer analysis of the crystal-chemical properties of compounds of
the ABO_3 type, where A stands for an alkali metal or Ag^+ or Cu^+, and
B for Nb, Ta, Sb or Bi, was carried out by Goodenaugh and Kafalas
(1973). Their analysis, apart from the size of the ions, also took into
consideration the Madelung electrostatic energy, the polarizability of the
ions, and the covalent component of the B–O bonds.

For alkali metal cations, the contribution of the covalent component
to the A–O bonding energy is relatively small. However, for a small
alkali ion (according to Goldschmidt (1926), the stability factor for a
cubic structure $t < 1$), the covalent component of the A–O bond may
distort the original perovskite cell to rhombic or hexagonal symmetry.

For Nb ions, which have an unfilled d shell, the contribution from the
covalent component to the Nb–O bond is rather substantial. According
to Meisner and Rez (1969), the degree of ionicity of the Nb–O is 0.445.
Figure 1.16 shows two possible variants of overlapping between the p
orbital of the anion and the t_2 orbital of the cation. Variant (*b*) is
attained in the perovskite structure, in which the octahedra are joined
by their vertices.

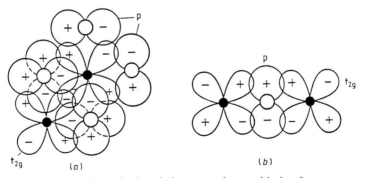

Figure 1.16. Formation of π bonds between the p orbitals of oxygen
and the t_{2g} orbital of a transition metal (Goodenaugh and Kafalas
1973); (*a*) angle $B\hat{O}B = 90°$; (*b*) angle $B\hat{O}B = 180°$.

On the other hand, the scheme of variant (*a*) may be realized for
distorted structures of alkali metal metaniobates, where the niobium ion
occupies the site of the alkali ion, or a vacant oxygen octahedron.

According to the data obtained by studying the phase diagrams for $Li_2O-Nb_2O_5$ systems, the chemical composition of litium metaniobate can only be described by the formula $LiNbO_3$ in a particular case. Real single crystals noticeably deviate from the stoichiometric composition without any change in their structure, or the separation of a second phase.

Violations of the stoichiometry of LN may occur as the Li/Nb ratio is changed or as the degree of oxidation of Nb changes. Solid solutions deficient in oxygen are produced by annealing LN in vacuum, or in an inert or reducing atmosphere (see Jorgensen and Barlett 1969, Wood *et al* 1974), and also by passing an electric current through a heated crystal. The reduced specimens are characterized by an intense colour and increased electric conductivity, which has the electron component. In most cases, the oxidizing annealing removes the deficit in oxygen (Fedulov *et al* 1965).

A change in chemical composition appreciably influences the physical properties of metaniobates. In this respect, LN is the best studied compound.

Table 1.8 presents the properties, which depend differently on the Li/Nb ratio in this crystal. From the table one can see that the conventional physico-chemical techniques for studying solid solutions, such as determinations of the density and lattice parameters, require an improved accuracy, because of the weak dependence of these properties upon the composition. At the same time, a study of those properties of solid solutions which strongly depend on composition is complicated by the necessity of having some reference curves. Such a method was employed by Carruthers *et al* (1971) when they determined the homogeneity range for LN by measuring the Curie temperature of ceramic samples of a given composition.

Table 1.8. Properties of $LiNbO_3$ sensitive to the Li/Nb ratio (according to Räuber 1978).

Dependence	Measured quantity
Strong	Curie temperature Birefringence Phase-matching SHG temperature NMR line broadening
Weak	Lattice parameters Density Dielectric constant Refractive index Electro-optic coefficient Nonlinear optic coefficient

A crystal-chemical analysis provides a basis for developing particular models for metaniobates of alkali metals (MAM). Their agreement with the experimental data is the accepted criterion of their validity.

For lithium niobate, the 'fault stacking' model of Nassau and Lines (1970) is such a model. It is described by the following formulae:

$$Li^+_{1-5x} Nb^{5+}_{1+x} (V^-_{Li})_{4x} O^{2-}_3 \qquad (Li/Nb < 1)$$

and

$$Li^+_{1+z} Nb^{5+}_{1-z} O^{2-}_{3-2z} (V^{2+}_0)_{2z} \qquad (Li/Nb > 1).$$

The composition of a congruently melted crystal (48.7 mol.% Li_2O), calculated by these formulae from the X-ray data and the results of density measurements, is in good agreement with the phase diagram of Lerner *et al* (1968) (49 mol.% Li_2O) and with the data of Ivanova *et al* (1980) (48.6 mol.% Li_2O).

The name of the model stresses the fact that in substituting, for example, niobium for lithium (Li/Nb < 1) the former does not occupy the place of the latter, but takes the vacant oxygen octahedron position, thereby causing a disturbance in the alternation of the layers in the cation sublattice as well as changes in the Nb–Nb and Li–Li spacings (Nassau and Lines 1970). The lithium (or oxygen, in the case of excessive lithium) vacancies resulting from the incorporation of Nb are distributed over the crystal volume, and not localized in sites of the cation substitution.

In the oxygen octahedron structure of LN (figure 1.9), the lithium ions occur at the common face of two octahedra, and are slightly displaced towards the centre of one of them. If the Nassau–Lines model, which best agrees with the experimental data, is adopted to describe LN-based solid solutions, it may be thought that the LN structure allows the niobium ion to be incorporated in a free octahedron. Simultaneously, there appear lithium vacancies so that the electroneutrality will be preserved.

Since the interstitial niobium has an octahedral anionic surrounding, typical of transition metals, its position inside the octahedron will be stable. Figure 1.14 shows an example of the overlapping p orbital of oxygen and t_{2g} orbital of niobium in that position. Owing to the stability of the niobium ions in the structure, the homogeneity region for compositions enriched in niobium is rather wide (figure 1.2) and the congruent melting point is shifted by 1.5 mol.% towards the excess of Nb_2O_5.

Application of the above model to solid solutions containing an excess of lithium in the crystal with respect to the stoichiometric composition of $LiNbO_3$ leads to a distortion of the oxygen–niobium octahedron and to the formation of two oxygen vacancies per excess lithium atom. In reality, a significant excess of lithium oxide in the melt (according to Lerner *et al* (1968) as much as 8 mol.%) is needed for the stoichiometric composition solution to be formed.

Jarzebski (1974) suggested that among the possible defects of the LN structure there should be free electrons, singly and doubly charged oxygen vacancies, and also Nb^{4+} ions in various crystallographic positions. The quantitative assessment of the contribution of each type of defect to the real structure of LN can only be made after an experimental study of the processes of diffusion and conduction of electricity in LN. An investigation of the spectral characteristics of the material (see chapters 5 and 6) is also necessary.

The measurements of electrical conductivity made by Jorgensen and Barlett (1969) at various oxygen pressures suggest the existence of free electrons and singly charged oxygen vacancies, which are formed in the following way:

$$O_0^* \rightleftarrows V_0^{\cdot} + e + \tfrac{1}{2}O_2 \text{ (gas)}.$$

As the oxygen pressure is raised, the number of free electrons rapidly decreases and, at an oxygen pressure of 1 atm, the electrical conductivity of LN is purely ionic, the lithium ions being the principal carriers of charge.

The problems related to the origin of structural defects (observed in LN crystals in the course of formation of a single domain) and to the determination of their charge state and mobility are most important. These will be discussed in chapter 4.

1.4 Phase formation in LN crystals

The peculiarities in the phase diagram of LN and the observed deviations from stoichiometry in crystal composition may, under certain conditions of heat treatment, result in separation of a second phase, viz. lithium triniobate $LiNb_3O_8$. The relatively scarce data available suggest that the phase formation should occur both in the bulk of LN crystals (Svaasend *et al* 1973, 1974, Scott and Burns 1972, Holman *et al* 1978) and on their surface (Armenise *et al* 1983, De Sario *et al* 1985, Jetschke and Hehl 1985, Rakova *et al* 1986).

The main results on the formation of $LiNb_3O_8$ within the volume of LN crystals have been obtained by Svaansend *et al* (1974), who used the data of X-ray phase analysis and measurements of the optical transmission coefficient to study the properties of LN crystals subjected to long annealing in air. The crystals were heat-treated for 100 to 1000 h in the temperature range from 600 to 1000 °C. After the LN crystals of various compositions had been cooled down to room temperature, one could observe a significant decrease of optical transmission due to the appearance of milky white opalescent regions. The transparence of the crystals decreased with the longer annealing time and the smaller Li_2O content in the original specimens. For instance, the crystals of LN grown from a melt with a Li_2O content less than 48 mol.% showed opalescence as early as after a 10 h annealing at 800 °C, whereas in crystals grown from melts with the increased content of Li_2O the opalescence only appeared after more than 500 h of annealing at the same temperature. Those authors have suggested that the change in transparency observed during annealing is due to the appearance of a second phase, $LiNb_3O_8$, which borders on $LiNbO_3$ on the side of Nb-rich compositions. This suggestion has been fully borne out by an X-ray phase analysis of the annealed crystals.

After a second annealing at temperatures above 1000 °C and a rapid cooling down to room temperatures, the scattering centres disappeared from the specimens containing regions of the second phase and the transparency of LN crystals was restored. X-ray patterns for such specimens contained reflections due solely to LN. The temperature above which the inverse transformation took place depended on the composition of specimens, being about 910 °C for congruently melted crystals. The measurements of the inverse transformation temperature for LN crystals of various composition have enabled Svaasend *et al* (1974) to ascertain both the behaviour of the curve describing the phase equilibrium for the $LiNb_3O_8$–$LiNbO_3$ system and the width of the solid solution region, and also to construct a phase diagram for temperatures below 1000 °C, discussed in §1.1 (figure 1.2). An important result which follows from the study just described is that at room temperature LN crystals are metastable, while under a long heat treatment they become unstable and, over a definite range of temperature, the $LiNb_3O_8$ phase may appear in them. However, the data described above can only give integrated characteristics of the process of phase formation.

The data on the concentration and localization of the $LiNb_3O_8$ phase within the crystal volume are rather scanty. Recently, techniques of optical microscopy and light scattering have been used to show that, under annealing in the two-phase region, sub-microscopic particles of the phase of length $\leqslant 10^{-5}$ cm across originate heterogeneously at the block boundaries and dislocations; moreover, along with the inclusions of particles of platinum and other impurities, they become centres of light scattering in LN crystals.

The precipitation rate of the second phase also depends on the rate of temperature lowering, because this determines the time during which the crystal will be in the temperature range where $LiNb_3O_8$ normally precipitates. Scott and Burns (1972) have noted that, as the cooling rate increases to 3 to 5 °C min^{-1}, the tendency of LN towards cracking decreases, as compared with that of crystals cooled at a rate of less than $1 \,°C \, min^{-1}$. Not excluding the contribution from other mechanisms, those authors believe that precipitates of the second phase may serve as nuclei for the origin and development of cracks in LN crystals. According to Holman *et al* (1978), for the precipitation of $LiNb_3O_8$ within LN crystals to be prevented, it is necessary that the cooling rate should be $\geqslant 20 \,°C \, min^{-1}$.

The first reports that the phase composition of the surface of LN crystal changes as a result of formation of lithium triniobate were made by Armenise *et al* (1983a,b) and De Sario *et al* (1985), who studied the diffusion of titanium into LN crystals which takes place during the fabrication of optical waveguides. The formation of $LiNb_3O_8$ on the surface of LN plates coated with a titanium layer occurred at annealing in the temperature range from 550 to 900 °C in an oxygen atmosphere. Viewed through a scanning electron microscope, lithium triniobate appeared as shapeless spots over $100 \, \mu m$ in diameter occurring within the layer of TiO_2. An analysis of their atomic composition has revealed that the titanium content within such spots is smaller, and the niobium content is greater, than within areas free from the phase. As the annealing temperature is raised above 900 °C, the decomposition of $LiNb_3O_8$ and the disappearance of the spots from the surface of LN plates can be observed.

When studying the surface of LN substrates, Armenise *et al* (1983a,b) have found that $LiNb_3O_8$ can also be formed in the absence of the titanium layer: i.e. phase formation on the surface of the crystals is peculiar to lithium niobate itself when it is annealed in the above mentioned temperature range. From Laue patterns taken in a sliding geometry and from the spectra of the Rutherford backscattering of helium ions, it has been found that the phase $LiNb_3O_8$ has an orientation relative to the (0001) and $(01\bar{1}0)$ LN substrates. The experiments have also shown that the above phase is precipitated on the crystal surface when the specimens are annealed not only in an oxygen atmosphere, but also in the air and in a flow of N_2 or argon. In this case, the introduction of water vapour into an annealing atmosphere precludes the precipitation of $LiNb_3O_8$ and causes the decomposition of the other phase if it is already present on the surface of the specimens. The decay of $LiNb_3O_8$ during annealing in a humid atmosphere was explained by those authors as the results of formation of the hydroxyl group OH^- and the $(Li_{1-y} H_y)NbO_3$ molecules due to the diffusion of protons into the crystals.

Another fact pertaining to the problem of phase formation on the surface of LN crystals has been obtained from a study of damage to them induced by radiation. Using the method of the Rutherford backscattering, Jetschke and Hehl (1985) have found that, at $T = 279 \,°C$, the phase composition of the surface of LN specimens irradiated with N^+ and P^+ ions undergoes a change. Due to an increased concentration of niobium in the sub-surface layer, a peak will then appear in the scattering spectra. The relation between structural disturbances in the surface layer of LN substrates and the appearance of phase precipitates in it has been noted by Gan'shin *et al* (1985, 1986), who recorded the $LiNb_3O_8$ phase which appeared upon annealing ($T = 450\,°C$, 3 h) of proton-exchange waveguides made on the faces (0001), (01$\bar{1}$0), (2$\bar{1}\bar{1}$0) and (01$\bar{1}$4) of lithium niobate. From the chemical formulae of the compounds of the phase and the die it follows that, in order that the $LiNbO_3 \rightleftarrows LiNb_3O_8$ transformation take place, it is necessary that the sub-surface layer of a crystal should contain an excess of niobium. There is no single view on the cause or the mechanism of production of such an excess. Thus Armenise *et al* (1983) and De Sario *et al* (1985) believe that it is the processes of diffusion or evaporation of Li or Li_2O which accompany heat treatment of LN crystals (see Carruthers *et al* 1974) that are responsible for the precipitation of $LiNb_3O_8$ on the surface of these specimens. The dependence of the growth rate of the phase upon the annealing time, $v \sim \tau^{1/2}$, which has been established from the intensity of reflections on Laue patterns, rather testifies to the process for which diffusion is a limiting stage. Another argument in favour of the model proposed is the annealing of LN in the presence of water vapour which, as is known, suppresses the diffusion of lithium from crystals and, as has been shown by De Sario *et al* (1985), simultaneously prevents the appearance of the second phase. The assumption made by Armenise *et al* (1983) and De Sario *et al* (1985) about the cause of precipitation of $LiNb_3O_8$ raises doubt. This is due to the absence of reliable data on the diffusion and evaporation of lithium during annealing of LN in the temperature range below 900 °C, typical of the formation of the monoclinic phase. Kotelyanski *et al* (1977), who employed a mass spectrometric technique, have recorded at $T = 300\,°C$ a peak of Li^+ above the surface of the LN crystals. This result, however, has a purely qualitative character.

A careful investigation of the evaporation kinetics of Li_2O during the annealing of LN in the temperature range from 930 to 1125 °C, which has been made by optical interferometric techniques, has made it possible to determine the values for the diffusion activation energy Q_D and the evaporation activation energy Q_v, and to find the values for the diffusion coefficient D, the evaporation flux I_v and the Langmuir pressure P_L of the vapour of Li_2O at various annealing temperatures (see table 1.10). The values for the lithium diffusion coefficient are listed in table 1.9. The extrapolation of the data of Carruthers *et al* (1974) into the low-temperature region, which has been made on the assumption that the values of Q_D and Q_v are constant, gives an infinitely small value for the evaporation flux $I_v \sim 2 \times 10^{-25}\,\mathrm{g\,cm^{-2}\,s^{-1}}$ at $T = 300\,°C$. This suggests that a deviation of the composition of the surface layer due to evaporation of Li_2O at $T < 900\,°C$ is rather unlikely.

Table 1.9. Diffusion coefficients for Li ions in $LiNbO_3$ crystals, D ($\mathrm{cm^2\,s^{-1}}$) (according to Carruthers *et al* 1974).

T (°C)	$Z^{\parallel}$	$Z^{\perp}$
195	4.5×10^{-12}	—
930	1.5×10^{-10}	1.3×10^{-10}
1000	6.5×10^{-10}	7×10^{-10}
1050	1.9×10^{-9}	1.8×10^{-9}
1100	4.3×10^{-9}	4.2×10^{-9}
1125	7×10^{-9}	6.5×10^{-9}

Neither can one explain, within the model of lithium evaporation, the decay of the second phase which takes place at $T > 900\,°C$. Indeed, as the annealing temperature is raised ($T > 900\,°C$), the lithium evaporation rate should also increase (Carruthers *et al* 1974), i.e. a further intense growth of the phase, and not its decay, should be observed, which contradicts the experimental data.

One may suppose that the $LiNbO_3 \rightleftarrows LiNb_3O_8$ transition under study belongs, by its nature, to solid phase transformations of the type of ordering and disordering of solid solutions, the process of precipitation of the monoclinic phase $LiNb_3O_8$ being then similar to the decay of over-saturated solid solutions. The inverse phase transition $LiNb_3O_8 \rightarrow LiNbO_3$ corresponds to the dissolution of the excess of one of the components (niobium) in a solid state. Phase formation in over-saturated solid solutions under heat treatment is best manifested in those cases when the solubility decreases with decreasing temperature, which is true of lithium niobate. If the cooling rate is rather low, the excessive component of the alloy may precipitate as a second phase even in the course of cooling; conversely a rapid cooling (tempering) would fix the metastable phase of a solid solution at room temperature in conditions of thermodynamic stability of the other phase.

Because of the physico-chemical peculiarities discussed above, crystals of lithium niobate of congruent composition are metastable at room temperature, and they contain point defects associated with a deficit of lithium whose concentration is higher than the equilibrium one. According to the phase diagram of figure 1.2, at temperatures below 900 °C the two phases, $LiNbO_3$ and $LiNb_3O_8$, may co-exist. The decrease of the

Table 1.10. Physico-chemical constants of $LiNbO_3$ crystals (according to Kuz'minov 1975).

Characteristic	Experimental data
Density of single crystals ($g\,cm^{-3}$)	4.612
Mohs' hardness	5
Melting point (°C)	1260
Curie point (°C)	1210
Parameters of a unit cell:	
Rhombohedral	
a (Å)	5.4920
Angle	55°53′
Hexagonal	
a (Å)	$5.148\,29 \pm 0.000\,02$
c (Å)	$13.863\,10 \pm 0.000\,04$
Number of formula units in cells	
Rhombohedral	2
Hexagonal	6
Thermal expansion coefficient	
a axis	16.7 ± 10^{-6}
c axis	2.0 ± 10^{-6}
Dielectric constant	$\varepsilon_{11}^s = 44 \quad \varepsilon_{11}^t = 84$
	$\varepsilon_{33}^s = 29 \quad \varepsilon_{33}^t = 30$
	$\varepsilon_{11}^s = 43 \quad \varepsilon_{11}^t = 78$
	$\varepsilon_{33}^s = 49 \quad \varepsilon_{33}^t = 32$
Refractive indices ($\lambda = 0.623\ \mu m$)	$n_o = 2.286 \quad n_e = 2.220$
Loss-angle tangent ($v = 1$ kHz)	less than 0.02
Specific resistance (Ω cm)	
200 °C	over 10^{14}
400 °C	5×10^8
1200 °C	140
Water solubility ($mol\,l^{-1}$)	
25 °C	2.8 ± 10^{-4}
50 °C	4.3 ± 10^{-4}
100 °C	7.4 ± 10^{-4}
Dissolution heat ($kcal\,mol^{-1}$)	6.2
Diffusion activation energy Q_D ($kcal\,mol^{-1}$)	
$Q_D^{\perp}$	68.21 ± 0.48
$Q_D^{\parallel}$	68.17 ± 1.24
($\perp, \parallel$ to c axis)	
Evaporation activation energy Q_v ($kcal\,mol^{-1}$)	
$Q_v^{\perp}$	70.6
$Q_v^{\parallel}$	59.0
($\perp, \parallel$ to c axis)	
Evaporation coefficient, α	$\leqslant 10^{-4}$
$\alpha_{\perp}/\alpha_{\parallel}$	3
Thermoelectric coefficient of the melt α_1 ($mV\,K^{-1}$)	-0.4
Thermoelectric coefficient of the crystal α_s ($mV\,K^{-1}$)	0.76 ± 0.02
Coefficient of crystallization emf α_v ($mV\,s\,\mu m^{-1}$)	1.25 ± 0.2

width of the homogeneity region observed with lowering temperature leads to a situation in which the annealing of metastable, non-stoichiometric crystals of lithium niobate in the temperature range from 300 to 900 °C is accompanied by the precipitation of $LiNb_3O_8$. As a result, the whole system passes over into an energetically more advantageous state, and the concentration of the point defects in the crystals decreases. The temperature range $T > 900$ °C corresponds to a single-phase system of LN, and it is characterized by a wide homogeneity region (up to ~ 6 mol.% Li_2O), within which the energy considerations allow the existence of LN with large variations in composition. As a result, at annealing temperatures above 900 °C, the reverse phase transformation $LiNb_3O_8 \rightarrow LiNbO_3$ occurs, the monoclinic phase decays and the surface of the specimens becomes single-phase.

Proceeding from the above considerations, Bocharova (1986) has estimated the amount of the lithium triniobate phase which can be precipitated in non-stoichiometric crystals of lithium niobate due to the excess of niobium present in them. The calculation of quantitative relationships was based on an account of the balance in lithium and niobium, given a complete decay of lithium niobate of the congruent composition $Li_{0.945}Nb_{1.0}O_3$ into the stoichiometric composition LN $Li_{1.0}Nb_{1.0}O_3$, and the phase $LiNb_3O_8$: $Li_{0.945}Nb_{1.0}O_3 = xLi_{1.0}Nb_{1.0}O_3 + yLiNb_3O_8$. The oxygen content in both compounds was assumed to be the same.

The estimates thus made show that, with a complete decay of the congruent LN, the amount of the phase $LiNb_3O_8$ precipitated because of the deviation of LN from the stoichiometric composition is maximum, constituting 7.15% of the crystal volume. If the initial specimen decays to the monoclinic phase and the non-stoichiometric LN, say, of composition $Li_{0.985}Nb_{1.0}O_3$, the total amount of lithium triniobate decreases to 5% of the initial crystal volume.

Let us evaluate the thickness of the continuous layer of lithium triniobate which may be formed upon a unit surface of the crystal given a uniform distribution of the entire phase precipitated under the complete decay of LN on the specimen surface. Evidently, the depth of the crystal at which the components are redistributed under annealing to form the surface phase with a higher niobium content is determined by the range of diffusion of the more mobile ion, i.e. lithium. Let the thickness of the initial crystal layer be equal to the lithium diffusion length in $LiNbO_3$, i.e. $l \sim (D\tau)^{1/2}$ (Carruthers *et al* 1974) where D is the diffusion coefficient at a given temperature (table 1.4), τ being the annealing time. Then, following a 4 h annealing at $T = 750$ °C, on the crystal surface there may be precipitated, from a $LiNbO_3$ layer with $l \sim 1.2 \times 10^3$ nm, a layer of $LiNb_3O_8$ about 90 nm thick. According to ellipsometric measurements, the experimental value of the thickness of an 'island' layer of lithium triniobate under the same annealing conditions is 30 to 40 nm. A comparison of the ellipsometric data, those of electron microscopy, and the above estimates leads one to conclude that in congruent composition crystals the deviation from the stoichiometric composition is enough to ensure the phase transformation $LiNbO_3 \rightleftarrows LiNb_3O_8$ due to the excess of niobium present in the original specimens. This also confirms the view expressed by Esdaile (1985).

Thus the phase transformation $LiNbO_3 \rightleftarrows LiNb_3O_8$ takes place in accordance with the phase diagram, the segregation of the lithium triniobate phase being due to the system's tendency to restore the thermodynamic equilibrium. Because of the small amount of lithium evaporated from a crystal at annealing temperatures below 900 °C, the contribution from the evaporation to the deviation from the stoichiometry and its role in the phase transformation are negligible. The suppression of phase formation on the surface of LN when the crystals are annealed in vapours of Li_2O seems to be due to the adsorption of Li_2O and the levelling off of the Li/Nb ratio in the sub-surface layer as a result of the lithium diffusion into the crystal.

Thus, taking into account the character of change in the phase composition of the surface of LN and the peculiarities of the phase diagram of the material, one should distinguish between the thermal annealing of the crystals in the two-phase ($T = 300$ to 900 °C) and single-phase ($T > 900$ °C) temperature ranges. The two-phase annealing is accompanied by the phase transformation $LiNbO_3 \rightarrow LiNb_3O_8$, which is the most intense within the sub-surface layer of crystals with disturbed crystal structure. Under the single-phase annealing, decay of the monoclinic phase $LiNb_3O_8 \rightarrow LiNbO_3$ and a complete recrystallization of the disturbed layer both occur.

Some of the physical and chemical constants of lithium metaniobate are listed in table 1.10.

2 Methods of Obtaining Single Crystals of Lithium Niobate

2.1 Peculiarities of growth of LN single crystals

Crystals of LN were first obtained by Matthias and Remeika (1949), who had grown them from a melt solution. A number of studies have shown that perfect single crystals of LN cannot be obtained by either the melt solution technique (Wood 1951) or the Stockbarger method (Foster 1969, Hill *et al* 1968, Nassau *et al* 1965).

Ever since it became obvious that LN is a highly promising material for quantum electronics applications (Giordmaine and Miller 1965, Miller *et al* 1965), interest in producing perfect single-crystal LN and the need for developing industrial technologies to ensure the fabrication of high-quality crystals with reproducible characteristics have both greatly increased.

The first relatively large crystals of LN were grown in 1965 by Fedulov and co-workers in the USSR and by Ballman in the USA. Somewhat later, Nassau *et al* (1965, 1966) and Ballmen *et al* (1967) investigated the conditions of growing LN crystals from a melt, determining their domain structure and electrophysical properties.

At the present time, single-crystal LN is mainly grown in air by the Czochralski technique. Most of the known procedures are based on the use of high-frequency heating of platinum crucibles. The use of iridium, rhodium and platinum–rhodium crucibles is undesirable, because these elements are capable of incorporating themselves into the LN crystal lattice. Nassau *et al* (1965, 1966) have recommended the growth of LN in an atmosphere of pure oxygen with a small admixture of argon. This should make for better diffusion of oxygen into the crystal. The procedure also reduces the number of oxygen vacancies, which affect the colour of the crystal. It is recommended that the melt should first be heated for 20 to 50 minutes at a temperature of 30 to 50 °C above the melting point to ensure its outgassing (Ballman 1965).

As has repeatedly been mentioned (e.g. Huber *et al* 1970), in growing perfect LN crystals it is necessary to avoid any sudden changes (however small) in thermal conditions at the crystallization front. A failure to do so will result in the emergence of defects and multidomain structure, a change in birefringence and, eventually, optical and structural imperfection of the specimen thus grown. To maintain the same thermal conditions at the various parts of the crystallization front and to keep the melt homogeneous, it is recommended that the crystal should be rotated, as it is grown, at a rate of 10 to 100 rpm. However, optimal conditions at the crystallization front cannot be achieved by this device alone, for these conditions are significantly dependent on the design of a crystallization chamber (CC). Zakharova and Kuz'minov (1969) have proposed a specially designed chamber for growing LN crystals, in which an additional resistance furnace is mounted above the crucible to decrease the vertical and radial temperature gradients and also to pre-anneal the single crystals. The requirements imposed on the stability of a thermal regime in the working volume of the additional furnace are quite stringent, because any temperature fluctuations in it would result in fluctuations in the growth rate and the emergence of variable-composition phases at the interface (Kuz'minov 1975). The use of the resistance furnace as an upper screen allows a preliminary annealing of the boule grown according to a given programme. This method, however, has not found a wide application in industry because of the possible contamination of the melt with the furnace lining material.

The screen may be shaped as a platinum sphere, as in Niizeki *et al* (1967). A crystallization chamber used for automated growth of LN crystals, with a platinum reflector, was described by Zidnik (1975). In this chamber crystals were grown whose length reached 70 mm and diameter 25 mm.

The use of a 'passive' thermal insulation lining in crystallization chambers has made it possible to grow crystals with a diameter up to 50 mm and a length up to 70 mm (Satch *et al* 1976). However, such insulation prevents the crystals from growing to any greater length because of the considerable axial temperature gradients, which in turn lead to thermal stresses in the crystal and its eventual cracking (Hausonne 1974). Somewhat better results have been achieved by using 'active' (heated) screens. For example, Kluev *et al* (1968) described chambers in which a platinum cylinder, heated by two or three extra coils of the inductor of an HF generator, serves as the thermal screen. Such a system of screening is convenient because it does not require an additional heater. Accordingly, the method has found application in the industrial technology of growing LN crystals.

However, the heating of the screen by additional inductor coils gives rise to the inhomogeneity of the thermal field of the screen. This is increased by the presence of an observation window. The axial and radial thermal gradients caused by these factors lead to the appearance in LN crystals of elastic stresses and cracking, and also the non-uniformity of their optical and electrophysical properties.

The temperature field of the melt as a function of the relative position of the crucible in the heater has been investigated by Konakov *et al* (1971). Lowering the crucible into the heater results in the lowered stability of the growth regime. With the crucible placed 5.0 to 20.0 mm below the heater's edge, it is rather difficult to grow crystals of a constant diameter. Even very small changes in the rotation rate and the lifting of the seed will affect the boule diameter. Placing the crucible 5.0 to 15.0 mm above the heater's edge renders the regime more stable and less sensitive to changes in the growth and ambient parameters.

Those authors, however, failed to take into account the effect of lowering the level of the melt in the crucible in the course of crystal pulling upon the stability of the growth regime. As the crystal grows, the character of convective fluxes in the melt changes. At the onset, it is determined by free convection, while it depends only slightly on the rotation of the crystal. Towards the end of the process, the free convection is suppressed by the forced one. As a result, the shape of the surface of the crystallization front changes from convex to concave, which has been associated with the lowered level of the melt in the crucible.

The effect of such initial factors as the position of the crucible in the inductor, the release of heat into the surrounding medium, the depth of the melt in the crucible, the rates of rotation and pulling upon the axial temperature gradient has been discussed by Laudise and Parker (1970).

Crystals of LN were grown at rates from 3 to 8 mm h^{-1}. The lower and upper limits of the pulling rate are determined by the purity of the original material. Many investigators believe that an initial axial gradient of 50 to 100 °C^{-1} and a rotation rate of 15 to 40 rpm, along with the above pulling, may ensure a flat interface and prevent the cellular growth of the crystal.

To prevent the crystal from cracking, the following devices are recommended.

(i) The crystal should be separated from the melt by slightly increasing the temperature of the latter without changing the pulling rate.

(ii) The crystal should be cooled gradually for 7 to 10 h to a temperature of 600 °C and then inertially down to room temperature.

(iii) The crystal should be removed from the chamber only after it is completely cooled, because on the surface of the crystal being cooled there arise, due to the pyro-effect, rather large static charges which flow away through the grounded platinum crystal holder or are neutralized by the ions in the surrounding medium (Gabrielyan 1978).

Practically all systems for growing LN crystals necessarily have a rather large (3 to 6 cm²) observation window, which is needed for controlling visually the diameter of the crystal being grown. The presence of such a window is, of course, a disadvantage because it significantly enhances the instability of thermal conditions at the crystallization front due to the temperature fluctuations of the gaseous medium above the melt, which may be as high as about 40 °C (Cockayne 1977). The thermal instability may sometimes lead to the cracking of the crystal being grown, because LN has a rather narrow range of plasticity (Brice 1977). Even if the crystal remains whole, the presence of a large window gives rise to a nonlinear temperature distribution along the pulling axis of the crystal, thereby increasing the effect of some uncontrolled factors upon the process of crystal growth. The temperature fluctuations at the crystallization front also affect significantly the chemical composition of the grown crystal. They may lead to a gradual change from crystalliza-tion with a stable flat surface to a facet growth. For example, a decrease of the temperature gradient above the melt in the course of crystal pulling may result in concentrational overcooling, formation of cells at the crystallization front and, in the long run, to the imperfection of the synthesized specimen.

The emergence of a cellular structure of the crystal due to the melt overcooling may be explained by the presence of impurities in the melt. The cellular structure in the LN crystal has also been observed in the absence of controlled impurities. The concentrational overcooling which occurs when LN crystals are being pulled from a nominally pure melt may also arise when crystals are grown from a melt whose composition differs from the congruent one (48.6 mol.% Li_2O), because the exces-sive components of the compound being crystallized will then play the part of an impurity.

Another very common defect closely related to temperature fluctua-tions at the crystallization front is the presence of so-called growth streaks, which, no matter what the crystallographic direction of crystal growth, are always normal to that direction. Their presence is attributed mainly to a lamellar distribution of the impurities, which arises due to the instability of the growth regime. The appearance of growth streaks in the absence of any impurities in LN may be caused by a deviation of the chemical composition at the crystallization front as a result of microscopic variations in the growth rate of the crystal.

Crystals without growth streaks can be obtained at relatively small temperature gradients above the melt and an almost constant melt temperature (fluctuations of not more than 0.2 or 0.3 °C). Using the congruent melting composition to grow single crystals of LN allows many adverse factors to be eliminated.

For a congruent melting, the compositions of the melt and of the crystal are the same, as the impurity distribution coefficient K is equal to unity. The solidified crystal has the composition corresponding to that of the melt. The composition of the crystal and the melt are independent of the solidified melt fraction. The composition of the crystal along the growth direction will therefore be relatively unaffected by small temperature fluctuations and changes in growth rate. For example, temperature fluctuations occurring at the time of growth may bring the crystal back into the melt and then crystallize the melt on the growth surface, but as the crystal composition is identical to the melt composition this will not affect the former. Similarly to that, a change in growth rate will not lead to that in composition. However, changes in crystal composition do take place when the melt composition is changed because of a preferential evaporation of one of the components. It has been established experimentally that the evaporation of lithium during the growth of a single boule is insignificant. This factor, however, should be taken into account if several boules are to be grown from the same melt or if some melt is added to that contained in the crucible to fill it up after a run of growing, instead of replacing the melt completely.

Crystals grown from a congruent melt are of a higher optical quality, and show smaller variations of birefringence along the growth direction, than those melted stoichiometrically.

2.2 Synthesis of the charge and preparation of the melt

To obtain polycrystalline LN, the solid-phase synthesis technique, which is very common in producing ceramic materials, is currently being used. If certain requirements are met, the method of solid-phase synthesis is quite suitable to synthesize a charge for growing LN crystals.

In what follows, we describe a method for charge preparation that has been used by D'yakov (1982) to grow crystals of alkali metal metaniobates.

To synthesize LN, the following initial reagents are used: high-grade lithium carbonate and high-grade niobium pentoxide. The solid-phase synthesis reaction is as follows:

$$Li_2CO_3 + Nb_2O_5 \rightarrow 2LiNbO_3 + CO_2 \uparrow .$$

When selecting regimes for drying the starting substances and firing the charge, the data of the differential-thermal, gravimetric and X-ray phase analyses have been used. The minimal temperature at which, according to the data of an X-ray phase analysis, a single-phase product is obtained after a four-hour firing and the loss of weight equals that calculated from the reaction equation, is taken to be the optimal synthesis temperature.

Carbonates of alkali metals were placed in platinum cups, dried in a muffle furnace at 250 to 300 °C (niobium pentoxide at 500 °C) for 5 to 8 h and were then cooled down to room temperature in a desiccator with P_2O_5 as a moisture absorber. The dried starting substances, weighed to within ± 0.05%, were mixed in an agate ball grinder for 10 min. The mixture was then shaken for 20 to 30 min in a polyethylene container.

The mixture prepared for firing was put in a platinum cup and into the muffle furnance, which was heated to the firing temperature at a rate of 200 °C h^{-1}.

For $LiNbO_3$, the firing temperature was 1000 °C. The temperature was maintained to within ± 10 °C; the isothermal conditions existed for 5 h, whereupon the furnace was inertially cooled for 11 to 12 h. The synthesized charge, cooled to 150–100 °C, was placed in the desiccator containing some P_2O_5.

The completeness of the reaction was controlled from the loss of the weight upon firing and the absence of the gas release during the preparation of the melt. The weights of the components were calculated so that the synthesized charge would be sufficient to fill the entire crucible. This precluded any disturbance of the melt composition due to an insufficient homogeneity of the charge.

The preparation of the melt prior to the process of growing is an important stage in producing single crystals of high quality.

From measurements of the kinematic viscosity and the surface tension of the melt as a function of the time of the isothermal regime, it has been found that there are, in the melt, some partially retained, directed covalent bonds, which are characteristic of the solid phase (Bol'shakov *et al* 1969). Therefore, as the temperature is changed, the melt remains for some time in a metastable state. This time depends on the temperature of the melt. At low temperatures of the melt, the time taken to reach stable values of the parameters measured is longer than that needed at high temperatures.

These observations are consistent with the existing ideas about the structure of a melt (Ubbelohde 1965); they also correspond with the practical recommendations for the preparation of a melt to obtain LN single crystals of a high quality offered by D'yakov *et al* (1981) and Ivanova *et al* (1980). These recommendations reduce to the desirability of overheating the melt and keeping it at the elevated temperature for some time. Should the melt be not overheated prior to the process of growth, there may appear macro-inclusions in the crystals thus grown.

In D'yakov's experiment, the melt was heated 100 to 200 °C above the melting temperature before the growth of single crystals started and, after keeping it under isothermal conditions for 15 to 20 min, it was cooled to the crystallization temperature at a rate not higher than $20 \,°C\,h^{-1}$.

The congruent melt to grow LN single crystals of a high optical homogeneity was prepared from a preliminarily recrystallized melt.

2.3 Choice of the optimum conditions for growth

In what follows, we describe the conditions for growing high-quality crystals of LN, determined by Rubinina (1976).

Single crystals of lithium metaniobate were Czochralski grown in the air in induction furnaces, where there were parts of crystallization set-ups 'Donets-1' and 'Donets-3'. The general layout of the experiment is shown in figures 2.1(*a*) and (*b*). The growth chamber is schematically

(*a*)   (*b*)

Figure 2.1. General view of installations for Czochralski crystal growing: (*a*) Donets-1; (*b*) Donets-3.

depicted in figure 2.2. The platinum crucible is placed in a ceramic vessel. The space between the walls of the crucible and those of the ceramic vessel is filled by highly pure aluminium oxide. The first ceramic glass containing the crucible is within another, much larger alundum glass and is sealed from above by a ceramic washer, the inner diameter of which corresponds to that of the platinum screen. The screen is shut by a ceramic glass with a hole to insert the crystal holder with the seed. The latter is attached to the crystal holder made of a sapphire rod by a platinum wire. The sapphire rod is fixed in a special holder in the growth installation. The growing crystal can be watched over through an observation slit in the vessel that shuts the platinum screen. Crystals were grown in platinum vessels 45 mm in diameter, 50 mm high and 1.6 mm thick.

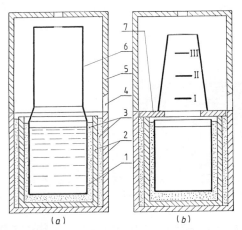

Figure 2.2. A schematic representation of the growth chamber (Rubinina 1976): (*a*) with a cylindrical screen; (*b*) with conical screen. I, II, III, are positions of the thermocouple in the crystal as the temperature gradients are measured. 1, crucible; 2, alumina glasses; 3, priming; 4, window; 5, alumina lid; 6, platinum screen; 7, ceramic washer.

The homogeneity of the composition and the optical perfection of Czochralski-grown crystals are determined by the following main factors.

(i) The homogeneity of the original melt and the constancy of the melt composition in the course of crystal growth.

(ii) The stability of the thermal conditions during the whole process of growth. A change in the temperature gradients near the growth surface leads to growth rate fluctuations, because the crystallization front tends to coincide with the isotherm corresponding to the melting point.

(iii) The shape of the growth isotherm which, in the optimum case, should be flat and parallel to the melt surface. It is the flat crystal–melt interface that ensures a low deficiency of the grown crystals and their high optical perfection.

(iv) The choice of optimal cooling conditions for the grown crystals with allowance for their thermoplastic properties and peculiarities of phase transition in the solid state (Laudise and Parker 1970).

In growing single crystals from non-congruent melts it is necessary to take into account the possibility of changing the crystal composition along the growth direction because of the gradual change in the liquid-phase composition, as the lithium oxide distribution coefficient will then be different from unity.

A change in Li_2O content in the melt during crystallization may be described, according to Midwinter (1968), as

$$c = c_0(1 - g)^{k-1} \qquad (2.1)$$

where c is the concentration (mol.%) of Li_2O in the melt at a time t, c_0 is the initial concentration (mol.%) of Li_2O, g is the fraction of the melt mass which has been crystallized at time t, and k is the distribution coefficient. Usually g is less than 0.1, and equation (2.1) may be written as

$$c - c_0 \simeq gc_0(1 - k). \qquad (2.2)$$

From equation (2.2) it follows that the melt composition, and therefore the composition of the crystal being grown, may be regarded as constant if $g \ll 1$.

Estimates made with the help of equation (2.2) have made it possible to determine the highest possible value of g at which the composition of a grown lithium metaniobate crystal may be considered to be practically constant. It has turned out that if the Li_2O content in the original melt deviates from the congruent one by 1 to 8 mol.% to either side, the crystal mass should be from 8 to 2% of the melt mass. The maximum change in the melt concentration, $\Delta c/c_0$, will not then exceed 0.2%.

To choose optimal regimes for crystal growth and cooling, the effect of the 'technological' parameters upon the temperature field near the crystallization front and upon the change in temperature gradients inside the crystal has been investigated. As the power of an HF generator is decreased, the temperature near the melt surface is found to depend on the extent to which the crucible was filled. Lowering the melt level in the crucible significantly affects the thermal conditions in the growth area (figure 2.3). Accordingly, crystals were grown so that the melt level would remain practically unchanged during a single run. This was achieved by growing small crystals from a crucible of a rather larger diameter. In all runs, crucibles were filled to the brim.

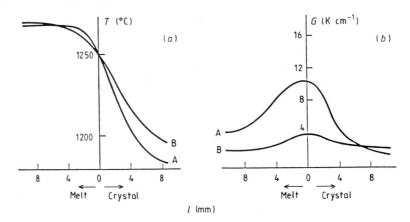

Figure 2.3. The effect of melt level on the temperature (*a*) and the radial temperature gradient (*b*) near the growth front (Rubinina 1976): curves A, melt level 5 mm below the crucible edge; curves B, crucible full.

Instantaneous temperature fluctuations occurring at the crystal–melt interface exert considerable influence on the homogeneity of the composition of the crystal being grown. This is especially noticeable when crystals are grown from non-congruent or doped melts. In this case, even small changes in temperature or growth rate may result in significant changes in the solid-phase composition. Then regions in the crystal appear which are parallel to the growth surface and which have different refractive indices and other characteristics. In lithium meta-niobate, these regions are often associated with the formation of oppositely charged domains, which can be established by etching.

Temperature fluctuations near the growth front may be caused by a poor thermal insulation of the chamber, an insufficient stabilization of the power of the heater, or the irregular convection of the melt. The power stabilization unit for the HF generator in the 'Donets-1' installation allows temperature fluctuations of not higher than $\pm 0.5\,°C$. However, real fluctuations observed near the growth interface are, as a rule, much higher, mainly because of an intense convection in the melt arising from high temperature gradients in the melt and above.

When growing crystals by the Czochralski technique temperature gradients can be lowered by using platinum screens mounted over the crucible with the melt. A vigorous stirring of the melt also makes for suppressing the irregular convection and lowering the amplitude of temperature fluctuations near the growth surface. Accordingly, the rotation rate of the seed used in growing LN crystals is usually 40 to 50 rpm.

Measurements have shown that, with good thermal insulation of the growth chamber, the use of a conical screen and a rapid (50 rpm) rotation of the crystal, temperature fluctuations near the growth front do not exceed $\pm 2\,°C$.

It has already been mentioned above that one of the conditions enabling one to grow a crystal of a homogeneous composition is to assure that the growth interface should be flat. In growing crystals by the Czochralski technique, the isotherms near the interface tend to curve. The shape of an isotherm is determined by the relationship between the heat flows from the melt, the walls of the crucible and the screen, and also the heat released in crystallization, and the heat that escapes through the crystal due to its thermal conduction and radiation.

Since the latent crystallization heat is proportional to the crystal mass, it is necessary to maintain a constant growth rate, which means that at a given pulling rate the crystal diameter should be kept constant. Because the curving of an isotherm increases as it approaches the crucible walls, the crystal diameter should be much less than that of the crucible.

Given a constant growth rate and a good stabilization of the heater's power, the shape of an isotherm depends on the conditions of heat release, which are determined by the radial temperature gradient. As has already been shown above, a decrease of the radial gradient can be achieved by vigorously stirring the melt and by using a conical platinum screen.

The optimum conditions for obtaining a flat growth front should be established experimentally, in each particular case, from the shape of the surface of separation of the crystal from the melt. The use of a conical screen enables one to obtain a flat isothermal growth surface in growing single crystals 8 to 12 mm in diameter from crucibles 45 to 50 mm in diameter at a rotation rate of 40 to 50 rpm.

In most works devoted to the study and growth of lithium meta-niobate, congruent composition single crystals are grown at a pulling rate of 5 to $8\,mm\,h^{-1}$ (Nassau *et al* 1965, Niizeki 1967, Parfitt and Robertson 1967). If crystals are grown from non-congruent melts, a concentrational overcooling arises near the crystal–melt interface, which gives rise to regions of inhomogeneous composition in the crystal (Kröger 1964). The degree of the concentrational overcooling increases as the difference in compositions of the melt and crystal increases. Concentrational overcooling can be reduced by increasing the temperature gradient in the melt near the interface. However, as has already been mentioned above, this gives rise to the more intense irregular convection in the melt, resulting in the formation of a streaky structure and the deterioration of the quality of the crystals thus grown.

With a low temperature gradient, it is possible to avoid an increase in concentrational overcooling due to a significant deviation of the melt composition from the congruent one by decreasing the growth rate. Figure 2.4 illustrates the minimum temperature gradient in the melt that assures the absence of concentrational overheating as a function of growth rate for lithium metaniobate melts of various compositions (Carruthers *et al* 1971). In the case of the crystallization conditions quoted above, the temperature gradient in the liquid phase near the growth front constitutes 15 to 20 K cm^{-1}. From figure 2.4 it is clear that, in these conditions, crystals of a homogeneous composition may be grown if the growth rate is decreased from $8\ \text{mm h}^{-1}$ to $0.5\ \text{mm h}^{-1}$, with a change in the Li_2O content in the melt from 50 mol.% to 58 mol.%.

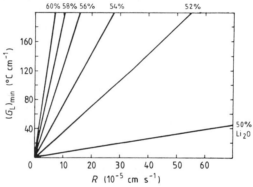

Figure 2.4. The minimum temperature gradient in the melt G_L needed to prevent constitutional overcooling for melts of various LN compositions and at various growth rates (Carruthers *et at* 1971).

In pulling single crystals from the melt the choice of the right cooling regime is of prime importance. With a rapid lowering of the temperature and large temperature gradient, there arise elastic stresses caused by a non-uniform plastic strain in the crystal. That the Curie point of lithium metaniobate is close to its melting point necessitates a careful control of the cooling rate immediately after the process of growing is at an end. Keeping the crystal for a long time at a temperature above T_c, at which an intense diffusion of ions take place, leads to the homogenization of the crystal composition and the dissolution of accumulations of impurity and intrinsic defects formed at the crystallization front as a result of the fluctuations of the temperature, growth rate and melt composition near the interface. It is these structural disturbances that often cause growth streaks and inner domains to appear in LN crystals.

In accordance with the currently adopted technology for producing LN single crystals, the cooling rate in the phase transition region should not exceed 30 to 35 K h^{-1}; the temperature is further lowered at a rate of 70 to 80 K h^{-1} down to room temperature. As has been mentioned above, the necessary condition for this is a small value of the temperature gradient in the crystal as it is cooled (Fedulov *et al* 1965, Fay and Dess 1968).

Rubinina (1976) adopted the following cooling procedure. After the crystal is separated from the melt, it is lifted into the screen at a rate of 3 to 6 mm h^{-1}, until it is completely inside the screen, which is in the field of the inductor. Lowering the power of the LN generator starts after the crystal has been lifted, and it continues according to a programme which envisages that the rate at which the temperature is lowered should be no less than 100 K h^{-1}.

The change in the temperature of the crystal being cooled is illustrated in figure 2.5, and the time dependences of the temperature gradients in the crystal with the lowering power of the generator are shown in figure 2.6. The maximum axial temperature gradient is observed at a distance of 5 to 12 mm from the growth front, constituting 60 to 70 K cm^{-1}. The radial gradient within this area does not exceed 20 to 22 K cm^{-1}.

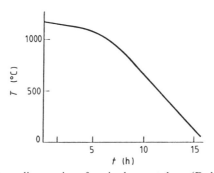

Figure 2.5. A cooling regime for single-crystal LN (Rubinina 1976).

After the crystal is drawn into the screen, a significant decrease of the axial gradient is observed (the lower edge of the screen is 10 cm above the melt surface). The difference in temperature along the growth axis is 35 to 40 K cm^{-1} in the lower part of the screen and 20 K cm^{-1} in the middle part, where the crystal is during cooling.

After the programme of lowering the power of the generator is switched on, the crystal temperature decreases at a rate of 25 to

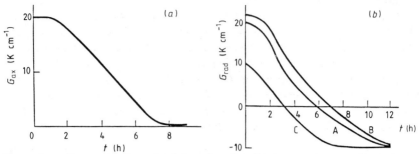

Figure 2.6. Change in the axial (*a*) and radial (*b*) temperature gradients in LN crystals being cooled; A, B, C correspond to positions I, II, III of the crystal in the screen shown in figure 2.2 (Rubinina 1976).

$30\,\mathrm{K\,h^{-1}}$ up to the moment of solidification of the melt, further cooling taking place at a rate of $\sim 90\,\mathrm{K\,h^{-1}}$ uniformly over the whole crystal length. Simultaneously, the temperature in the crystal volume is levelled off, which is suggested by the gradual decrease of the temperature gradients (figure 2.6).

The sign reversal of the radial gradient in the course of crystal cooling is evidently due to the fact that, in the initial stage of cooling, the crystal surface is heated by the radiation of the platinum screen, and the temperature at the crystal interior turns out to be lower. As the crystal is cooled, the amount of heat released through the side surfaces increases, the temperature inside the crystal becoming higher than that near the surface.

As mentioned above, the quality of lithium metaniobate crystals significantly depends on the cooling rate near the phase transition point and the duration of crystal annealing in the paraelectric phase. From this viewpoint the duration of cooling for crystals of various compositions can be estimated. As the content of Li_2O in a crystal increases from 47 to 50 mol.%, the Curie temperature increases from $1080\,^\circ\mathrm{C}$ to $1180\,^\circ\mathrm{C}$ (Carruthers *et al* 1971). Crystals with a low content of Li_2O pass through the phase boundary after they are completely drawn into the screen, when the power of the LN generator is lowered (figure 2.5). The cooling rate in the T_c range is 30 to $35\,\mathrm{K\,h^{-1}}$. The Curie isotherm for crystals whose composition is close to the stoichiometric one passes at a distance of 10 to 12 mm from the melt surface. The temperature gradient along the growth direction in this range is about $50\,\mathrm{K\,cm^{-1}}$. Since the pulling rate for crystals of this composition did not exceed $2\,\mathrm{mm\,h^{-1}}$, the cooling rate of the crystal that passed through the phase boundary was then about $20\,\mathrm{K\,h^{-1}}$. The duration of crystal annealing at temperatures above the ferroelectric phase transition was not less than 5 to 7 h.

The constancy of the chemical composition of the crystals obtained is confirmed by the uniformity of the refractive indices of the optical elements made out of these crystals, which has been found from the temperature dependence of the SHG of a YAG:Nd laser (see chapter 8).

It should also be mentioned that in these specimens there are practically no internal domains typical of those crystals which have not undergone a long annealing above T_c. To obtain homogeneous perfect crystals of lithium metaniobate, Rubinina recommends that the following basic conditions should be fulfilled.

(i) The crystal mass should not exceed 8% of the melt mass at the Li_2O concentration in the melt, deviating by not more than ± 1 mol.% from the congruent composition. At the maximum possible excess of Li_2O in the melt (~ 8 mol.%), the solid-phase mass should not exceed 2% of the melt mass.

(ii) The crystal diameter should be much shorter than the crucible diameter. From a crucible 45 to 50 mm in diameter, crystals of not more than 10 to 12 mm in thickness may be grown.

(iii) The pulling rate should decrease with increasing Li_2O content in the original melt from 4 to 6 mm h^{-1} for the congruent composition down to 0.5 to 1 mm h^{-1} at a Li_2O content higher than 56 mol.%.

(iv) The cooling regime for grown crystals should include a long (at least 5 to 7 h) annealing of the crystals at temperatures above the Curie point.

2.4 Growth conditions for LN crystals of a constant radius

2.4.1 Factors affecting the crystal radius

To control crystal structure and properties, one should know the main factors that influence the conditions of their growth. Timan and Burachas (1978) have established the relations between the various physical parameters that determine the conditions of crystal growth, in particular the size of the crucible, the size of the crystal and the thermal conditions.

Figure 2.7 schematically illustrates the Czochralski technique. The crystal is pulled at a rate V_b along the z axis, the melt lowering at a rate V_L. Therefore the growth rate is given by

$$V_s = V_b + V_L. \tag{2.3}$$

From the balance of the change in the mass of the crystal and in that of the melt it follows that

$$\rho_L R_L^2 V_L = V_s \rho_s R_s^2 \tag{2.4}$$

whence, making use of equation (2.3), one obtains

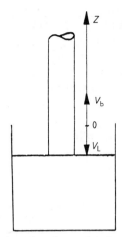

Figure 2.7. A scheme for crystal growing by the Czochralski technique.

$$V_s = V_b/[1 - \rho_s R_s^2/(\rho_L R_L^2)]. \tag{2.5}$$

Equation (2.5) is the relation between the growth rate, the pulling rate and the ratio of the squared radii of the crystal grown and the crucible at a given ratio of the densities of the crystal and the melt.

Crystal growth at a fixed rate is determined by thermal conditions, which ensure the heat balance at the crystallization front. When there is no convection in the melt the condition may be written as

$$V_s q \rho_s = \lambda_s(\partial T_s/\partial z) - \lambda_L(\partial T_L/\partial z). \tag{2.6}$$

If, however, heat is transferred through the melt by convection, the condition will have the following form:

$$V_s q \rho_s = \lambda(\partial T_s/\partial z) - \alpha(T_L - T_m) \tag{2.7}$$

where q is the latent crystallization heat, λ_s and λ_L are the heat conduction coefficient for the crystal and the melt, T_m is the melting point, $\alpha = \lambda_L/\delta$ is the heat transfer coefficient at the phase boundary, and δ is the thickness of the boundary layer in the melt near the crystallization front.

Eliminating the quantity V_s from the equation (2.6), or from equations (2.7) and (2.5), one can obtain the relationship between the thermal conditions of growth and the geometric size of the crucible and the crystal there. For example, one can derive a relation between the radius of the crystal being grown, that of the crucible, heat fluxes at the crystallization front and the pulling rate. Such relations have been obtained by Timan (1977) and Burachas in the form

$$R_s = R_L\left[\frac{\rho_L}{\rho_s}\left(1 - \frac{q\rho_s V_b}{\lambda_s(\partial T_s/\partial z) - \lambda_L(\partial T_L/\partial z)}\right)\right]^{1/2} \qquad (2.8)$$

in the absence of convection in the melt and

$$R_s = R_L\left[\frac{\rho_L}{\rho_s}\left(1 - \frac{q\rho_s V_b}{\lambda_s(\partial T_s/\partial z) - \alpha(T_L - T_m)}\right)\right]^{1/2} \qquad (2.9)$$

if convection is operative. Equation (2.9) corresponds to such thermal conditions in the melt that the temperature of the lower layers is higher than that of the upper, which favours free convection. One should note that a change in the character of convection in the melt may lead to a change in the value of α, as a result of which the heat flux from the melt to the crystal may change even if the melt temperature remains constant. Equation (2.9) is also applicable in the case of forced convection. The heat transfer coefficient will then depend on the rotation speed of the crystal relative to the melt, because α is directly proportional to $\sqrt{\omega}$ (ω being the rotation speed of the crystal: see Cochran 1934).

Occasionally, to ensure crystal growth, it may be necessary to overcool the melt, which is especially true of crystals with a relatively low heat conductivity. There may be two possibilities here. In the first case, overcooling at the crystallization front may occur. Then the temperature gradients in the crystal and the melt will be of the same sign and the directions of the heat flows in the solid and liquid phases will coincide. The other possibility is associated with the release of heat from the crystallization front via both the solid and liquid phases. The temperature gradients in these phases will then be of opposite sign.

An analysis of equations (2.8) and (2.9) reveals that the radius of the crystal being grown may be regulated by changing either the thermal conditions of growth or the pulling rate. To assure the constancy of the radius of the crystal being grown at a constant pulling rate, it is necessary that either the difference between the amounts of heat flowing in the same direction between the solid phases or the sum of the heat flows directed away from the crystallization front to the liquid and solid phases (but not in the same direction) should be kept constant. In reality, the direction of heat flow depends on a number of factors, such as changes in the thermal conditions of growth, the position of the heaters, the position of the crucible in the heater, the release of heat to the surrounding medium, the depth of the melt in the crucible and the rates of crystal pulling and rotation.

When crystals are Czochralski grown without adding to the melt the positions of the level of the melt and the crystal relative to the crucible walls change. This leads either to an increase or to a decrease in the amount of heat flowing away from the surface of the crystal being grown, which results, other factors being equal, in an increase or a decrease the temperature gradient in the crystal. A decrease or an increase in the amount of heat flowing away from the crystallization front is associated with changes in the temperature gradient above the melt, the heat conduction of the crystal, the ratio of the radii of the crystal and the crucible, the pulling rate etc, on which, as follows from equations (2.8) and (2.9), the radius of the growing crystal depends. A change in the radius of the growing crystal may be avoided by either decreasing or increasing the term responsible for the heat transfer from the liquid phase to the solid one, i.e. by adjusting the power of the heater. However, it is not always possible to ensure the growth of a crystal of a strictly constant radius by simply varying the power of the heater, because of the thermal inertia of the process.

The crystal radius may also be controlled by varying the pulling rate. This factor is almost free of inertia. Equations (2.8) and (2.9) thus reveal the main factors affecting the radius of a growing crystal and evaluate their respective roles.

2.4.2 *The effect of thermal conditions upon the growth of* LN *crystals*

As mentioned above, thermal conditions of growth significantly influence the quality of grown crystals, because these conditions may give rise to thermally induced elastic stresses, which lead to structural defects and cracking. Changes in growth rate also affect the homogeneity of a crystal and its properties. Thermal conditions and changes in them during growth are, in turn, influenced by the heat insulation and the relative position of the units of the crystallization chamber. This influence should be taken care of in designing a chamber so that optimal conditions may be chosen for automated growth of LN crystals.

It is known that LN crystals have a relatively low strength, and tend to crack. A non-uniform temperature distribution in the area of the growth and cooling of the crystal gives rise to elastic stresses, whose value may exceed that of the strength of the material.

By setting up heat screens above the crucible (these may be either active, with an autonomous or a common heating by the same source, as is used for heating the crucible, or passive) one can obtain such temperature distribution in the above mentioned areas which will not lead to any significant elastic stresses. As has been pointed out by Kuz'minov (1975), the use of a platinum cylinder, heated by two or three extra coils of the inductor of an HF generator, as a heat screen may technologically be the best device for growing LN crystals in induction heating installations.

As mentioned in §2.1, the local heating of the screen, as well as the use of a large (3 to 5 cm^2) observation window to control the crystal diameter, lead to a non-uniform temperature distribution over the melt in the screened volume of the crystallization chamber (figure 2.8 (*a*)), which induces considerable elastic stresses and often results in the cracking of the boules.

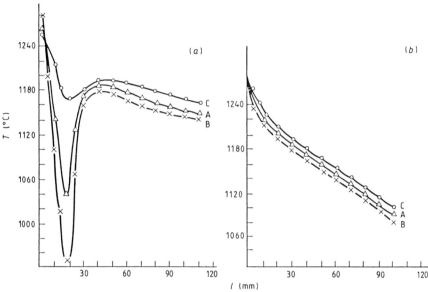

Figure 2.8. Temperature distribution over the melt surface and in the growth chamber: (*a*) non-monotonic; (*b*) monotonic. Curve A, at the centre of the crucible; curve B, 25 mm from the centre of the crucible (on the side of the window); curve C, 25 mm from the centre of the crucible (to the right of the window) (Burachas *et al* 1978).

Burachas *et al* (1978) succeeded in decreasing the thermally induced elastic stresses to values at which LN crystals remain intact by creating a uniform, almost linear, temperature distribution above the melt surface. The decisive factor for assuring the wholeness of the crystal is the constancy of the axial temperature gradient along the pulling direction, the absolute value of the axial gradient being no less important.

A steady temperature variation along the pulling axis may be achieved by choosing the relative position of the extra coils of the inductor, their number, the ratio of the geometric size of the platinum screen to that of the crucible, the size of the observation window, and also by using passive screens. It has experimentally been shown that the platinum screen should be a fraction of 0.75 to 1 in diameter and 1.2 to 2 in height of the crucible diameter (d_c), and should be placed at a distance of 0.15 to 0.35 d_c from the edge of the crucible. An extra coil should be removed from its neighbour by 0.35 to 0.6 d_c and the diameter of the window should not exceed 0.12 d_c.

Figure 2.9 shows the design of a crystallization chamber which assures a steady temperature variation above the melt along the pulling axis (see figure 2.8).

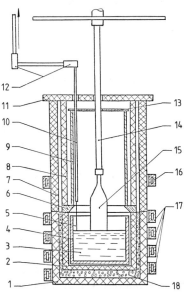

Figure 2.9. Design of the growth chamber. 1, 2, alumina glasses; 3, melt; 4, crucible; 5, probe; 6, alumina ring; 7, 8, alumina cylinders; 9, screen; 10, alumina padding; 11, lid; 12, cantilever; 13, platinum lid; 14, crystal support; 15, crystal; 16, extra wind of the inductor; 17, inductor; 18, refractory priming (Burachas *et al* 1978).

To study the effect of the relative position of the individual elements of the chamber upon the character of change in thermal conditions in the growth area, Burachas simultaneously measured the temperature of the surface of the melt T_{ms}, the temperature above the melt T_{am} and the variation in the power of the HF generator ΔP. The crystal diameter was automatically kept constant by means of an electric contact gauge of the melt level (Timan and Burachas 1981). One thermoelectric couple was in contact with the melt surface; the other was at a distance of 10 mm above the melt surface and 8 to 9 mm from the crucible wall. Using a melt depth gauge, the thermoelectric couples were synchronously shifted downwards as the depth of the melt in the crucible changed, while their position with respect to the melt surface remained unchanged during growth (position 10, figure 2.9). In all experimental runs, the surface of the melt was initially 2 or 3 mm lower than the edge of the crucible.

LN crystals 24 mm in diameter were pulled from a platinum crucible

60 mm in diameter and 60 mm high at a rate of 6 mm h^{-1}. The rotation rate was 30 rpm. In run 1, the crystal was grown under the following starting conditions. The crucible was placed at the level of the upper coil of the inductor. The extra coil was at a distance of 26 mm from the upper coil. An observation window 7 mm in diameter to ensure visual control of the seed was 25 mm away from the edge of the crucible. After the required diameter was reached, the crystal grew without changing its diameter for 4.5 h. A small change in diameter took place when the upper part of the crystal was shifted, in the area opposite to the observation window. During the tenth hour of growth the system that automatically controlled the crystal diameter went over into an unstable regime, as a result of which oscillations of crystal diameter arose. The level of the melt in this run dropped by 15 mm. The crystallization front remained flat throughout.

The conditions of run 2 differed from those of run 1 in that the extra coil of the inductor had been displaced towards the other coils (the screen was not then heated). The crucible was mounted at the level of the upper coil of the inductor. In that experiment, the crystal diameter control system automatically maintained the diameter within 0.7%. Small deviations of the diameter occurred as the observation window was overlapped by the crystal. The melt level was again lowered by 15 mm, the crystallization front having been slightly convex.

In contrast to run 2, in run 3 the crucible was mounted in the inductor 10 mm lower than the top coil. After 10 h of growth, the system of automated diameter control went over to an unstable regime, giving rise to substantial periodic changes in crystal diameter relative to its average value. The crystallization front in this run was flat, the melt level lowering by 14 mm.

In run 4, the crucible was mounted 7 mm above the upper coil of the inductor. The screen was not heated. There were not any significant warps on the crystal surface, but the diameter slightly tapered as the crystal was pulled. At the very end of the process, there appeared some instability in the operation of the control system. A peculiarity of this run was that after 5 h of growth the melt temperature suddenly began to drop. The crystallization front was slightly concave. The melt level fell by 14 mm.

The results of the measurements made in these four runs are presented in figure 2.10 (*a*), (*b*), (*c*), and (*d*), showing respectively the temporal dependences of the melt surface temperature, the temperature above the melt (relative to its initial value in run 4), the difference in temperature at the melt surface and 10 mm above the surface, and finally the power of the HF generator.

A comparison of the curves which correspond to the various initial conditions of growth enables one to elucidate some general regularities governing changes in thermal conditions caused by pulling the crystals from the melt.

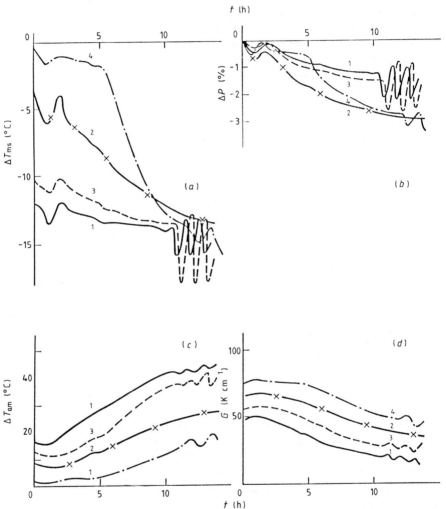

Figure 2.10. Variation with time of: (*a*) melt surface temperature; (*b*) the power of the HF generator relative to the initial value (see text) (the data of Burachas *et al* 1978); (*c*) the temperature 1 cm away from the melt surface (relative to the initial value in run 4); (*d*) temperature difference between the melt surface and 1 cm away from it.

From the curves shown in figures 2.10 (*a*) and (*b*) it can be seen that as a constant-diameter crystal grows the temperature in and above the melt changes differently. However, the relative change in temperature gradient above the melt invariably remains about the same (figure 2.10 (*d*)), which is indicative of an identical heat outflow from the crystal, irrespective of the initial gradient. From figure 2.10 (*a*) it can also be seen that in all cases the melt temperature is lowered down to almost the same value during crystal growth. This seems to be due to the fact that crystallization takes place in a melt whose temperature is close to the melting point of the substance being crystallized. The lowering of the melt temperature is caused by the decreased heat release from the surface of the growing crystal.

As follows from equation (2.9), one observes about the same change in the amount of heat flowing away from the crystal at different changes in melt temperature ($\alpha = \lambda_L/\delta$ is constant). Accordingly, maintaining the crystal radius constant should lead to a change in the boundary layer thickness δ. The latter should increase to compensate for the decreased heat flow from the crystal. Furthermore it increases as the gradient above the melt and the gradient in the melt in front of the crystallization surface decrease. The observed variations in thermal conditions lead to a change in character of convection in the melt as the crystal grows. A similar phenomenon was observed by other investigators who used the Czochralski technique.

As can be seen from figure 2.10 (*a*) and (*b*), the magnitude and character of the change in the power of the HF generator needed to keep the crystal diameter constant depend on the specific thermal conditions in the growth area and the cooling of the crystal. They are determined by changes in the melt temperature. Thus, from the character and magnitude of the changes in the HF generator's power required for maintaining a constant crystal diameter, one may obtain a qualitative picture of the changes in the thermal conditions in the growth area and crystal cooling.

As the melt temperature decreases due to the decreasing heat flow from the growing crystal, which is caused by the screening effect exerted by the crucible walls, the stability of the control unit is apt to break down at some temperature characteristic of the particular conditions. This instability arises because of the decreased temperature gradient above the melt.

Attention is also drawn to the fact that, at relatively low initial gradients near the melt surface, the melt surface temperature and the power of the HF generator change comparatively little as the crystal is grown. Crystals may be grown without a significant change in thermal conditions also from an 'overheated' melt (conditions of run 4, see above), by using a crucible in which the level of the melt changes only slightly relative to the initial one. As the level is lowered by 5 mm in the course of growth, the melt temperature changes relatively little, but with a further lowering of the melt level it decreases abruptly.

The tapering of the crystal diameter observed in that experimental run is explained by a change in the melt density resulting from a substantial decrease of the melt temperature (by 15 °C).

The effect of the position of the crucible in the inductor, the heat insulation of the crucible bottom, and some other factors concerning the magnitude and character of the change in the power of the HF generator maintaining a constant crystal radius, may be elucidated from an analysis of the registered 'power-grams' taken during an automated growth of LN crystals.

If a crucible with an uninsulated bottom is used, the growth of a crystal of a constant radius occurs, from some moment, at a constant power of the HF generator (figure 2.11, curve 1), other factors being equal. In this experiment, small changes in power occurred as the foremost cylindrical part of the crystal passed, as the crystal was pulled, through the region opposite to the observation window.

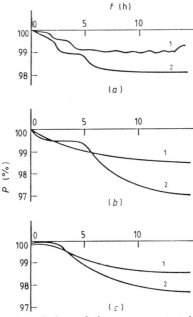

Figure 2.11. Power variation of the HF generator in the course of growing a constant-diameter LN crystal (see text).

Figure 2.11 (*b*) illustrates the curves showing how the power of the HF generator is changed to ensure a constant radius when the crucible in the crystallization chamber is placed 10 mm above the upper coil of the inductor, the observation window (5 by 10 mm) being just above the edge of the crucible. Curve 1 corresponds to conditions in which the initial level of the melt in the crucible was 11 mm above the edge, and curve 2 corresponds to when it was 2 mm above the edge.

The curves of power variations in time shown in figure 2.11 (*c*) correspond to conditions which differ from those for run 1 only, in that the observation window is placed immediately above the edge of the crucible. When the first crystal was grown in that run, the initial level of the melt was 8 mm below the edge (curve 1). For the second crystal, the distance was 1 mm (curve 2). These experimental results suggest that the initial level of the melt should considerably affect the character of change in thermal conditions during the growth. The lower the initial melt level, the smaller the necessary reduction of the HF power (other factors being the same) to keep the crystal diameter constant.

One should note that even a relatively small (e.g. 7 mm in diameter) observation window with quartz glass destabilizes the automatic control of crystal growth. As the cylindrical part of the crystal is lifted into the region opposite to the window and then taken out of this region, the amount of heat flowing away from the crystallization front drastically decreases, causing the crystal to melt. To keep the diameter constant, the automatic control unit (ACU) at these moments reduces the power. And the greater the area of the window the greater the power reduction (figure 2.11, curve 2). When the window is just above the edge of the crucible, and its area is as small as possible, its negative effect is less pronounced (figure 2.11 (*b*) and (*c*)). The window is then completely shut out by the cylindrical part of the crystal, and it does not exert a significant influence on the change in the heat flow away from the growing crystal.

To sum up we can say that, in pulling a Czochralski-grown crystal, the thermal conditions in the growth region may change differently, depending on the initial conditions. The range of conditions that are optimal for crystal growth is limited, both on the side of high initial temperatures of the melt and on that of low initial temperatures. In the case of high initial temperatures, the melt temperature may be changed greatly, which can cause undesirable alterations in the composition and structure of the crystal. Moreover, in this case temperature fluctuations ($\simeq 8$ to 13 °C) caused by irregular convection will arise in the melt. These fluctuations will in turn lead to a change in the distribution coefficient for the constituent components of the crystal. In the case of low initial temperatures, oscillations of the position of the crystallization front (growth rate) will appear, leading to the introduction of impurities and the emergence of growth substructures due to the concentrational overcooling of the melt.

To control crystal growth automatically, it is necessary that in the course of pulling, the thermal conditions in the growth area change smoothly, and not discontinuously. Otherwise the ACU, which compensates for the changing thermal conditions in the growth region by varying the generator's power alone, may cause fluctuations of the melt temperature to appear.

A change in the screening effect of the walls of the crucible, which occurs as the crystal grows and heat is released from it as its end departs from the melt mirror, is relatively smooth. At the chosen initial temperature distribution above the melt surface, which is close to linear, this change can readily be compensated for by a smooth change in the heater's power. By judiciously choosing the size and position of the observation window one can also reduce the undesirable power fluctuations.

2.4.3 The effect of power fluctuations on the level of the melt and the radius of the crystal

The radius of a lithium niobate crystal is usually regulated by varying the power of the heater at which a change in melt temperature is brought about. The change in the melt temperature leads not only to a deviation of the crystal radius from the norm, but also to a change in the melt volume due to its thermal expansion.

In the beginning of this sub-section we will give a theoretical evaluation of the effect of the thermal expansion of the melt upon the parameters used to control the process of crystal growth. It will be shown that the effect may be quite significant.

To control the process of growth, it is important to establish the relationship between the change in the radius of the growing crystal and the change in the power of the heater, both in the static regime, when the crystallization front is immobile and the radius does not change, and in the dynamic one, when the radius undergoes a change. An increase in crystal radius due to a change in power may be found from a time dependence of the melt depth in the crucible as the power changes discontinuously.

The measurements were made in the crystallization chamber shown in figure 2.9. The melt depth and the melt temperature were measured simultaneously.

Figure 2.12 illustrates the time dependences of the change in temperature and depth of the melt upon a discontinuous change in power when there is no contact between the crystal and the melt. As can be seen from the figure, the change in melt depth practically follows the change in melt temperature when the power is either raised or lowered. The change in melt depth is then caused by that in melt density. An increase in the time of setting-in a new melt temperature as a result of a lowered power is associated with the lining effect. When the power is lowered discontinuously, the time constant τ_2 for the transient process involving the change in the melt depth is larger than that in the case when the power is raised τ_1 (figure 2.12).

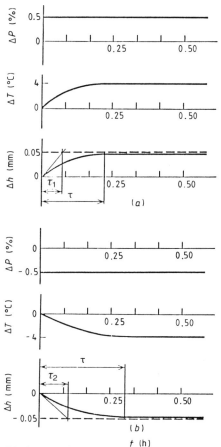

Figure 2.12. Variation of temperature and melt level as the heater power increases (*a*) or decreases (*b*) discontinuously (Timan and Burachas 1977).

Figure 2.13 presents the experimental dependences of the change in the melt depth on the change in the power when the crystal is in contact with the melt. Curves shown in figure 2.13(*a*) have been obtained under the stationary conditions when the crystal is not pulled and its cross section remains unchanged. The power was raised discontinuously by 0.5% and in 20 min was returned to its original value. Curves of figure 2.13(*b*) correspond to the condition that the crystal expands but is not pulled. Here again the power was raised abruptly by 0.5%, and then it was kept constant for 30 min. Such a change in power resulted in a change in melt temperature by 4 °C.

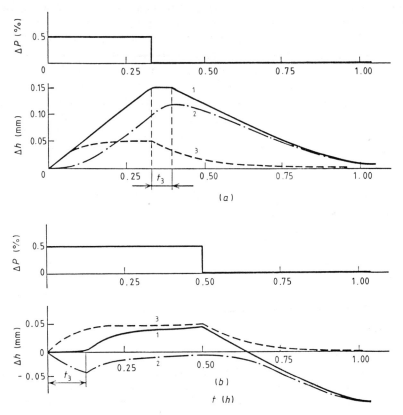

Figure 2.13. Variation in melt level due to a discontinuous increase in the heater's power. The crystal is in contact with the melt (1), due to a phase transformation (2) and a change in melt density (3) (Timan and Burachas, 1977): (*a*) the crystal is unchanged; (*b*) the crystal is expanding.

By graphically subtracting appropriate quantities it was possible to separate the effect of change in melt depth due to the phase transition (curve 2).

From figure 2.13 it follows that the change in the melt depth due to crystallization accompanies that in power with a lag of about 8 min.

If the crystal radius increases, a decrease in power also occurs with a lag. For a while, the radius continues to increase, which is evidenced by the resulting curve 2 in figure 2.13(*b*). The delay time t_3, i.e. the time which passes from the moment of the power 'jump' to that of the beginning of the change in crystal radius in the desired direction, depends on the expansion rate of the crystal in the initial period.

Thus there is a delay time during which the process of change in crystal radius, already in operation, cannot be stopped by changing the power of the heater. The radius of a growing crystal can be regulated more easily in the stationary conditions of growth, because the shorter the delay time the greater the accuracy of regulation. To control the growth of a crystal of a constant diameter, one should be able to forecast a change in diameter over the time t_3. Only after this period can the adjustment of the power made affect the quantity being regulated (i.e. the crystal diameter). Thus to ensure the constancy of the crystal radius, one should make use of the available information about the melt temperature, and this will enable the elimination of random disturbances which might lead to changes in crystal radius during the time t_3.

The experimental results suggest that a melt-depth gauge used to control the radius of a growing crystal 'feels' the change in melt temperature long before (~ 8 min) the change in the radius itself actually takes place. Therefore, using this information, one may prevent the change in radius by adjusting the power. In so doing, one should try to eliminate all possible factors affecting an instantaneous change in radius, such as a sudden change in the heat outflow from the crystallization front, fluctuations of the rate of pulling and rotation and also of the melt temperature.

2.4.4 *The emergence of instability in the control unit*
One of the most important requirements that a unit intended for the automatic control of crystal growth should meet is the stability of its operation during the whole process of crystal pulling. However, as shown above, a unit properly adjusted at the beginning of the process may, towards the close of the process, go over to an oscillating or unstable regime of operation. To rule out such an undesirable possibility and to ensure a stable control of the process during the entire period of growth, one usually varies the parameters of the regulator (Green *et al* 1977).

The experimental results presented above testify that an instability in the automatic control unit mainly arises from a significant change in the thermal conditions in the growth area—due to an increased screening effect exerted by the walls of the crucible. To maintain a constant radius, one should then lower the melt temperature and therefore increase the temperature gradient in front of the crystallization surface in the melt.

To establish the nature of such an instability, it is necessary to determine the regularities governing the variations of certain parameters due to change in thermal conditions in the growth area. To this end, LN crystals 24 mm in diameter were experimentally grown in a 60 mm diameter crucible, the pulling rate being 6 mm h^{-1} and the rotation rate

62   *Physics and Chemistry of Crystalline Lithium Niobate*

30 rpm. The parameters of the melt-depth regulator were chosen such that the deviation of the crystal radius in the initial period from the assigned value would be within ~ 0.3%, and would not be adjusted during the whole period of growth.

The experimental data thus obtained have made it possible to perform a qualitative analysis of the time variation of some individual parameters of growth. Figure 2.14 presents the averaged experimental curves (*a*), (*b*), (*c*), (*d*) showing respectively changes in the differential signal, crystal diameter, melt temperature and power of the HF generator; the curves (*e*), (*f*) and (*g*) constructed from the above data are also given in the figure.

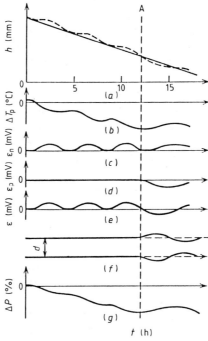

Figure 2.14. Variation of control parameters during a stationary growth. (*a*) Melt level; (*b*) melt temperature; (*c*) signal $\varepsilon_n(t)$ due to a melt density variation; (*d*) signal $\varepsilon_p(t)$ due to crystal diameter variation; (*e*) total signal $\varepsilon(t)$; (*f*) crystal diameter; (*g*) heater power (Burachas *et al* 1978).

Figure 2.14(*a*) shows both the programmed and the actual (broken curve) variations of the parameter under control in the course of growth which correspond to the experimental change in differential signal (curve (*a*)). The broken curve shows the crystal diameter assigned by

the programme; the full curve represents the actual diameter (curve
(*e*)). The moment at which diameter oscillation began is marked by line
A. Up to that moment, the differential signal had been due entirely to
the change in melt density caused by temperature variations.

Curves (*c*) and (*d*) illustrate variations in the differential signal arising
in the course of control. These variations are due to the thermal
expansion of the melt and the change in crystal diameter respectively.
The sum of the ordinates of these curves is identical to the experimental
curve for the differential signal.

The rate at which the melt temperature decreases gradually slows
down and, from the moment shown by A, it remains practically
unchanged. The power of the HF generator also reaches a constant level.
At the same time, the temperature in the medium that surrounds the
crystal continues to rise (see figure 2.15), as a result of which the heat
outflow from the crystal decreases. If the thermal balance at the
crystallization front is retained, this should lead, given a constant radius
of the crystal and a constant rate of its growth, to a decreased heat flow
from the melt to the crystal, i.e. to a decreased temperature gradient at
the crystallization front in the melt, which is described as $(T_L - T_m)/\delta$.
From these experimental considerations it follows that T_L remains
practically constant. The temperature gradient may therefore decrease
solely at the expense of an increase in the boundary layer thickness δ in
front of the crystallization surface in the melt, which depends on the

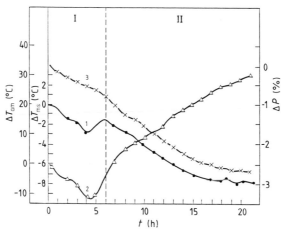

Figure 2.15. Time dependence (relative to the initial value) of the
melt surface temperature (1), temperature 10 mm above the melt
surface (2) and heater power (3). (The initial axial temperature
gradient above the melt is 60 °C cm^{-1}; at the end of the growth it is
22 °C cm^{-1}). I, growth of the upper cone of the crystal; II, growth of
the cylindrical part of the crystal.

intensity of convection in the melt. Therefore, in the stage of crystal growth being considered, there occurs a significant change in the character of convection in the melt. As the melt depth decreases (due to crystal pulling) the bottom of the crucible, through which the heat flows out, begins to play an increasingly important role in the heat exchange with the surrounding medium, and the free-convection conditions are violated. That the character of the convection in the melt changes for these reasons is confirmed by the established effect of the relative positions of the heater and the crucible with the melt upon the temperature distribution in the melt (Konakov *et al* 1971).

The lowered intensity of the convection in the melt leads to a situation in which an adjustment of the power affects the crystal radius and the growth rate through changing the thickness of the boundary liquid layer at the crystallization front. Such a change in the object of regulation leads to a lack of correspondence with the parameters of the regulator which ensure a constant diameter. The regulating unit does not receive preliminary information about changes in the object being regulated and therefore it is unable to compensate for these changes in advance.

The capillary forces arising when a crystal is Czochralski grown also make for the instability of the crystal radius and the position of the crystallization front (Hurle 1977). In the case of relatively high temperature gradients in the melt at the crystallization front, this instability is suppressed (Tatarchenko and Brener 1976), which is the reason why crystals can be grown by the Czochralski method. At small gradients at the crystallization front in the melt there arise such conditions that even insignificant disturbances may give rise to the instability of the growth. This manifests itself in that once a small increase in crystal radius and a change in the position of the crystallization front occur, they may only increase further. If the level of the melt in the crucible is low and the average melt temperature is close to the melting point, the regulating changes in melt temperature may go beyond the isotherm of the melting temperature, thereby giving rise to sharp oscillations of growth rate.

In the regulating unit, this will in turn lead to an enhancement of the differential signal. Compensating the signal $\varepsilon(t)$ by adjusting the power of the heater ΔP introduces a perturbation in the object of regulation, which exceeds the original perturbation in magnitude and has the opposite sign.

As a result, the conditions of growth periodically change with time, giving rise to warps on the surface of the growing crystal. To eliminate this, it is necessary either to decrease the value of the heat transfer coefficient K or to change the regulating law of the regulator or to compensate for a part of the differential signal $\varepsilon(t)$, as the temperature gradient in the melt decreases. A method for compensating for the part of $\varepsilon(t)$ was first proposed by Burachas *et al* (1978). A photograph of single crystals of lithium metaniobate, grown with the use of an automatic control of crystal diameter, is shown in figure 2.16.

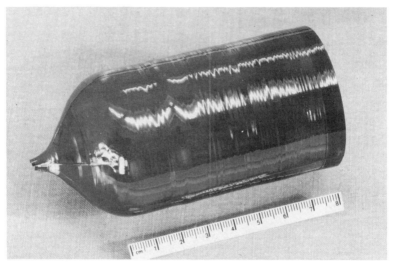

Figure 2.16. Photograph of crystals grown on the Donets-1 installation furnished with a unit for automatic control of crystal diameter.

2.5 High-temperature annealing and formation of single domains in crystals

To diminish the optical inhomogeneity of crystals, use is made of the phenomenon of directed diffusion of ions in an electric field. The electric field is applied in such a way that the positive electrode is situated on the side of an increased concentration of lithium oxide in a crystal, which is found from the deviation of the synchronism angle (see chapter 9).

The high-temperature diffusive annealing of single crystals is carried out in a furnace with a temperature gradient of less than $0.3\,\mathrm{K\,cm^{-1}}$. A crystal of LN is placed in a corundum ceramic cell. Figure 2.17 shows schematically how the electric field is applied and the electrolysis is controlled.

To diminish the effect of an increased electric current density at points of contact with the platinum electrode, polished discs made of the same single crystal were inserted between the crystal and the electrode. The cell containing the crystal is placed in the furnace which is heated, at a rate of $300\,\mathrm{^\circ C\,h^{-1}}$, up to 1240 or 1245 °C.

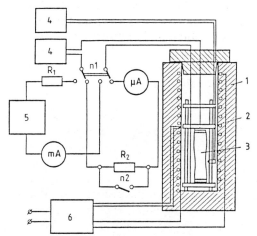

Figure 2.17. Scheme of an installation for polarization and high-temperature electro-diffusion annealing of LN single crystals (D'yakov 1982): 1, low temperature gradient furnace; 2, alumina crystal support; 3, LN crystal; 4 current and temperature recording devices; 5, constant voltage source; 6, programmed temperature regulator; R_1, operation switcher; R_2, measuring resistance shunt.

At these temperatures, the processes of diffusion are so greatly accelerated that in an hour all cracks present in a crystal are completely healed, the crystal surface is melted as a result of a change in the surface layer composition and the electrical conductivity reaches $1 \times 10^{-2} \, \Omega^{-1} \, cm^{-1}$. These observations are in good agreement with the previously obtained data on changes in composition (Holman *et al* 1978) and refractive indices of the surface layer (Kaminov and Carruthers 1973), due to annealing at high temperatures.

Under the influence of an electric current the distribution of the components in a crystal changes, and its optical inhomogeneity decreases in length by about a factor of ten, as is suggested by the measurement of deviation of the synchronism angle after a high-temperature electro-diffusion annealing (D'yakov *et al* 1981).

Simultaneously with the decrease in optical inhomogeneity, a single domain is formed. For LN crystals, this usually occurs as a result of annealing at 1160 to 1210 °C. Under these conditions, an increase in current density to $50 \, mA \, cm^{-2}$ or an increase in the time of treatment results in the appearance at the negative electrode of a brown coloration which tends to spread throughout the crystal. This is associated with the reduction of niobium ions and the emergence of charged defects in the crystal.

An increase in annealing temperature reverses the process of decay of the solid solution under the influence of the electric current: the free diffusion of oxygen into the crystal volume prevents the reduction of niobium. The decay of the solid solution occurs at the positive electrode. X-ray diffraction patterns of the surface layer at the positive electrode show distinct lines due to $LiNb_3O_8$, which suggest that LN crystals decay with isolation of the phase enriched in Nb (Lundberg 1971).

The change in composition over the crystal length, which an X-ray phase analysis fails to disclose, can be calculated from the data of nonlinear optical measurements or from those on refractive indices as functions of crystal composition (see chapters 6 and 7). The experimentally observed change in optical inhomogeneity corresponds to the change in the difference between the concentrations of lithium oxide at the ends of the crystals by about 0.01 mol.% Li_2O. With a mass of 120 g of the crystal being annealed, such a change in composition corresponds to the transport of 1.22×10^{-3} g of Li_2O in the crystal.

During the electro-diffusion annealing, which lasted for 20 min, about 80 C of electric charge passed through the crystal (the average current being 70 mA), whereas the electric transport of 1.22×10^{-3} g of Li_2O requires only about 4 C.

Therefore, less than 5% of the electric current passed goes to make the crystal composition more homogeneous, while most of it is spent on the electrochemical decomposition of lithium niobate near the electrodes, and is transferred by oxygen vacancies and by ions in the surface layer.

The curves for the synchronism angle deviation $\Delta\varphi$ as a function of crystal length l (see figure 9.19) show irregular zigzags, reflecting local changes in composition. In these parts of the crystal, the concentration gradient of Li_2O may be of opposite signs at neighbouring points. If, during annealing, the current passing through the crystal is stabilized, then the specimen will acquire a neutral grey hue, without any pronounced absorption peaks in the visible and near IR regions of the spectrum, due to the presence of sub-microscopic inclusions. When the crystal is illuminated sideways by a focused beam from a helium–neon laser one can observed a luminous cone: the Tindale effect. The presence of the scattering centres in the crystal is probably due to accumulation of lithium ions in areas where the concentration gradient of Li_2O has the opposite sign. If the voltage applied to the crystal is stabilized during annealing at a constant temperature, no scattering centres are formed.

The high-temperature electro-diffusion annealing of Nd-doped single crystals of LN has made it possible to produce single-domain specimens of these crystals. At the lower temperatures this could not be achieved, because ferroelectric domains 'stick fast' within the growth bands with an elevated impurity concentration present in the crystal. Under the influence of the electric field at a temperature close to the melting point the growth bands largely disappear, and the crystal becomes a single domain.

2.6 The Stepanov technique

Growing single crystals by the Stepanov technique has some advantages over the commonly used method due to Czochralski. For a variety of technological applications, in particular for the holographic recording of information and for the manufacture of acousto-optical and acousto-electronic devices, single-crystal plates are required. At the present time, such plates are cut out of Czochralski-grown massive crystals. But growing large single crystals means considerably lower rates of pulling boules from the melt, as well as an increased annealing time in the growth chamber. There are, however, some fundamental difficulties, which cannot be overcome if the Czochralski technique is used. Here, for example, the thermal conditions at the crystallization surface are apt to change uncontrollably due to the lowered level of the melt as the boule is being pulled out of the crucible. If the melt is non-congruent, a change in composition will occur over the length of the crystalline boule and rotational growth streaks will also appear, which significantly deteriorate the optical quality of the crystal. In the Stepanov technique, the crystal is not rotated and the rotational growth streaks are therefore absent.

The level of the melt in the die, out of which a crystal plate is pulled, remains constant and therefore the temperature gradients also remain constant, which is highly important for the stability of growth conditions. In the Stepanov technique, as a rule, the crystallization rate is much higher and the effective coefficient of the impurity distribution is close to unity. Therefore impurities are evenly distributed in crystalline plates. An important advantage of Stepanov's method is also the possibility of growing several crystalline plates at once. Furthermore, shaped crystals of LN in the form of pipes, rods, polyhedrons etc open up great possibilities for developing novel instruments in opto- and acousto-electronics.

Successful growth of LN crystals by the EFG technique (a modification of Stepanov's method) was first reported in 1975 (Fukuda and Hiraro 1975). Shaped LN crystals were grown in an installation designed for the Czochralski technique. A platinum crucible 50 ml in volume contained a platinum die with a capillary gap of 0.05 cm. A platinum screen was placed over the crucible to lower the temperature gradient. This made it possible to obtain a gradient of about $100\,°C\,cm^{-1}$ near the melt.

Ribbon-like crystals 20 by 1 and 5 by 40 mm in size of various orientations were grown at rates from 0.5 to 10 mm min^{-1}. Whole crystals were only obtained at rates less than 3 mm min^{-1}, while at the higher rates the crystals were already destroyed in the course of their growth. The physical properties of the crystals were as good as those of Czochralski-grown crystals. Nevertheless, in spite of the fact that shaped crystals were in such demand and their production was highly promising, very few publications devoted to the problems of growth of high-quality LN crystals in the form of ribbons (Matsumura and Fukuda 1976, Fukuda and Hirano 1976) and pipes (Red'kin *et al* 1983) have appeared since that first report.

Shaped LN crystals containing Cr^{+3} as an impurity and crystals of LiTa$_x$Nb$_{1-x}$O$_5$ with $x = 0.05$ and 0.2 were grown by a technique similar to that described by Fukuda and Hirano (1976). The crystals thus grown were 20 by 1 mm, 5 by 40 mm and 15 by 1 by 40 mm respectively.

The feasibility of growing long ribbon-like crystals of LN with length up to 200 mm and cross section up to 20 by 3 mm was investigated in China by Shi *et al* (1980). The crystals were grown along the *c* axis, while the (1010) plane was the plane of the ribbons. The growth rate ranged from 20 to 40 mm h^{-1}. The temperature gradient above the die varied between 50 and 80 °C cm^{-1}. When very long crystals are grown, overcooling occurs at the crystallization front, which is due both to the loss of heat by the already grown part of the crystal and to the lowered level of the melt. To prevent the undesirable overcooling, the temperature was gradually raised as the crystal was grown.

The thermal unit developed by Shi *et al* (1980) was also used to obtain LN crystals in the form of pipes with an outer diameter of 9 mm and inner diameter of 5 mm (Red'kin *et al* 1983). The length of the pipes thus grown reached 100 mm. The procedure for growing ribbon-shaped crystals of various complex oxide compounds, including LiNbO$_3$, was described by Ivleva *et al* (1983), who used a conventional installation of the Donets-1 type with HF heating and improved stabilization of the power generated. The temperature fluctuations of the melt were about 0.3 °C. Plane-type dies were made out of a platinum sheet 0.5 mm thick, the gap width being 0.5 and 1.0 mm and the lengths 25, 35 and 60 mm.

The upper edge of the die was ground, and it was placed in the crucible in such a manner that its forming edge would be at the same level with the upper edge of the crucible. A crystallization unit for growing LN plates is schematically shown in figure 2.18. The temperature gradients were varied by means of replaceable platinum and ceramic screens. The windows in the crystallization chamber were symmetrically placed relative to the pulling axis. The non-symmetry of the thermal field gave rise to cracking in the process of crystal growth.

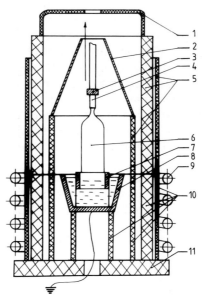

Figure 2.18. Scheme of a crystallization assembly for growing LN plates: 1, ceramic lid; 2, platinum screen; 3, platinum crystal support; 4, seed; 5, ceramic heat screens; 6, crystal plate; 7, platinum crystal form; 8, platinum crucible; 9, inductor; 10, ceramic heat screens; 11, ceramic stand (Ivleva *et al* 1983).

Some technological parameters characterizing the growth of single-crystal LN plates are listed in table 2.1. These parameters ensure that crystalline plates without cracks or gas inclusions are obtained. Photographs of such plates and pipes are shown in figure 2.19. An analysis of the defect structure of crystalline plates, depending on the conditions of crystallization, has revealed the following regularities. The main defects present in single-crystal LN plates are cracks and micro-inclusions. Cracks arose when on the growing surface of a plate there had appeared defects in the form of growth ridges situated at angles of 60° and 30° to the growth axis, thus reflecting the crystal symmetry. Plates grown along the [010] direction are more resistant to cracking. The formation of micro-inclusions in the form of gas bubbles is possibly due to the dissociation under local overheating of the melt and presence of non-isomorphic impurities in it. Lithium niobate plates free from micro-inclusions were obtained either at low pulling rates of 12 to 14 mm h^{-1} or with the use of an over-crystallized melt. In the latter case, the plates had no inclusions at pulling rates as high as 110 mm h^{-1}.

To sum up, single-crystal plates of the above mentioned compositions,

Table 2.1. Some technological parameters for single-crystal LN plates grown by the Stepanov technique (Ivleva et al 1983).

No	Growth direction	V (mm h^{-1})	Temperature gradient (°C cm^{-1})	Annealing time (h)	Size (mm)	Colour	References
1	[001], (010)	110	80	4–8	105 × 20 × 1.5	colourless	Ivleva (1983)
2	[0001], (01$\bar{1}$0) [11$\bar{2}$0], (0001)	6–30		6	50 × 20 × 2 70 × 12 × 2		Red'kin et al (1983)
3	[001], (100) [210], (001) 35° [001] 131° [001]	3–600 200	100		40 × 20 ×1.5		Fukuda and Hirano (1975)

grown by the Stepanov technique, are more homogeneous, both optically and chemically, than plates cut out of Czochralski-grown massive crystals.

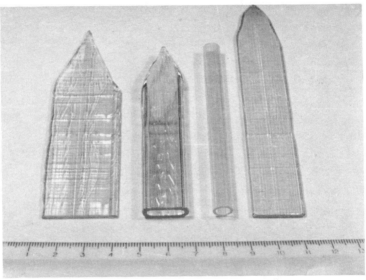

Figure 2.19. Specimens of profiled LN crystals (Red'kin *et al* 1987).

2.6.1 Physical constants characterizing a melt of lithium niobate

In order to choose optimal conditions for crystal growth by either Czochralski's or Stepanov's techniques, one should know such physical constants of the melt as its density, coefficient of thermal expansion, magnitude of surface tension, capillary constant, angle of wetting of the material of the crucible and the die, and also the value for the growth angle. In spite of the evident importance of knowing these constants, there have been few determinations of them for lithium niobate melts.

The surface tension of LN was first measured by using the molten drop technique at the boundary of the melt with a vacuum at the melting point (Bol'shakov *et al* 1969). These measurements have shown that the value of surface tension is in the range 50–150 dyn cm^{-1}.

Satunkin *et al* (1985), who made use of a platinum rod equipped with a mobile gauge, determined both the density and the surface tension of LN. The procedure of the density measurements is presented in figure 2.20(a). A certain known amount of LN was melted in a crucible with a known capacity and shape. The change in the melt level Δh was then found, which made it possible to calculate the density. This was found at

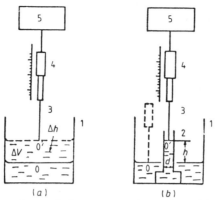

Figure 2.20. Experimental lay-out to determine (*a*) density and (*b*) surface tension of LN melt (Red'kin *et al* 1983): 1, platinum crucible; 2, slit between parallel platinum plates; 3, platinum probe; 4, mobile measuring device; 5, hoister.

two values of the temperature: at $T = 1270\,°C$, $\rho = 3.57\,\mathrm{g\,cm^{-3}}$; and at $T = 1300\,°C$, $\rho = 3.42\,\mathrm{g\,cm^{-3}}$. To determine the surface tension value, the height h was measured to which the melt rose through a narrow capillary slit, of diameter d (see figure 2.20(*b*)). Then using the formula

$$h = \sigma\, 2\cos\theta/d\rho_{m}g \qquad (2.10)$$

where θ is the wetting angle, σ is the surface tension strength, ρ_{m} is the melt density and g is the acceleration due to gravity, the value for surface tension was calculated on the assumption that a lithium niobate melt completely wets platinum, i.e. $\theta = 0$. The surface tension value was also found at two temperature values: at $T = 1270\,°C$, $\sigma = 204\,\mathrm{dyn\,cm^{-1}}$ and at $T = 1300\,°C$, $\sigma = 194\,\mathrm{dyn\,cm^{-1}}$. The capillary constant for a lithium niobate melt was found from the formula

$$a^{2} = dh \qquad (2.11)$$

where d is the slit width and h is the height of capillary rise. Again, at $T = 1270\,°C$, $a^{2} = 11.7\,\mathrm{mm^{2}}$ and at $T = 1300\,°C$, $a^{2} = 11.5\,\mathrm{mm^{2}}$.

The thermal expansion coefficient for a LN melt was measured by Timan and Burachas (1978), who used a level gauge in their measurements. It has been found that the coefficient is $10^{-3}\,°C^{-1}$.

Satunkin *et al* (1985) and Red'kin *et al* (1985) refined the methods of determining the characteristics of the melt, the accuracy of determination having been raised. Table 2.2 presents the results of the determination of the characteristics of the melt.

Table 2.2. Physico-chemical constants of the melt of lithium niobate (according to Red'kin *et al* 1985).

Density	$3.8 \pm 0.05 \ \mathrm{g\,cm^{-3}}$
Thermal extension coefficient	$10^{-4} \ \mathrm{K^{-1}}$
Capillary tension	$297 \pm 12 \ \mathrm{dyn\,cm^{-1}}$
Capillary constant	$4.0 \pm 0.01 \ \mathrm{mm}$
Angle of damping of platinum	$0°$
Angle of growth	$0°$

2.7 Electric phenomena arising in crystallization of LN

D'yakov *et al* (1985) recorded the EMF that appeared in the course of crystallization of LN from the melt at a temperature of 1520 K, at which both crystal and melt are ionic conductors. Under conditions close to equilibrium, it has also been possible to observe, along with the electric phenomena associated with the thermal EMF of the crystal and the melt, the EMF that is proportional to the rate of crystallization. This may be caused by the neutralization of the charge trapped by the growing crystal. Lithium niobate crystals were grown from the congruent melt by both the Czochralski and the Stepanov technique. The EMF was measured by means of two platinum–rhodium thermocouples, which also served as the electrodes. One thermocouple was immersed in the melt, the other was inserted in the crystal. The effects of a change in the pulling rate, of interruptions in pulling and of sudden displacements of the growing crystal were investigated. The temperature of the crystal and the melt, as well as the potential difference, were measured by the compensation technique to within 0.01 mV. Figure 2.21 illustrates the dependence of potential difference between the two thermocouples on the temperature T_1 of a Czochralski-grown crystal at a constant temperature T_2 of the melt. Point 1 corresponds to the initial moment of pulling, when the inserted thermocouple, which measures temperature T_1, is at a depth of 30 mm in the melt. Point 2 corresponds to the moment when the soldered point of the thermocouple was close to the crystallization front. Point 3 corresponds to a constant pulling rate; point 4 marks an interruption of pulling; point 5 denotes the initial moment of lowering the crystal; point 6 corresponds to lowering the crystal into the melt at a constant rate; point 7 denotes the melting of the electrode thermocouple out of the crystal. The observed $\varphi(T_1)$ is due to EMF proportional to the crystallization rate. The dependence obtained is indicative of the contributions to φ from the thermal EMF of the melt (thermoelectric coefficient d_1) and the crystal (α_s) and the EMF proportional to the growth rate v. On segment 1–2, when both thermocouples are in the melt,

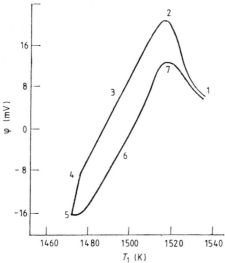

Figure 2.21. The EMF of the crystal melt system φ as a function of the temperature of the electrode inserted in the crystal. The pulling rate is $4.4\ \mu\mathrm{m\,s}^{-1}$, the rotation rate 40 rpm and the crystal radius 7 mm. The numbers refer to growth stages (D'yakov *et al* 1985).

$$\varphi = \alpha_1(T_1 - T_2). \tag{2.12}$$

The value of α_1, which is found from two independent experiments, is equal to $-0.4\ \mathrm{mV\,K}^{-1}$. On segment 3–6, when one thermocouple is in the crystal,

$$\varphi = \alpha_s(T_1 - T_m) + \alpha_1(T_m - T_2) + \alpha_v v \tag{2.13}$$

where T_m is the melting point and α_v is the crystallization EMF coefficient. On segment 3–4, when T_2 and the growth rate are kept constant, the slope of the curve is given by $d\varphi/dT_1 = \alpha_s$. It has been found (from over 30 experimental runs) that $\alpha_s = (0.76 \pm 0.02)\ \mathrm{mV\,K}^{-1}$. The opposite signs of α_s and α_1 cause a peak to appear on the curve for $\varphi(T_1)$ in the vicinity of point 2. The value of α_v, found from hysteresis loops, is $\alpha_v = 1.25 \pm 0.2\ \mathrm{mV\,s\,\mu m}^{-1}$. An interruption in pulling leads to relaxation of the growth rate to zero. A very simple thermal model, which takes into account linear dependences of the growth rate and of the temperature gradient in the crystal upon the height of meniscus h at a constant crystal diameter, yields an exponential relation for the growth rate

$$v = v_0 \exp(-t/\tau). \tag{2.14}$$

If the temperature in the crystal obeys the law

$$T_1 = T_m - \operatorname{grad} T_s \cdot \int_0^t v \, dt \qquad (2.15)$$

then the temperature gradient is given by

$$\operatorname{grad} T_s = \operatorname{grad} T_{s0} + \frac{\partial \operatorname{grad} T_s}{\partial h}(h - h_0) \qquad (2.16)$$

where $\operatorname{grad} T_{s0}$ stands for the stationary gradient at $v = v_0$ and $(\partial \operatorname{grad} T_s/\partial h)(h - h_0)$ is the change in the crystal temperature gradient due to a change in the height of the meniscus, as compared with the stationary value h_0 for $v = v_0$. The curve for $\varphi(T_1)$ should have a salient point in accordance with the relation

$$d\varphi/dT_1 = \alpha_s + \alpha_v/\varphi \operatorname{grad} T_{s0}. \qquad (2.17)$$

This salient point is shown in figure 2.22, and it enables one to find the coefficient α_v at known values of τ and $\operatorname{grad} T_{s0}$, which in the experiment just mentioned are $\tau = 300 \text{ s}$ and $\operatorname{grad} T_{s0} = 100 \text{ K cm}^{-1}$ for a crystal with $R = 7 \text{ mm}$. This yields $\alpha_v = 1.1 \pm 0.2 \text{ mV s } \mu\text{m}^{-1}$. Knowing the values of α_v and α_s allows one to control the crystallization rate, as well as to determine its sudden increase ΔV_0 due to a rapid rise of the height of the meniscus by Δh_0. The values of $\Delta v_0/\Delta h_0$, calculated for the Czochralski technique, lie in the range $2.5 \times 10^{-3} \text{ s}^{-1}$ to $1.25 \times 10^{-3} \text{ s}^{-1}$ for various rotation rates and crystal radii, the corres-

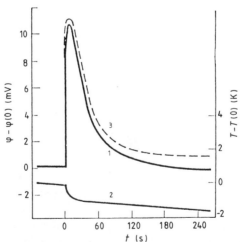

Figure 2.22. The dependence of the EMF $\varphi(1)$, the temperature of the electrode in the crystal T_1 (2) and the crystallization EMF (3), calculated by formula (1) in the case of a sudden increase in the meniscus height $\Delta h_0 = 0.34 \text{ m}$. The pulling rate is 2.2 $\mu\text{m s}^{-1}$ (Stepanov technique: a many-capillary crystal form) (D'yakov *et al* 1985).

ponding value for the Stepanov technique being about $2.5 \times 10^{-2}\,\text{s}^{-1}$ (for $R = 6\,\text{mm}$). The higher values of Δv_0 obtained in the Stepanov technique suggest that the crystallization front should thermally hitch to the mold.

The crystallization-induced EMF proportional to the growth rate is apparently due to disturbances arising in the double electric layer at the interface. The observed phenomena appear to be typical of a broad class of oxide crystals.

3 Defect Structure of Single-crystal Lithium Niobate

3.1 Morphology and macro-defects

The degree of perfection of Czochralski-grown LN crystals depends on the composition of the charge and the crucible material, the crystallographic direction of the growth axis, quality of the seed and the conditions of growth and cooling. The above factors affect the formation of macro- and microdefects in crystals. It has been found experimentally (Niizeki *et al* 1967) that the relative growth rates of the crystallographic faces of lithium metaniobate obey the following relation: $V_{(100)} \simeq V_{(120)} > V_{(001)}$, which causes a circular or an oval cross section of the crystalline boule in the direction perpendicular to that of growth.

Another observed feature of LN crystals is the appearance of growth ridges on the surface of a crystalline boule. Their configuration also depends on the direction in which the crystal is pulled. These questions have been taken up by Nassau *et al* (1966). Figure 3.1 illustrates ridges on the surface of crystalline boules pulled in various directions. It also shows the cross section normal to the pulling axis. In figure 2.16, a growth ridge can be distinctly seen on the surface of a crystal pulled along the *c* axis, in the [001] direction.

The number of growth ridges on the cylindrical part of the crystal differs from that on the conical part. Some ridges, observable on the conical part, disappear when a stable state (cylindrical part) is reached in crystal growth.

Figure 3.2 schematically depicts the cross section of the ridges in the vertical plane. From this figure, it may be assumed that growth ridges result from the fact that a certain crystallographic plane fails to develop as the crystal is pulled.

The formation of ridges can be explained if we assume a mechanism of growth of a LN crystal from the melt such that the crystallographic

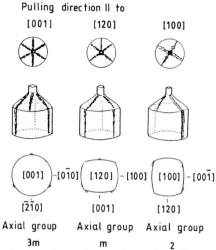

Figure 3.1. Drawings to show the configuration of growth ridges on the surface of LiNbO$_3$ crystals. Top row, ridges on the shoulder viewed from the seed. Middle row: perspective view. Bottom row, cross section normal to the pulling direction (Niizeki *et al* 1967).

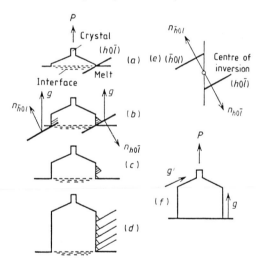

Figure 3.2. Mechanism of gradual formation of growth ridges (for explanation see the text) (Niizeki *et al* 1967).

plane ($h0l$) grows fastest. Separate stages of the proposed mechanism of ridge formation are illustrated in figure 3.2. The plane causing ridges to appear is shown as normal to vector n (see figure 3.2).

The growth vector g lies in the vertical plane, containing the vector in the direction of drawing P and parallel to the crystal surface. Once the crystal has reached a constant diameter, the directions of vectors g and

P coincide in the initial stage of growth. As the crystal diameter becomes larger, the vector *g* makes an acute angle with the melt surface. Niizeki *et al* (1967) have considered two relations between the vectors *n* and *g*.

(i) The angle between the vectors is obtuse. Having been once generated on the interface, the plane (*h0l*) has a high tangential growth rate. If the crystal is continuously pulled from the melt, the growing (*h0l̄*) planes will retain a small portion of the melt until the balance between the viscosity, surface tension and specific gravity of the melt is broken. The melt surface will then be lowered to the initial level, but a crystallized part of the melt will remain on the crystal surface (figure 3.2(*c*)). As the crystal is pulled, the above process will be repeated periodically and growth ridges will appear on the boule surface, as is shown in figure 3.2(*d*).

(ii) The angle between the vectors is acute. The (*h̄0l*) planes will then grow inside the melt, as is shown in figure 3.2(*c*), and ridges will not be formed. From figure 3.2(*e*) it can be seen that, if in crystals with a symmetry centre the plane (*h0l̄*) causes ridges to appear, the inverse plane (*h̄0l*) does not.

As mentioned above, the direction of the vector *g* changes as the expansion of the crystal stops and it continues to grow with a constant diameter. If the vectors *n* and *g* form an obtuse angle in the expansion stage, then ridges will appear on the conical part. These will disappear as the crystal goes over to the cylindrical part of the boule.

Planes $\{h0l̄\}$, responsible for ridge formation, can be indentified if the angle between the vector *n* and the direction of the *c* axis is known. On the surface of separation of the crystal from the melt there are numerous trihedrons, whose planes are oriented parallel to those of growth ridges. Measurements made on these pyramids yield an angle of about 57° between the *X* axis and the plane normal to the $\{10\bar{2}\}$ plane in the hexagonal packing. The greater growth rate of the rhombohedral faces $\{10\bar{2}\}$ is also evidenced by the formation of these faces as the melt solidifies in the crucible. By making use of the symmetry elements of the ridges on the surface of the crystal, one can determine its point symmetry group above the Curie point.

It is known that the symmetry elements of a crystal subjected to some external influence are obtained as the general symmetry elements of the crystal prior to the application of the external force plus the symmetry elements of the action itself. In the Czochralski technique, the external action has the symmetry of the polar axial group $C_{\infty,V}$. The symmetry of the crystal then corresponds to the symmetry at the growth temperature. Accordingly, the operation of pulling leaves only those symmetry elements of the crystal which are oriented parallel to the pulling direction. And it is to these elements that the configuration and shape of the growth ridges on the boule surface corresponds.

The axial symmetry group for the axis along which the crystal is pulled may therefore be determined from the symmetry elements present in the ridge pattern. By combining the latter, one can find the point group of the crystal at the growth temperature.

High-temperature X-ray powder patterns of lithium metaniobate have led Abrahams (1966a,b,c) to suggest the point symmetry group C_{3i}, whereas Megaw (1954), who commented on that work, proposed the group D_{3d}. This uncertainty cannot be resolved by X-ray diffraction methods alone, but it can be settled by observing growth ridges.

As can be seen from figure 3.1, the axial groups for the pulling directions [001], [120], and [100] will be 3m, m and 2 respectively. If these groups are combined, the high-temperature phase point group is found to be $\bar{3}2/m$ (D_{3d}). This non-polar group loses, at the Curie point, the horizontal two-fold reflection axis $\bar{2}$, becoming in the low-temperature phase the polar group 3m.

The number and orientation of growth ridges on the boule surface may be found by means of stereographic projections, which is illustrated in figure 3.3. The upper part of the figure shows the projected $\bar{3}2/m$

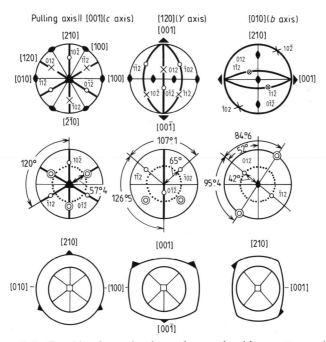

Figure 3.3. Graphic determination of growth ridge patterns (for explanation see the text) (Niizeki *et al* 1967).

symmetry elements and also six equivalent normals to the $\{10\bar{2}\}$ planes. The polar directions in the stereograms are parallel to the crystal pulling axis. The lower part of the figure only shows the symmetry elements parallel to the pulling axis, representing the axial groups for particular pulling axes. In order to determine the number of ridges on the boule surface, one should take into consideration the angular dependence of the growth vector upon the normals to the planes. When a stable growth stage is reached and g is normal to the melt surface, the above angle is obtuse relative to the normals to the planes on the lower hemisphere of the projection, and the growth ridges should appear on the cylindrical part of the crystal. If the vector g makes a small angle with the melt surface, then all normals to the planes giving rise to ridges beyond the line of equal width, shown by the broken circumference in the second (lower) row of the figure, will make an obtuse angle with g. The small double and single circles stand for the normals to the faces producing ridges which satisfy the conditions described above. These normals correspond to the ridges formed on the conical part of the crystal. The relative orientation of these ridges can be found from the known position of the axes; the results of such a calculation are shown in the lower part of the figure.

To elucidate the genetic features of crystal morphology, overcooling conditions were chosen such that the crystallization front would be almost flat, and the growth fragments on it well developed. The growth surfaces of LN crystals grown along the $[000\bar{1}]$, $[10\bar{1}0]$ and $[10\bar{1}2]$ are shown in figure 3.4.

On the crystallization front of LN crystals grown in the $[000\bar{1}]$ direction, one can see fragments in the form of trigonal pyramids (figure 3.5(a)). Both on the crystal surface and on the growth front, ridges end in flat parts.

The fragments on the growth front of LN crystals arose from the development and intergrowth of pyramids on the $[101\bar{2}]$ faces. The tangential development of the elementary rhombic fragments of a $[10\bar{1}2]$ growth pyramid leads to an edge shape for the $[10\bar{1}0]$ and the top one for the $[000\bar{1}]$ (figure 3.5). At high growth rates (overcooling of several °C), crystals approach in shape a double trihedron (figure 3.5(c)), while the $\{10\bar{1}2\}$ planes also develop on the growth front.

Microscopic studies have revealed that the $\{10\bar{1}2\}$ faces grow in layers (Honigmann 1958). Occasionally in the centre of the face one can notice a growth spiral corresponding to its symmetry. Thus ridges on the surface of Czochralski-grown crystals should be considered as a sign of under-development of growth pyramids.

The morphology of the side surface of crystals depends on the direction of growth, while that of the growth front depends on its curvature. Figure 3.4 shows the growth surface of a crystal pulled in the

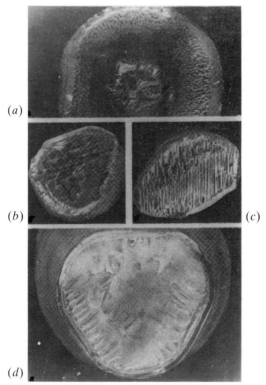

Figure 3.4. Surfaces of a highly concave crystallization front of LN crystals grown in various directions, ~ five-fold magnification: (*a*) [000$\bar{1}$]; (*b*) [10$\bar{1}$2]; (*c*) [10$\bar{1}$0]; (*d*) [000$\bar{1}$] (Garmash *et al* 1972).

[000$\bar{1}$] direction, with a highly concave crystallization front. In the centre of the surface, one can see trihedral pyramidal juts, as in crystals pulled in the same direction with a flat crystallization front. Going over from the centre to the periphery, one can observe a change in morphology. In the intermediate part, it is similar to the morphology of [10$\bar{1}$2] pulled crystals, while on the side surfaces it is similar to [10$\bar{1}$0] pulled ones.

In the central part of Czochralski-grown crystals, macroscopic cavities can be encountered. These are either rounded macroscopic pores from 1 to 100 μm in size or dendrite-like negative crystals. Their cross section is of a closed flat edge shape with re-entrant angles.

The origin of cavities is usually a change in growth regime. As these cavities expand, negative crystals may develop. When cavities suddenly narrow pores appear with a rounded surface. In changing from expansion to contraction both pore shapes can be seen simultaneously. If the material to be crystallized contains from 0.05 to 0.1% of impurities, then negative crystals will be formed without a change in growth regime.

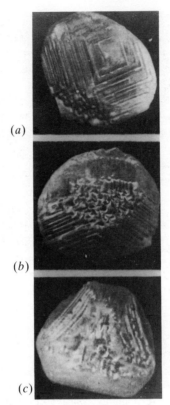

(a)

(b)

(c)

Figure 3.5. Surfaces of the crystallization front of LN crystals at different rates and directions of growth: (a) low growth rate along $[10\bar{1}2]$; (b) the same along $[000\bar{1}]$; (c) high rate along $[000\bar{1}]$ (Garmash *et al* 1972).

Among macro-defects in LN crystals, one can mention gas inclusions (Garmash 1972), cracks (Tsinzerling 1976, Vere 1968), twins (Deshmukh 1972, 1973, Nosova 1977) and streaky structure due to growth (Dubovik *et al* 1973).

Cracks in crystals arise because of a significant temperature gradient in the region immediately above the melt or a rapid cooling of the just grown crystal. As a rule, LN crystals are apt to crack along the planes of the pyramid $\{\bar{1}012\}$; less frequently, they may crack in the planes of the $\{10\bar{1}0\}$ prism and the (0001) base. The degree of pyramid cleavage varies from perfect to medium, the prism cleavage being imperfect. The origination of cracks in the planes of the $\{10\bar{1}0\}$ prism is associated (Vere 1968) with the intersection of twins. LN crystals pulled along the $[10\bar{1}0]$ direction are more apt to crack in the pyramid planes, when cooled upon growing, than are the crystals pulled along the *c* axis. It has been noted that certain impurities, e.g. Zn and Al, make for a sharp increase in the stress within the crystal and therefore make its destruction during cooling more probable. To avoid cracking, crystals are doped with Mg in amounts from 0.02 wt.% to 5 wt.% (Carruthers *et al* 1971). Twins represent another type of defect in LN crystals. Twinning is associated with thermal stresses (Blistanov *et al* 1975). Morphologic (Anghert and Garmash 1973) and X-ray diffraction studies have shown that twinning occurs in the $\{\bar{1}0\bar{1}2\}$ planes. The angle between the three-fold axes of the twins is $65°\,30'$. The number of ways in which twinning may occur depends on the pulling direction. Twinning in LN crystals is most often encountered in the $[10\bar{1}2]$, $[10\bar{1}4]$, and $[11\bar{2}0]$ directions.

It has been found (Azarbayejani 1970) that direct current affects the formation of twins in LN crystals pulled along the c axis. If a current of 5 to $10\,\text{mA cm}^{-2}$ is allowed to pass through a crystal, twins can only arise when the current is directed from the melt to the seed, whereas if the electric current is passed in the opposite direction no twinning is observed. Those authors have attempted an explanation of the influence exerted by an electric current on twinning on the basis of the kinetics of crystal growth. However, twins in LN crystals may also appear in the course of cooling of the grown crystal. In such a case, twins may evidently be due to thermal stresses. Twins in the $\{\bar{1}101\}$ planes or $35°$ domains of LN crystals have also been reported (Deshmukh 1972). Polarized light investigations of crystals with such defects have shown that the extinction of the matrix crystal and that of the twin differ in position by $35°$. Regularities governing the origin of these defects and their structure have not yet been investigated.

A characteristic structural imperfection of Czochralski-grown crystals is their growth streakiness, which arises due to rotation of the crystal and to melt-temperature fluctuations. Growth streaks in LN crystals repeat the shape of the crystallization front and determine the position of the boundaries of 180° ferroelectric domains. A method of eliminating this defect, consisting of annealing the crystal in a constant electric field at a temperature close to the Curie point, has been proposed by Carruthers *et al* (1971).

Linear defects in LN have been studied by the etching technique. It has been found that the structure of dislocations is largely determined by the seed crystal parameters. Thus the density of dislocations in a crystal may be reduced by more than an order of magnitude by a gradual narrowing, followed by a gradual expansion, of the crystal. A sudden separation of the crystal from the melt results in a great deterioration of the lower part of the crystal due to generation of dislocations by a thermal shock. On the other hand, a decrease of the pulling rate from $15\,\text{mm h}^{-1}$ down to $5\,\text{mm h}^{-1}$ leads to a drastic decrease of the number of dislocations (from 6×10^4 cm^{-1} to 10^3 cm^{-1}).

X-ray topogram studies (Sugii *et al* 1973) have revealed that in LN crystals grown along the [0001] direction the lines of dislocations run in the [01$\bar{1}$0] and [$\bar{1}$2$\bar{1}$0] directions. The absence of interactions between dislocations and the accumulations of defects near the domain boundaries suggest that the dislocation structure of the crystals had appeared before the formation of the domains, i.e. at temperatures above the Curie point. It has experimentally been shown that dislocations in LN crystals are formed at temperatures not lower than 1270 K (Blistanov *et al* 1976).

An etching study of LN crystals with the growth axis [10$\bar{1}$0] has shown that in the (10$\bar{1}$0) plane the preferable direction of dislocation lines is [1$\bar{2}$10]. In the ($\bar{1}$012) plane, the straight parts of dislocation lines run in the [$\bar{1}$01$\bar{1}$] direction, while dislocation segments in the [1$\bar{2}$10] direction are distorted, intersecting those in the [$\bar{1}$01$\bar{1}$]. In the (1$\bar{2}$10) plane, dislocation lines form a complicated geometric pattern, in which one can distinguish segments directed along the [10$\bar{1}$1] and [10$\bar{1}$0] directions (see Levinstein *et al* 1967).

The number of growth pyramids developing at the crystallization front (with crystals being pulled in various directions) and the average density of dislocations have been found (Sheibaidakova *et al* 1976) to correlate with one another: the greater the former, the higher the latter. The average density of dislocations is different for different growth directions: [0001], $\sigma \sim 5 \times 10^4$; [1$\bar{2}$10], $\sigma \sim 6 \times 10^5$; [$\bar{1}$012], $\sigma \sim 5 \times 10^4$; [$\bar{1}$104], $\sigma \sim 10^5$; [1010], $\sigma \sim 10^5$ cm^{-2}. Dislocation density topograms are illustrated in figures 3.6 and 3.7. The dislocation distribution in LN crystals has a different symmetry for different growth directions: for the [0001], it obeys the c_3 axis; for the [$\bar{1}$104], there is a symmetry plane coinciding with the ($\bar{1}\bar{1}$20); for the [10$\bar{1}$2], the symmetry plane coincides with (1$\bar{2}$10).

The correspondence between the symmetry of the dislocation distribution and that of growth direction can be explained by proceeding from a simple form of crystal growth. When LN crystals are grown along the [0001] direction, on the crystallization front three growth pyramids $\langle 10\bar{1}2 \rangle$, corresponding to the three faces of a rhombohedron, may develop. When the growth is along the [10$\bar{1}$0] direction, only two such pyramids will develop, and for the [10$\bar{1}$2] direction there will be only one such pyramid. The intergrowth of these pyramids leads to structure distortions which stimulate the development of dislocations. For the convex crystallization front, the most perfect regions lie in the vicinity of [10$\bar{1}$0] ridges. There the dislocation density is the lowest. For crystals with a concave crystallization front, the regions of minimum dislocation density are rotated through 60°. This shift is explained by the fact that, in the former case, a system of $\langle 10\bar{1}2 \rangle$ pyramids is formed in the crystal,

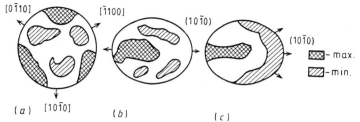

Figure 3.6. Distribution of dislocation density in LN crystals grown along: (*a*) [0001]; (*b*) [1$\bar{1}$04]; (*c*) [10$\bar{1}$2] (Sheibaidakova *et al* 1976).

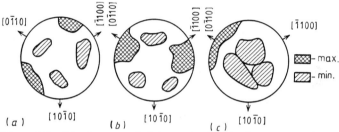

Figure 3.7. Distribution of dislocation density in LN crystals grown in the [0001] direction: (*a*) convex crystallization front; (*b*) concave front; (*c*) flat front. (Sheibaidakova *et al* 1976).

while in the latter a $\langle \bar{1}01\bar{2} \rangle$ system rotated at 60° with respect to the first one is formed. On the (0001) plane, dislocations are distributed more evenly, the regions of minimal dislocation density occupying a large area. Heat treatment of LN crystals does not change the symmetry observed in the topograms.

By using decorating and X-ray topography techniques it has been found that Czochralski-grown LN crystals may have small-angle boundaries. Upon annealing, best crystals are characterized by a small-angle disorientation of about 0.1°.

To summarize, LN crystals may contain the following defects: electric domains, cracks, dislocations and small-angle boundaries. Twins may also occur in these crystals. Electric domains are dealt with in chapter 4.

3.2 Point defects in LN

This compound has a rather wide range of homogeneity: from 48 to 50% Li_2O at room temperature. From a practical point of view, of greatest interest is the composition of 48.6 mol.% Li_2O and 51.4 mol.% Nb_2O_5, which corresponds to the congruently melted compound. Since the latter is non-stoichiometric, the number of point defects in such crystals is determined by deviation from stoichiometry. There are different views concerning structure defects present in non-stoichiometric LN, which rest on two models: the one proposed by Lerner *et al* (1968) and the other by Fay *et al* (1968). The first model assumes that, in crystals with $R = \text{Li}/\text{Nb} < 1$, excess Nb^{5+} ions are substituted for Li^+ in their sites. To satisfy the electroneutrality condition, there will then arise lithium vacancies. In order to improve the model in the sense of lowering the defect energy, Nassau and Lines (1970) have suggested that there may be partial defects of packing in the structure, i.e. the violation of the succession of the cations in stacks perpendicular to the oxygen layers. Using NMR techniques, it has been shown that, in congruently melted crystals, 6% of Nb^{5+} are in a position which is different from the usual one (Peterson and Carnevale 1972). In accordance with the above model (see Lerner *et al* 1968), these were interpreted as substitutional lithium ions. However, a quantitative disparity has been pointed out, because the model yields only 1% of the ions in the new positions. To improve the model's agreement with experiment, it had to be modified by proposing the possibility of formation of niobium vacancies, thus yielding 5% of Nb in new positions. The presence of Nb^{5+} in uncommon positions of the lattice, supposedly in Li sites, was also observed in luminescence studies on LN (Krol *et al* 1980). However, the corresponding luminescence peak has not been confirmed by other authors (Arizmendi *et al* 1981).

A second model assumes that, in niobium-enriched crystals, some lithium sites remain vacant, the electroneutrality requirement being fulfilled at the expense of oxygen-ion vacancies which arise in the process, i.e. $\text{Li}^+_{1-\delta}(V_{\text{Li}^+})'_\delta \text{Nb}^{5+}\text{O}^{2-}_{3-(\delta/2)}(V_{\text{O}^{2-}})^{\cdot\cdot}_{\delta/2}$ (Fay *et al* 1968).

Changes in the properties of LN due to deviation from stoichiometry are accounted for in different manners, depending on the model adopted (Redfield and Burke 1984, Földvari *et al* 1984). For example, the shift of the UV edge of proper absorption towards the long-wavelength region that is observed with decreasing Li/Nb ratio may be ascribed either to the effect of the excessive Nb^{5+} placed in Li sites and interstitial positions or to the influence exerted by the micro-electric field set up by lithium and oxygen vacancies. Similarly the reverse shift, observed in congruently melted crystals doped with MgO (2.7 mol.%), towards the short-wavelength region is explained, within the first model, by the expulsion of the excessive niobium by Mg^{2+} from lithium sites. According to the other model, this shift is due to the complete occupation of oxygen sites in the lattice (Bollman 1983), which follows from the electroneutrality requirement when the Mg^{2+} are implanted in vacant Li sites.

A unified model for the defect structure of non-stoichiometric LN has been attempted by Jarzebski (1974). In the general case, at a high temperature, in a state of thermodynamic equilibrium with the surrounding medium LN may contain the vacancies of lithium, niobium and oxygen ions, interstitial ions Li^+, Nb^{5+} and O^{2-}, as well as antistructural defects (i.e. Nb^{5+} in lithium sites and Li^+ in niobium sites) and quasifree electrons. The concentration of these defects is determined by the Li/Nb ratio and by the oxygen pressure in the surrounding medium. In particular, for the congruently melted composition and for normal oxygen pressure, the electroneutrality equation is written (Jarzebski 1974) as

$$2[V_O^{2+}] + [Li^+] + 5[Nb_i^{5+}] + 4[Nb_{Li}^{4+}] = [V_{Li}^-] + 5[V_{Nb}^{5-}] + 4[Li_{Nb}^{4-}] + 2[O_i^{2-}].$$

$$(3.1)$$

At the same time, it has been pointed out that it is rather hard to speak of the concentration values for these defects.

A comparison of the two models and their agreement with various experimental data has led to the conclusion (Bollman 1983) that it is the vacancies of Li^+ and O^{2-} that play the dominant role, while packing defects, if present at all, are much less important. The absorption spectra of congruently melted LN crystals annealed in steam show two absorption lines, at 2.865 and 2.875 μm, at OH^- contents as low as 1.2×10^{21} cm^{-3} (see figure 3.8) (Bollman and Stöhr 1977). This is in good agreement with the second model because, having occupied all vacant oxygen sites, the OH^- ions begin, as soon as their concentration exceeds 1.05×10^{21} cm^{-3}, to be inserted into the interstitial positions, whereas the first model assumes that it is the interstitial sites that are first occupied and that at OH^- concentrations above 2.19×10^{21} cm^{-3} these ions may appear in oxygen sites.

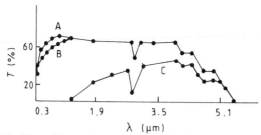

Figure 3.8. The effect of annealing in hydrogen upon the spectral transmission of LiNbO$_3$: A, transmission before annealing; B, transmission after 5 min annealing at 500 °C; C, transmission after 15 min annealing at 800 °C (Guseva *et al* 1967).

The calculated density value of a congruently melted crystal depends on its defect-structure model. In the case of lithium and oxygen vacancies ($Li_{0.944}NbO_{2.972}$), it ranges from 4.163 to 4.602 $g\,cm^{-1}$, depending on the unit cell volume chosen. This is in good agreement with the experimental data: $4.617 \pm 0.02\,g\,cm^{-3}$ (Redfield and Burke 1984) and 4.612 $g\,cm^{-3}$ (Kuz'minov *et al* 1966). The model that assumes Nb^{5+} to be in Li^+ sites, with a complete filling of the cation sites and with the 'excess' oxygen in the interstitial positions ($Li_{0.971}Nb_{1.029}O_{3.058}$), yields densities from 4.746 to 4.736 $g\,cm^{-3}$, i.e. it is at variance with experiment. If the excess Nb^{5+} ions are in interstitial positions, the calculated density will be still greater (from 4.874 to 4.888 $g\,cm^{-3}$). However, one should also consider the model (Nassau and Lines 1970) which assumes the incorporation of Nb^{5+} into Li^+ sites, leading to formation of four lithium vacancies ($Li_{0.952}Nb_{1.009}O_3$). In this case, the calculated density lies within 4.657–4.646 $g\,cm^{-3}$, i.e. the agreement with experiment is worse than for the model with the V_{Li} and V_O. Although one can find other values for LN in the literature, e.g. 4.628 $g\,cm^{-3}$ (Abrahams *et al* 1966a,b,c), 4.64 $g\,cm^{-3}$ (Nassau *et al* 1966) and 4.7 $g\,cm^{-3}$ (Ballman 1965), the authors failed to give the crystal compositions for which the above densities had been obtained.

NMR data indicating 6% of Nb^{5+} in uncommon surroundings are highly consistent with the model assuming the vacancies of lithium and oxygen if the observed change in the surroundings of Nb^{5+} is due to lithium vacancies. Nassau *et al* (1966) have reported that the results of their electron spectroscopic study of congruently melted LN support the validity of the vacancy model.

It may therefore be believed that in LN with a composition of 48.6 mol.% Li_2O and 51.4 mol.% Nb_2O_5 the prevailing defects are the lithium and oxygen vacancies.

3.3 Formation of F centres in LN crystals

The band structure of oxygen-octahedral dielectrics ABO_3, in particular $SrTiO_3$, has been studied rather well (Michel-Calendini *et al* 1980).

It is mainly determined by octahedrons BO_6 (B being a transition metal ion). Considering that the NbO_6 octahedron also is the principal structural element in LN, one may suppose that the chief peculiarities of the band structure of $SrTiO_3$ will also be retained in LN (Di Domenico and Wemple 1969). The upper, completely filled, band is formed by oxygen electron states (2p), while the lower conduction band is formed by the d orbitals of the niobium ion, the conduction band being split into two sub-bands $d\varepsilon$ and $d\gamma$ separated from each other by 5 eV. The width of the forbidden band (E_g) for LN varies, according to different authors, from 3.7 eV (Clark *et al* 1973) to 3.9 eV (Ionov 1973). There has been observed an anisotropy of the edge of the proper absorption at various orientations of the light wave polarization vector E, viz. $E_g = 3.825$ eV at $E \| c$ and $E_g = 3.840$ eV at $E \perp c$ (Kozyreva *et al* 1971). Although no dichroism has been observed by Redfield and Burke (1984), Bityurin *et al* (1978) found a dichroism of about 0.1 eV. The shift of the proper absorption edge yields a temperature coefficient for the forbidden bandwidth, $\Delta E_g / \Delta T$, equal to -8.4×10^{-4} eV °C^{-1} and -6.5×10^{-4} eV °C^{-1}. The optical absorption edge is exponential in character, obeying the modified Urbach rule down to $T = -187$ °C (or 90 K) (Mamedov *et al* 1984). At still lower temperatures, the edge becomes sharper, and non-direct optical transitions can be observed. The forbidden bandwidth for such transitions at $T = -267$ °C ($T = 6$ K) is 4.19 eV.

The basic information on how point defects affect the electronic subsystem of the crystal has been obtained by measuring the spectra of optical absorption, the spectra of electron spin resonance, and the temperature dependence of electrical conductivity.

A typical spectrum of light absorption in a lithium niobate crystal is presented in figure 6.1. In the wavelength range between 320 to 350 and 5500 to 7000 nm (Guseva *et al* 1967), LN is transparent. The absorption line at 2870 nm, observed in all specimens, is attributed to OH$^-$ ions incorporated in LN structure (Herrington *et al* 1973). Heat treatment in various atmospheres significantly affects the concentration and character of intrinsic point defects in LN. High-temperature annealing in a reducing atmosphere (vacuum or an inert gas) gives LN crystals a yellowish brown or even black colour, depending on the degree of reduction (Nassau *et al* 1966). In the spectrum, one can then observe a broad absorption band with a peak in the wavelength range from 450 to 510 nm (Ketchum *et al* 1983, Sweeney and Halliburton 1983). An additional absorption peak has been observed in the short-wavelength part of the main band. The complicated character of this band has been interpreted as being the sum of two major peaks of 2.6 eV (at 477 nm) and 3.2 eV (at 387 nm) (see Arizmendi *et al* 1984).

The absorption that occurs during reduction is associated with the fact that, as the equilibrium between the crystal and its environment depleted in oxygen sets in, there arises a deficiency in the oxygen crystal sublattice (Pareja *et al* 1984). Even with a low degree of reduction, a decrease of micro-hardness of the surface is observable, confirming the suggestion that some oxygen has left the crystal. Using photoelectron spectroscopy (Courths *et al* 1980), observation was made of the increased concentration of oxygen and lithium vacancies as LN was reduced by heating the specimen with an electron beam ($E = 2.5 \times 10^3$ eV, $T \simeq 600$ to 700 °C). Oxygen vacancies, with an effective charge of $+2$, can trap one or two electrons, producing F$^+$ or F centres which may be responsible for the coloration of LN due to its reduction.

Annealing in an oxygen atmosphere results in the disappearance of that absorption band and the discoloration of the crystal.

There exists the view (Bollman and Gernand 1972) that it is the ions Nb^{4+} arising from the above reduction that are responsible for the appearance of the 500 nm (2.5 eV) absorption in LN. The ESR data have revealed, however, that there are rather few Nb^{4+} ions in reduced LN crystals cooled to liquid nitrogen temperature, and that only after the crystals are optically discoloured at $T = -196\,°C$ (77 K) does the number of these ions drastically increase (Ketchum *et al* 1983, Sweeney and Halliburton 1983). The ESR spectrum of the Nb^{4+} ion is thermally unstable: when the crystal is heated to room temperature, the spectrum vanishes. Photoelectron spectroscopy studies of reduced LN have shown that Nb^{4+} ions are present even at room temperature (Feng *et al* 1985). The concentration of Nb^{4+} can be as high as 15% of the total number of Nb ions when LN is reduced in a vacuum at 1000 °C for an hour. The absence of an ESR spectrum of the Nb^{4+} ion is accounted for by formation of the Nb^{4+}–F^+ complex. The close values of the g factor for the Nb^{4+} and the F^+ centre favour the expansion of the resulting ESR spectrum to a wide range of wavelengths, and therefore the peaks due to Nb^{4+} are poorly resolved.

The optical discoloration does not increase the absorption at 500 nm, but rather suppresses it. Therefore, Nb^{4+} ions cannot be responsible for this absorption. The weakening of the absorption at 500 nm (2.5 eV) or at 477 nm (2.6 eV) and 387 nm (3.2 eV) as a result of the optical discoloration of the reduced LN at $T = -196\,°C$ (77 K) is accompanied by the appearance of a new absorption band with a peak at 760 nm (1.6 eV).

The explanation of the above phenomenon offered by Ketchum *et al* (1983) and Sweeney and Halliburton (1983) differs from that given by Arizmendi. The former is based on the assumption that in the process of reduction in the crystal there arise oxygen vacancies, each of which traps two electrons. It is exactly these defects that are responsible for the 500 nm absorption line. Optical discoloration at low temperature removes one electron form the F centre, transferring it to the F^+ centre. The liberated electron is meanwhile 'self-trapped' on the Nb^{5+} ion, giving rise to a minor polaron Nb^{4+}. The oxygen vacancies with one electron, i.e. the F^+ centres, are responsible for the 760 nm absorption.

A setback of the model under discussion is that it associates with the F^+ centre a lower (1.6 eV) level in the forbidden band than it does with the F centre (2.5 eV). It is known, however, that the removal of each successive electron from a multi-charged centre requires more and more energy.

The elucidation of the complex structure of absorption of the reduced LN, consisting of two bands—one at 477 nm (2.6 eV) and the other at

387 nm (3.2 eV)—has enabled Arizmendi *et al* (1984) to suggest that immediately after the reduction both F(2.6 eV) and F^+(3.2 eV) centres may exist in the crystal. The 760 nm (1.6 eV) absorption has been ascribed to the polaron states of Nb^{4+} produced at a low temperature. This is quite consistent with the experimental value for the activation energy of the drifting mobility of minor polarons (0.4 eV) (see Nagels 1980), which according to the theory of Austin and Mott (1969) should be four times as low as the energy of optical destruction of the polaron state. Irradiation of the crystal with light quanta of 1.96 eV (or 2.2 to 2.5 eV) results in the appearance of a polaron absorption band at 775 nm (1.6 eV) and in the decreased absorption of the F and F^+ centres, the relative amount of the latter somewhat increasing. This corresponds to the reaction

$$F + h\nu \rightarrow F^+ + e^-$$

and to the self-capture of the freed electrons on to Nb^{5+}. An increase in the radiation dose in the F band leads to saturation of the effect. A further increase in polaron absorption can be obtained if the crystal is irradiated with 3.1 to 3.5 eV light quanta, which is enough to ionize F^+ centres.

A theoretical calculation of the energy levels of the F^+ centre in lithium niobate (Ho 1981) has yielded a value of 3.11 eV for the transition between the ground E_{1s} and the first excited state E_{2p} of the electron in an oxygen vacancy.

As a result of the calculation, which made use of the hydrogen and helium models, levels have been obtained which are separated from the bottom of the conduction band by 0.8 eV for the F centre and by 1.9 eV for the F^+ centre (Ohmori *et al* 1975).

The experimental absorption data have failed, however, to confirm the presence of such energy levels in the forbidden band of LN. An exception is perhaps the 1200 nm (1 eV) absorption, which was arbitrarily ascribed by Bollman and Gernand (1972) to the F centre. This could be based on the fact that absorption in that band increased upon annealing in a vacuum but decreased upon annealing in oxygen. The irradiation of the crystal with light waves of 1200 nm failed, however, to weaken that absorption band, which runs counter to the author's assumption, because the F centre should then be ionised.

Two absorption bands, 424 nm (2.9 eV) and 1925 nm (0.64 eV), have been attributed to F^+ centres; and the other two, 482 nm (2.6 eV) and 1200 nm (1 eV), to F centres (Soroka *et al* 1974). According to other authors, the energy levels due to these centres should be placed at 2.9 and 2.6 eV from the conduction-band bottom. Then the absorption bands 0.64 and 1.0 eV should correspond to the electron capture from the valence band to these levels. It seems wrong, however, to believe that both F and F^+ centres could be associated with the two absorption bands, because the F centre is incapable of capturing an electron.

Therefore the model of the energy levels of oxygen vacancies and Nb^{4+} ions proposed by Arizmendi *et al* (1984) appears to be the best substantiated one; furthermore, the conclusions reached on its basis are consistent with the results of a number of experimental investigators (Rubinina 1976, Anghert *et al* 1972). In accordance with this model, the energy levels belonging to the F and F^+ centres lie in the forbidden band at depths of 2.4–2.7 eV and 3.2 eV respectively, while the 1.6 eV level corresponds to the Nb^{4+} ion.

A wide absorption band (from 350 to 700 nm), similar to that observed in reduced LN, may be produced by irradiating the crystal with γ quanta (Rosa *et al* 1982), X-rays (Schirmer and Von der Linde 1978), neutrons (Rosa *et al* 1982) or fast electrons (Pareja *et al* 1984). A complicated structure of this band has been noted by several authors.

The ionising radiation produces free electrons and holes, which subsequently may recombine or be trapped. The formation of minor polarons by capturing electrons (by Nb^{5+} ions) and holes (by O^{2-} ions or lithium vacancies) has been observed in the ESR spectra of LN irradiated with X-rays at $T = -250\,°C$ (20 K). All absorption bands observed in LN are explained by Bernhardt (1979) within the bound-polaron model. The position of the peak of an absorption band and its temperature stability are determined by the polaron binding energy.

In the absorption spectrum of LN irradiated with X-rays at $T = -263\,°C$ (10 K), three main absorption peaks, at 760 nm (1.6 eV), 500 nm (2.5 eV) and 387 nm (3.2 eV), have been isolated by Arizmendi *et al* (1984). The 760 nm absorption band is similar to that observed after the optical discoloration of reduced crystals; it is associated with the polaron state of the electrons (Nb^{4+}). The absorption at 500 nm (2.5 eV) and that at 387 nm (3.2 eV) have both been ascribed to holes trapped by lithium vacancies, the same authors believing these absorption bands in reduced LN crystals to be due to the F and F^+ centres respectively. The same temperature and the same destruction rate of these absorption bands have led the authors to conclude that both bands are determined by transitions inside one and the same centre. The activation energy needed for a thermal escape of a hole from a given trap, which has been found from the annealing kinetics of the absorption bands, is 0.54 eV. This value coincides with the distance between the acceptor level of a lithium vacancy from the valence band, which has been estimated within the hydrogen model as 0.5 eV.

The point defect model described above fails, however, to take into account the presence of oxygen vacancies in congruently melted crystals, although they, along with lithium vacancies, are the dominant point defects in such crystals. A change in the relative concentration of oxygen vacancies in the various charge states may be responsible for the emergence of certain absorption bands upon irradiation with γ quanta. This is evidenced by the absence of saturation of the 475 nm absorption with increasing γ radiation dose (Karaseva *et al* 1977). Proceeding from these facts, one may attribute the absorption observed in the irradiated LN crystals with a peak in the range from 420 to 510 nm, just as the similar absorption in the reduced crystals is attributed, to the presence of oxygen vacancies.

Lithium niobate is readily doped with rare earth and transition metal ions. Ions of iron, manganese and nickel have been found even in crystals grown without special doping. Cation impurities are capable of replacing Li^+ and Nb^{5+} in LN structure, or they may occupy vacant octahedrons. The preservation of electroneutrality is then assured by formation of the necessary number of vacancies in the cation or anion sublattices.

Mg^{2+} ions tend to occupy lithium vacancies, but as the concentration of Mg^{2+} increases above that of V_{Li} these ions seem to be placed in other positions, which is suggested by the emergence of an absorption peak at 1200 nm (Sweeney *et al* 1984). Subsequently, the authors have ascribed this absorption to the formation of a complex representing the Nb^{4+} polaron state bound with the impurity ion Mg^{2+}, when the electron alternately spends a considerable time on the manganese ion and on the neighbouring niobium ion (Sweeney *et al* 1985).

The splitting of corresponding optical absorption bands is indicative of the fact that the ions Nd^{3+} and Eu^{3+} occupy in LN two structurally non-equivalent positions, supposedly those of Li^+ and Nb^{5+} (Valyashko *et al* 1974, Arizmendi *et al* 1984).

Most of the relevant studies have been devoted to doping LN with transition metals. Nickel enters the LN structure in the divalent state (Arizmendi *et al* 1980). From the results of EPR studies and by analogy with α $LiIO_3$ it has been suggested that the ion Ni^{2+} occupies the position of Li^+ (Mirzakhanayan 1981), whereas the ESR data indicate that nickel takes the niobium sites (Rosa *et al* 1982).

The EPR studies have shown that the ions Mg^{2+} and Cu^{2+} replace Nb^{5+} (Malovichko *et al* 1983, Kobayashi *et al* 1979), while the ions Cr^{3+} are found in the positions of Li^+. Taking into account the Jahn–Teller effect in an interpretation of the EPR spectra has allowed a new model of Cu^{2+} substitution in LN to be proposed, namely, Cu^{2+} in Li^+ sites (Petrosyan *et al* 1984). The elucidation of the point symmetry of Cr^{3+} ions in LN from the temperature dependences of the EPR spectra has shown that they may be found either in Li or Nb sites, but there should necessarily be present an additional defect on the three-fold axis in the most immediate surroundings (Grachev and Malovichko 1985). The spectrum of a LN crystal doped with copper is illustrated in figure 3.9.

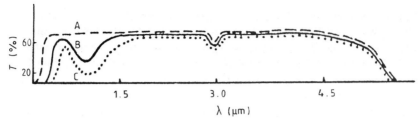

Figure 3.9. Optical transmission of $LiNbO_3$: A, pure crystal; B, 0.1 wt.% CuO; C, 0.5 wt.% CuO (Sirota and Yakunishev 1974).

Different authors have determined the different positions of Fe^{2+} and Fe^{3+} in LN. According to Vladimirtsev *et al* (1983), each of these ions may occupy only one position in the LN lattice, viz. Fe^{2+} is substituted for Li^+ and Fe^{3+} for Nb^{5+}. However, there is another view that both Fe^{2+} and Fe^{3+} may be found in the Li^+ and Nb^{5+} sites, the main portion of the iron ions occupying the neighbouring Li and Nb octahedrons and forming paired defects. A paired defect formed by Fe^{3+} ions is electrically neutral, whereas that formed by Fe^{2+} is accompanied by the emergence of an oxygen vacancy. The ions Fe^{2+} and Fe^{3+} can also occur in Li sites, while Fe^{3+} ions alone may occupy single Nb sites. When the crystal composition deviates from the stoichiometric one towards the lithium deficiency, the relative content of Fe^{2+} ions increasingly exceeds that of Fe^{3+}.

The ratios Fe^{2+}/Fe^{3+}, Mn^{2+}/Mn^{3+} and Cu^+/Cu^{2+} are known to vary as a result of irradiation or reduction annealing of LN crystals (Belogurov *et al* 1976). The reduction tends to increase the concentration of ions of lower valency, i.e. Fe^{2+}, Cu^+ and Mn^{2+}. The capture of the free electrons produced by X-rays also transfers some of the Cu^{2+} ions to Cu^+, which is accompanied by the absorption band shifting from 1.2 eV to 3.1 eV.

Most absorption bands observed in doped LN crystals are associated with electron transitions to the higher lying energy levels of the impurity centres, rather than to the conduction band. Accordingly, data on the position of the impurity levels in the forbidden band are rather scarce.

The absorption in the band 2.6–2.7 eV is associated with the intervalent transition $Fe^{2+} \rightarrow Nb^{5+}$ (Kurz *et al* 1977), and is therefore indicative of the depth of the Fe^{2+} level in the forbidden band. An electron transition from the valence band to the Fe^{3+} level is accompanied by absorption in the 3.1 eV band, indicating that the level lies at a depth of 0.6 eV from the bottom of the conduction band (Zulberstein 1976). The energy of the impurity levels in LN evaluated within the hydrogen and helium models is 1.9, 0.8 and 0.5 eV below the conduction band for the ions Fe^{3+}, Fe^{2+} and Cu^{2+} in lithium sites respectively.

However, these results have not yet been confirmed experimentally.

From the above discussion one may conclude that the existing data on energy levels and specific point defects corresponding to them are rather controversial (see figure 1.12). This is true, for example, of the levels of the oxygen vacancies, the Nb^{4+} level, and the levels due to Fe^{2+} and Fe^{3+}.

From the results of optical absorption studies, one can regard as reliable the inference that there is a level of 2.4 to 2.7 eV, which is most likely due to two-electron oxygen vacancies (if we confine ourselves to considering intrinsic point defects alone). Moreover, there seem to exist the (1.5–1.6 eV) level due to the (Nb^{4+}) electron–polaron state and the (2.9–3.2 eV) level associated with singly ionised oxygen vacancies.

The structure of the energy levels in the forbidden band determines the electrical properties of LN at temperatures far from the melting point.

The energy separating the valence band from the level responsible for the optical absorption in the band with a peak at 2.4–2.7 eV, associated with the oxygen deficiency, is equal to 1 or 1.3 eV. That this value coincides with the activation energy of the electrical conductivity of LN in the temperature range from 100 to 400 °C has enabled Maksimov *et al* (1984) to advance the idea of hole-type conductivity in LN at these temperatures. In accordance with the best substantiated model, however, the energy level belongs to two-electron oxygen vacancies, and therefore no electron may come to it from the valence band. Furthermore, the thermal EMF (Jorgensen and Bartlett 1969) and Hall effect (Josch *et al* 1978) data indicate an electron-type conductivity in LN (see chapter 5).

Uncontrolled impurities or oxygen vacancies, which are present in large amounts in non-stoichiometric (Li/Nb < 1) crystals, may serve as donor centres, and they determine the electron conductivity in LN.

The electron conductivity activation energy (1 to 1.3 eV) and the position of the oxygen vacancy level in the forbidden band determined from optical absorption data (2.4 to 2.7 eV) may probably be related to each other by assuming that the activation energy H is equal to half the ionization energy of the donor centres. This is valid for a single type of centre or for a weakly compensated semiconductor at temperatures above T^* defined as:

$$kT^* = (E_g - \Delta E_d)/(\ln N_a - \ln 2N_c) \qquad (3.2)$$

where k is the Boltzmann constant, ΔE_d the depth of the donor level, N_a the concentration of the compensating acceptor impurity and N_c the density of states in the conduction band.

The level 0.6 eV or 0.68 eV, found from the data on optically induced birefringence and the electric conductivity activation energy in the range

50 °C < T < 170 °C (H = 0.34 eV), is generally attributed to oxygen vacancies. This seems to be hardly justified, as it has never been confirmed by optical data.

Suggested by the hologram survival time, the hole states with a depth of 0.64 eV may evidently be ascribed to singly ionized oxygen vacancies (0.5 to 0.64 eV from the top of the valence band, as found from the optical absorption data) or to lithium vacancies (0.5 to 0.54 eV from the valence band).

The 1 eV level assumed from the time of build-up and erasure of holograms has not been observed in the overwhelming majority of optical absorption studies. Exceptional in this respect are the works of Sweeney *et al* (1985), who observed the 1200 nm (1 eV) absorption for reduced and γ-irradiated LN crystals and for specimens doped with Mg with a concentration above 5%. In Rh-doped specimens, a peak of thermal luminescence has been observed, which corresponds to the 1.1 eV level (Ohmori *et al* 1975). It seems that the activation energy of 1 eV that was measured by Pashkov *et al* (1979a,b) is an effective value of H, and is due to the superposition of the various mechanisms of conductivity.

No energy level with a depth of less than 0.5 eV has been found in LN. Therefore the conductivity below 100 °C with a low activation energy is determined by a mechanism other than the carrier transition between a defect level and the allowed band. The conductivity is possibly associated with the hopping motion of the carriers through deep local levels due to defects (Blistanov *et al* 1981) or with the polaron charge transfer mechanism.

The disparity existing between the experimental data of the various authors precludes an unambiguous interpretation of the energy level structure due to point defects in the forbidden band of LN. The depth of those levels, regarded as rather reliably determined, differs from the conductivity activation energy at temperatures far from the melting point, which should hold for a compensated semiconductor if charge transfer is effected by excitation of electrons from donor centres. The absorption bands and the various types of structural defects observed in LN crystals are presented in tables 3.1 and 3.2. A scheme for supposed energy levels in the forbidden band of LN is shown in figure 3.10.

3.4 Twinning in LN crystals

Under the influence of mechanical stresses and also heat stresses arising during the growth, annealing and the subsequent formation of single domains in LN crystals, there may appear elastic stresses, accompanied by deformation of the crystals and the emergence of twins. Such crystals

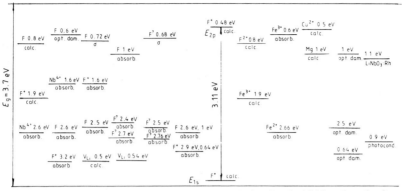

Figure 3.10. Scheme of electron energy levels in the forbidden band of LN (Stepanova 1986).

Table 3.1. Light absorption bands, possible types of structural defects and their activation energy in $LiNbO_3$ (according to Bollman and Gernad 1972).

Absorption band (nm)	Activation energy (eV)	Structural defect
482	0.5–0.6	Nb^{5+} Nb^{4+}
1200	0.2–0.3	$2e^-$
1925	1.1	O_i^{2-}
2870	0.9	OH_O^-
	1.3	V_O^{2+}

Table 3.2. Light absorption bands, corresponding colour centres, charge carriers in $LiNbO_3$ and their activation energy (according to Khromova 1975).

Absorption band (nm)	Colour centre	Type of defect	Activation energy (eV)
$\begin{cases} 1430 \\ 1925 \end{cases}$	F	V_O^{2+}	1.31
$\begin{cases} 480 \\ 1200 \end{cases}$	F'	V_O^+	1.10
700	V	O_i^{2-}	0.3–0.4
2870		OH_O^-	0.8–0.9
		Li^+	1.87

are useless in electro-optics and in nonlinear optics. From this viewpoint twins may be regarded as macroscopic defects in the crystal lattice of LN, although it would be more correct to classify them under mechanical properties and treat them as a reaction of the crystal upon a mechanical action. Plasticity of LN has been investigated by Vere (1968) and Blistanov *et al* (1976). It has been found that compression of polydomain LN crystals along the [0001] and [01$\bar{1}$0] axes at temperatures above 1433 K leads to twinning according to the [$\bar{1}$011] (10$\bar{1}$2) system. The destruction of strained specimens takes place at a plastic strain of about 1 per cent. Below 1433 K specimens are destroyed without a trace of plastic strains. Cracks appear in the planes {0$\bar{1}$10} at the intersection of twinning interlayers. The boundaries of these interlayers remain undistorted in going over to the region of the host crystal with a different direction of spontaneous polarization. Etching experiments have led to the conclusion that the spontaneous polarization direction in the twin crystal is opposite to that in the matrix one. Actually in LN crystals the angle between these directions differs from 180 °C (Blistanov *et al* 1976).

Twins have been observed in LN near scratches made at room temperature on the prism faces {1120} (see Katrich *et al* 1975).

The Moh's hardness of LN crystals is equal to 5. The material is brittle and, as a Knoop pyramid is being indented into the crystal, the destruction already begins at a load of 5×10^{-2} N (Nassau *et al* 1966).

The micro-hardness of the faces of the base (0001) and prisms (1100), (1$\bar{1}$20) at room temperature was estimated from the results of indentation of a tetrahedral pyramid (Katrich *et al* 1975) and a Knoop pyramid (Noda and Ida 1972, Brown *et al* 1975). These results are listed in table 3.3.

Table 3.3. Micro-hardness of LN and LT crystals calculated from Knoop pyramid indentation data of Brown *et al* (1975).

Direction of long edge of indentor	Plane	Micro-hardness according to Knoop $10^{-7}H$ (Pa)	
		LiNbO$_3$	LiTaO$_3$
[11$\bar{2}$0]	(01$\bar{1}$0)	530 ± 10	660 ± 20
	(0001)	560 ± 5	780 ± 10
[01$\bar{1}$0]	(1$\bar{1}$20)	570 ± 10	775 ± 5
	(0001)	570 ± 5	760 ± 15
[0001]	(1$\bar{1}$20)	650 ± 15	920 ± 5
	(0$\bar{1}$10)	780 ± 10	970 ± 20

An analysis of the brittleness characteristics of LN crystals has revealed that the faces of the (0001) base are more brittle and less strong than those of the prisms. This can be explained by the more favourable position of the cleavage planes {10$\bar{1}$2} relative to the basal plane (Noda and Ida 1972).

Micro-hardness and micro-brittleness studies, carried out by Katrich *et al* (1975) on the faces of the monohedrons (0001) and (000$\bar{1}$) of LN, have failed to establish any differences in the mechanical properties of these faces. Nosova (1977) has examined the micro-mechanical properties of the opposite faces of the pyramids (10$\bar{1}$2) and ($\bar{1}$01$\bar{2}$), which belong to the various simple forms, and also the characteristics of the faces of the pyramids ($\bar{1}$012) and (10$\bar{1}\bar{2}$). No differences between the properties of the opposite faces have been found, which is consistent with the result given by Katrich *et al* (1975). The anisotropy of micro-hardness in the basal plane of LN crystals is absent.

From the scratching data, the thickness of the surface layer of LN crystals, which can be damaged by mechanical processing, needed for various purity grades, has been determined. The results of the thickness determination are listed in table 3.4.

Table 3.4. Thickness of the damaged surface layer of LN crystals (Data of Katrich *et al* 1975).

Purity grade	Damaged layer thickness (μm)		
	Face (0001) Random grinding direction	Face (1120) Grinding direction [0001]	Grinding direction [10$\bar{1}$0]
7a	22.0	14.5	22.0
7b	17.5	11.0	18.0
7v	14.0	9.0	14.0
8a	11.0	7.0	11.0
8b	9.0	5.5	9.0
8v	7.0	4.6	7.0
9a	5.5	3.5	5.5
9b	4.5	2.8	4.5
9v	3.5	2.2	3.5

The study of mechanical properties has thus revealed that LN crystals are not very plastic. When LN crystals are uniaxially compressed, their plastic deformation has the form of twinning in the planes {10$\bar{1}$2} at temperatures above 1430 K. At room temperature twins in LN arise near scratches made on the crystal surface. The presence of electric domains of opposite signs does not affect the process of twinning in the crystal. The interaction of twins gives rise to cracks.

Mechanical stresses set up in loaded crystals will relax in the same way as will thermal stresses, i.e. by twinning and crack formation. This allows one to study the regularities governing the formation of defects by setting mechanical stresses in the crystal by means of controlled loading.

A complete description of twins in LN crystals covers the following aspects:

(i) the change in the shape of the crystal region which becomes a twinning interlayer;
(ii) the relationship between the crystallographic orientations of the lattices in the twinning interlayer and in the host crystal;
(iii) possible ways of atomic displacement in the process of twinning.

A practically important task is the search for conditions hindering LN twinning under mechanical and thermal stresses.

3.4.1 Crystal strain due to unaxial compression

Specimens strained by compression along the $[01\bar{1}1]$ and $[0001]$ directions in the temperature range from 290 to 620 K were destroyed without apparent traces of a plastic deformation.

On the faces of specimens strained at 670 K and above that temperature, relief streaks can be seen through a microscope (figures 3.11 and 3.12). The width and shape of the strain-induced streaks depend on the temperature at which the strain occurred.

In the temperature range up to 970 K, strain often results in the formation of narrow wedge-shaped streaks, which start from an edge of the specimen and end at some distance from the edge. Signs of destruction are also observable. As the temperature is raised, strain-induced clefts and cracks vanish, while both the length and the breadth of the above mentioned streaks increase.

At strain temperatures above 970 K wide streaks with parallel boundaries, which cross the whole specimen, appear in the crystal. Some of them may be as wide as 100 μm. The formation of such a wide twinning interlayer shortens the crystal in the direction of the applied force by 10 to 15 μm, which constitutes about 0.2% of the specimen length. Where strain-induced streaks intersect, breaks of the crystal wholeness are often visible (see figures 3.11 and 3.12).

On the faces of a specimen, there are relief streaks due to the emergence of one and the same interlayer to the crystal surface. Plane-parallel interlayers may intersect two pairs of opposite side faces or the side faces and the top (bottom) end of the crystal.

Interlayers arising from compression along the $[01\bar{1}1]$ direction can be divided, from their position relative to the edges and faces of the crystal, into four types (see figures 3.11 and 3.12 and table 3.5) and those arising due to compression along the $[0001]$ direction into six types (table 3.6).

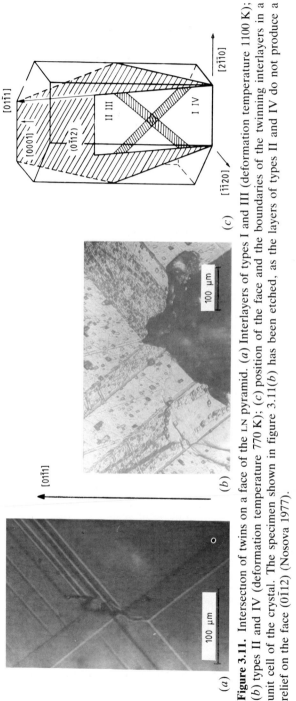

Figure 3.11. Intersection of twins on a face of the LN pyramid. (*a*) Interlayers of types I and III (deformation temperature 1100 K); (*b*) types II and IV (deformation temperature 770 K); (*c*) position of the face and the boundaries of the twinning interlayers in a unit cell of the crystal. The specimen shown in figure 3.11(*b*) has been etched, as the layers of types II and IV do not produce a relief on the face ($0\bar{1}12$) (Nosova 1977).

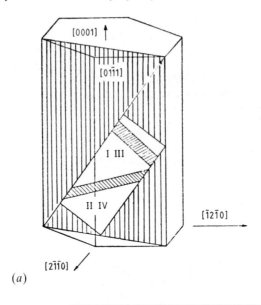

(a)

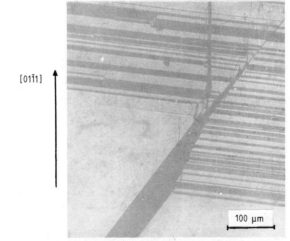

(b)

Figure 3.12. Intersection of twins on the face of the LN prism
$(2\bar{1}\bar{1}0)$: (a) specimen strained at 1100 K; (b) position of the face and
the boundaries of the twinning interlayers in a unit cell of the crystal
(Nosova 1977).

Interlayers of types I, III, and V may appear over the whole
temperature range of crystal plasticity. As the strain temperature
increases, a change in the shape of strain-induced streaks (from narrow
wedge-like to wide plane-parallel) is distinctly observable. As a rule,
there are many interlayers of these types in a single specimen (see figure
3.12).

Table 3.5. Geometry of strain-induced interlayers in a specimen with faces $(0\bar{1}12)$ and $(\bar{2}110)$ (according to Nosova 1977).

Type of interlayer	Angles of interlayer boundaries with edge $[01\bar{1}1]$ on specimen face (deg)		Angles θ of interlayer surfaces with specimen faces (deg)	
	$(01\bar{1}2)$	$(\bar{2}110)$	$(01\bar{1}2)$	$(\bar{2}110)$
I	137	85	8	4
II	43	144	0	7
III	43	85	8	4
IV	137	144	0	7

Table 3.6. Angles of boundaries of strain-induced interlayers with the edge [0001] in specimens with faces $(\bar{3}4\bar{1}0)$ and $(52\bar{7}0)$ (according to Nosova 1977).

Face of specimen	Angles with the edge [0001] of the boundaries of twinning interlayers (deg)					
	I	II	III	IV	V	VI
$(\bar{3}4\bar{1}0)$	34	53	70	80	41	60
$(52\bar{7}0)$	67	78	34	52	42	61

Interlayers of types II, IV, and VI appear as a result of the strain applied to the crystal at temperatures below 1270 K. These interlayers tend to be plane-parallel even at minimum plasticity temperatures. As a rule, several interlayers of the above types can be formed in a single specimen (see figure 3.11(b)).

A comparison of X-ray diffraction photographs, taken by the Laue technique, suggests that in the strained crystal a totality of unsmeared, ordered spots is superimposed on the pattern of X-ray reflections from the planes of the unstrained crystal. One may suppose that the interlayers that have arisen possess the crystal structure of a different orientation from the host crystal, and represent a twin.

Surfaces of strained specimens upon etching are illustrated in figure 3.13. The different shapes of the etch figures seen in the host region and in the twinning interlayers on the face of the crystal which is the plane of the $(\bar{2}110)$ prism of the unstrained crystal suggest that the surfaces being compared represent different crystallographic planes (see figure 3.13).

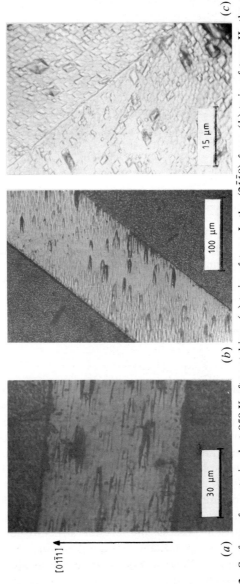

[01$\bar{1}$1]

(a)   30 µm

(b)   100 µm

(c)   15 µm

Figure 3.13. Surface of LN strained at 950 K after etching: *(a)* twin of type I, the (2$\bar{1}$10) face; *(b)* twin of type II, the (01$\bar{1}$2) face; *(c)* twin of type II, the (2$\bar{1}$10) face. The positions of the faces of the specimen and the boundaries of the twinning interlayers in a unit cell of the crystal are shown in figures 10.1(c) and 10.2(b) (Nosova 1977).

The etch figures in the region of the twinning interlayers of types II and IV on the face which is the plane of the $(01\bar{1}2)$ pyramid and in the region of the contiguous surface of the unstrained crystal are the same, but have different orientations (see figures 3.11(*b*) and 3.13(*c*)). The surfaces of the host crystal and the twin will then represent faces of the same, simple shapes; they are, as it were, rotated relative to each other around a common normal.

The twinning interlayers of types I and III on the face of the $(0\bar{1}12)$ pyramid are etched in the same way as the opposite face of the $(01\bar{1}2)$ specimen, which corresponds to the projection of the negative end of a domain (figure 3.14). The long diagonals of the etch figures on the $(01\bar{1}2)$ face in the host crystal region are parallel to the edge of the $[01\bar{1}1]$ specimen. On the opposite face of the $(0\bar{1}12)$ crystal, in the twin region, the long diagonals of the etch figures make an angle of 87° with the same edge.

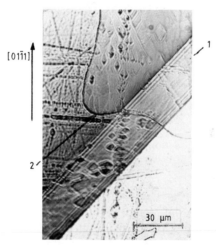

$[01\bar{1}1]$

30 μm

Figure 3.14. Intersection of a twinning interlayer (1) of type I with the domain boundary (2) on a face of the prism of a LN crystal strained at 1270 K, as seen in natural transmitted light. One can see interference bands parallel to the twinning boundary and etch figures (Nosova 1977).

Figure 3.14 shows how the twin crosses the boundary of a 180° electric domain in a LN specimen. Etching has failed to reveal any disturbances at the point of intersection, which agrees with the results reported by Vere (1968).

Short relief streaks may sometimes be observed near the intersection of twinning interlayers of types I and III on the face which is the plane of the pyramid (figure 3.15). Traces of these streaks are paralled to the edge of the [$2\bar{1}\bar{1}0$] specimen. The angle θ which the surface of the relief streaks makes with the ($0\bar{1}12$) crystal face is 6°. The above streaks seem to represent accommodation twins. Their appearance is evidently due to relaxation of the mechanical stresses which arise in the original crystal when twins of types I and III intersect (Blistanov *et al* 1975).

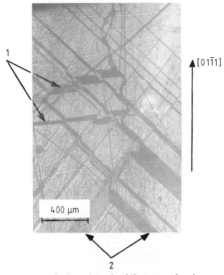

Figure 3.15. Accomodation bands (1) near the intersection of twinning interlayers (2) of types I and III on a face of the pyramid of a LN crystal strained at 1270 K (Nosova 1977).

Straining LN crystals by compression along the [$01\bar{1}1$] and [0001] directions at temperatures below 670 K thus leads to their destruction, leaving no visible traces of a plastic deformation.

Such traces, due to twinning, have been observed in LN specimens strained at temperatures above 670 K. The twinning nature of defects produced as a result of straining has been ascertained by X-ray diffraction techniques, as well as by etching crystal surfaces.

In the temperature range from 670 to 970 K, plastic straining by twinning and the destruction of LN crystals along the [$01\bar{1}1$] and [0001] compression axes occur independently of each other. At temperatures above 970 K, breaks of continuity only arise when twinning interlayers intersect.

Indices of the twinning planes may be established from the position, on the faces of the crystal, of the twinning interlayer boundaries, which are the traces of the twinning planes (Klassen-Neklyudova 1960).

The study of the relative position of twins in LN crystals has shown that the planes of the pyramids $\{\bar{1}01\bar{2}\}$ and $\{10\bar{1}2\}$, $\{10\bar{1}4\}$ and $\{\bar{1}014\}$ may serve as the twinning planes. It is impossibe, however, to distinguish the opposite plane, which in the structure of the ferroelectric phase represents faces of the various simple forms, merely by studying the geometry of twinning interlayers.

In twinning on the above mentioned planes, shear plane corresponds to the plane of the prism $\{1\bar{2}10\}$, as a symmetry plane perpendicular to the twinning planes.

The crystallographic symbols of the shear directions caused by twinning are given by the cross multiplication of the indices of the twinning planes K_I and those of the shear planes S. As a result, the indices for the possible shear direction, $\langle\bar{1}011\rangle$ and $\langle1011\rangle$, $\langle20\bar{2}1\rangle$ and $\langle\bar{2}02\bar{1}\rangle$, have been found.

The polarity of the ferroelectric phase structure imposes a limitation upon the number of twinning systems in operation. Experiments devoted to the study of the electric field upon the formation of twins in LN have made it possible to isolate 6 operative systems out of the possible 12. The indices of the twinning elements K_I and the shear directions η_1, are listed in table 3.7.

Table 3.7. Crystallographic indices of active twinning planes and shear directions of the twinning interlayers whose geometry is described in tables 3.5 and 3.6 (according to Nosova 1977).

Twinning elements	Crystallographic indices of active twinning planes and shear directions for twinning interlayers of type:					
	I	III	V	II	IV	VI
K_I	$(\bar{1}01\bar{2})$	$(1\bar{1}0\bar{2})$	$(01\bar{1}\bar{2})$	$(10\bar{1}4)$	$(\bar{1}104)$	$(0\bar{1}14)$
η_1	$[\bar{1}011]$	$[1\bar{1}01]$	$[01\bar{1}1]$	$[20\bar{2}1]$	$[\bar{2}201]$	$[0\bar{2}21]$

The boundaries of accommodation twins coincide with the traces of the planes $(0\bar{1}1Z)$, $(01\bar{1}\bar{Z})$, $(01\bar{1}Z)$, and $(0\bar{1}1\bar{Z})$, the indices along the Z axis remaining unknown.

The value for the specific shear g due to twinning in LN crystals in the systems $[\bar{1}011]$ $(\bar{1}01\bar{2})$ and $[20\bar{2}1]$ $(10\bar{1}4)$, found from an experimental study of the change in the shape of the specimens caused by twinning interlayer formation, is given by $g_1 = g_2 = 0.18$ (Nosova 1977).

It has thus been established that the twinning system $\langle\bar{1}011\rangle$ $\{\bar{1}01\bar{2}\}$ manifests itself when LN crystals are uniaxially compressed over the whole temperature range of plasticity from 670 to 1370 K. The twinning system $\langle20\bar{2}1\rangle$ $\{10\bar{1}4\}$ appears, due to uniaxial compression, in the temperature range between 670 and 1220 K.

3.4.2 Deformaiton of LN crystals by gliding

When the crystal is compressed along the [01$\bar{1}$1] direction, substantial shearing stresses are set up in the gliding system [0$\bar{1}$10] (0001). At the same time, the orientation of the crystal allows twinning in the above system. No traces of gliding have been found in the investigated specimens, the plastic deformation having been due to twinning.

When the crystal is compressed in the direction of the normal to the plane of the pyramid (0$\bar{1}$12), deformation by twinning is forbidden. Simultaneously, considerable shearing stresses are set up in the gliding system [0$\bar{1}$10] (0001). Straining the specimens in the range from room temperature to 1370 K leads to their rupture without any traces of a plastic deformation. The compression at 1370 K also leads to rupturing, but this time narrow relief streaks near the crystal ends are observable. Upon etching, etch figures appear in their places. The emergence of such sequences as a result of compression allows us: (i) to conclude that in the given case a plastic deformation due to gliding took place; (ii) to interpret the etch figures on the pyramid faces as dislocations. From the position of the gliding streaks, the indices of the gliding plane (0001) have been determined. The most probable direction of gliding is [0$\bar{1}$10], which is the direction of the shortest translation in the basal plane.

From the results described above one may suppose that the critical shear stress for twinning is less than that for gliding in the basal plane. Accordingly, mechanical stresses in LN crystals partially relax at the expense of a gliding deformation in that plane only in those cases when twinning is forbidden. At the same time, gliding cannot result in high-degree deformations, and along with the gliding streaks the traces of rupture are visible in the crystal.

LN specimens compressed along the [$\bar{3}$4$\bar{1}$0] are ruptured without noticeable traces of a plastic deformation. Since in these specimens the maximum shearing stresses are set up in the supposed gliding system [0$\bar{1}$10] ($\bar{2}$110), one may conclude that there is no gliding in the prism plane in LN crystals.

3.4.3 Mechanical stresses

The optical polarization technique (see chapter 9) has been used to determine the stresses in the plane normal to the optical axis of the crystal.

Under the influence of mechanical stresses (σ) a uniaxial LN crystal becomes a biaxial one, the plane of its optical axes being perpendicular to the direction of compression (σ_2) and parallel to that of dilatation (σ_1) if the stresses are applied normally to the optical axis of the crystal.

The relation between the presence of two optical axes in the crystal and its birefringence is dealt with in chapter 9. For crystals of the point symmetry group 3m, to which the LN crystal belongs, the relation between the stresses applied to the crystal and its birefringence is given by Indenbom and Tomilovsky (1958):

$$\Delta = c(\sigma_1 - \sigma_2) = c\Delta\sigma \tag{3.3}$$

$$c = n_o^3(\pi_{11} \times \pi_{12}) \tag{3.4}$$

where σ_1 and σ_2 are the principal stresses in the plane normal to the optical axis, π_{12} and π_{11} are the piezo-optical constants, and Δ is the birefringence induced by a mechanical stress. Hence

$$\Delta\sigma = \frac{2\Delta}{n_o^3(\pi_{11} - \pi_{12})}. \tag{3.5}$$

For stresses applied along the X axis and for the light propagation along the Z axis, we have

$$\pi_{11} - \pi_{12} = 1.02 \times 10^{-13} \text{ cm}^2 \text{ dyn}^{-1}$$

while for stresses along the Y axis and the light propagation along the X axis (Spencer *et al* 1967)

$$\pi_{11} - \pi_{12} = 6.92 \times 10^{-14} \text{ cm}^2 \text{ dyn}^{-1}.$$

The X, Y and Z axes coincide with the crystallographic ones. The values of Δ along the X and Y direction, which have been measured on the investigated specimens of LN, are listed in table 10.7. The birefringence Δ reaches its maximum value in the centre of the crystal and decreases towards the periphery. The stresses measured along the X and Y axes coincide. The asymmetry seen in the experimental curves can be explained by displacing the element away from the geometrical axis of the crystal boule.

Using the maximum values of Δ, stresses in the crystal, also listed in table 3.8, have been calculated. It should be mentioned that the specimens had a dislocation density of about 10^4 cm^{-2}.

Table 3.8 Birefringence and thermal stress in LN (according to Kuz'minov 1975).

Number of crystal specimen	Δ (10^{-6})	σ_x (kg cm^{-2})	σ_y (kg cm^{-2})
1	5.28	8.7	12.8
2	13.21	21.8	26.0
3	21.14	34.8	51.4

4 The Domain Structure of Lithium Metaniobate Crystals

4.1 Ferroelectric domains in lithium metaniobate

Lithium metaniobate is a ferroelectric whose Curie temperature is rather high ($T_c \sim 1200\,°C$), being only several tens of degrees below its melting point. When grown from the melt in the para-phase, the crystal at room temperature turns out to be divided into domains, and a polydomain crystal cannot, of course, be utilized in either field of its application with the same efficiency as a monodomain crystal. Accordingly, a number of techniques have been developed to turn many-domain crystals into single-domain ones (Levinstein *et al* 1967). At the same time, it has remained obscure which factors cause the specific distinctions from the standard schemes to appear in the growth domain structure of LN crystals, and what mechanism is responsible for its change under the influence of heat treatment or an electric field.

An elucidation of these problems seems important from two points of view. First, the study of the mechanism by which the growth domain structure is formed may open up a possibility of controlling this process; in particular, it may allow single-domain crystals to be grown directly from the melt. This should lead to an improved optical homogeneity of the crystals. It is also clear that the development of a rational technology for producing single-domain (i.e. polarized) crystals is impossible without understanding the mechanisms of the processes involved.

Secondly, lithium metaniobate has a most unusual type of domain structure, in which oppositely directed polarization vectors occur at the domain boundaries. Theoretical ideas about the possibility that an equilibrium domain structure of this type may be formed have not been advanced until quite recently (Selynk 1973). It has more recently been

112

found that this type of domain structure is also present in a number of other more complex ferroelectric crystals, for example in niobates with the structure of the tetragonal tungsten bronze (Kuz'minov 1975).

Czochralski-grown single crystals of lithium metaniobate are, as a rule, many-domain. There are several ways of changing them into single-domain ones (Nassau *et al* 1965):

(i) growing a crystal in an electric field;
(ii) application of an electric field to a crystal at temperatures close to the Curie point;
(iii) addition of 0.5 at.% MoO_3 to the melt.

The third method has not gained much popularity because of the poor reproducibility of its results.

The first method is to superimpose an electric field upon a growing crystal, a platinum wire on the seed and the crucible serving as the electrodes. Near the phase transition region, the electric field is kept constant in spite of the increasing length of the crystal. This is achieved by maintaining the current flowing through the crystal constant, assuming that the resistance of the crystallization region remains constant throughout the crystal growth. Although the degree of single-domain formation by such a method is rather high (in fact it is almost 100%), the applied electric field gives rise to electrolysis near the growth surface. As a result, defect regions (cavities) of a macroscopic size are formed in the crystal. For this reason, the second method has recently been employed much more frequently. It consists of applying an electric field of several volts per cm to the grown crystal at temperatures close to the Curie point for 0.5 to 1 h. Platinum serves as the electrodes; the whole process takes tens of minutes. The method simultaneously makes it possible to get rid of the coloration of the crystal which arises, as Reisman and Holtzberg (1958b) believe, due to formation of oxygen vacancies and the lowering of the niobium valency to 4. Moreover, this regime of single-domain formation reduces the residual tension, although it cannot as a rule be removed completely. Among the disadvantages of the second method, one should mention the diffusion of platinum into the near-electrode crystal layers and the cracking of the near-electrode regions of the specimen. To avoid such undesirable effects, ceramic electrodes made of LN are placed between the platinum electrodes and the crystal (Yamada *et al* 1968).

Lithium metaniobate is a uniaxial ferroelectric; therefore, in conformity with the symmetry change in the phase transition $\bar{3}m \rightarrow 3m$, only 180° domains may exist in the crystals.

Ferroelectric domains in lithium metaniobate crystals were first discovered by Nassau and Levinstein (1965). Their investigation was carried out using one of the first Czochralski-grown crystals. Later, an attempt was made to establish the regularity governing the arrangement of the domains in the crystal and to determine the orientation of the boundaries between them.

In order to visualize domains, use was made of two phenomena. First, it has turned out that a relief on the polished crystal surface can be observed in the reflected light. The relief is due to the different hardness of the crystal on either side of the domain boundary (Carruthers *et al* 1971).

Secondly, it has been found that some etchants differently affect the surfaces of the domains of opposite polarity. There have been tested such procedures as the etching in KOH melt at 400 °C for 15 to 60 s; etching in a mixture consisting of two parts of 30% hydrogen peroxide and one part of sodium hydroxide at 50 °C for 15 min; and 10 min etching by a mixture of one part of hydrofluoric acid and two parts of nitric acid at the boiling point (110 °C). Of all the etchants listed, the latter was found to be the most convenient, because unlike the other ones it was capable of revealing domains not only on the surface normal to the polar axis but also on the surfaces parallel to that axis.

That the regions thus isolated are indeed domains of the opposite sign can be proved by the fact that, when locally heated by a focused laser beam, these regions, because of the pyro-effect, are polarized with the opposite sign. The domain nature of these regions is also confirmed by experiments on repolarization in the external field. Upon formation of a single domain the crystal cross sections at both electrodes are etched uniformly, though at different rates. The characteristic size of a single-domain region is about 0.1 mm.

Nassau *et al* (1966) have found that, in crystals grown along the polar axis, domains form a system of axial cylinders oriented along the growth direction, which gives rise to a series of concentric rings appearing on the etched (0001) surface. When crystals are grown in the direction perpendicular to the polar axis, domains form a sandwich-type structure parallel to the Z axis.

The domain structure of Czochralski-grown crystals has been investigated at greater length by Parfitt and Robertson (1967), who studied in particular how it may be affected by various growth conditions, e.g. the pulling rate (from 5 to 12.5 cm h^{-1}), the rotation speed (0 to 100 rpm) and the seed orientation (the $\langle 000\bar{1} \rangle$ and $\langle 2\bar{1}\bar{1}0 \rangle$ directions). The grown crystals were annealed in an oxygen atmosphere at 1100 °C for 24 h. A solution of KHF$_2$ in concentrated nitric acid served as an etchant (a 5 min etching at 100 °C).

Crystals grown from a very pure raw material along the polar direction $\langle 0001 \rangle$ only had a thin (less than 1 mm) periphery envelope of a parallel domain, whereas those obtained from a less pure material had the domain structure in the form of rings and antiparallel layers lying parallel to the growth surface. In crystals grown along the $\langle 2\bar{1}\bar{1}0 \rangle$ two domains were formed, the boundary between them passing along the principal axis of ellipsoidal crystalline boules.

As the pulling rate increases to $1.7\ cm\ h^{-1}$, there arises a many-domain structure, whatever the crystallographic orientation of the specimen. A change in rotation speed does not affect the geometry of the domains in crystals grown along the $\langle 2\bar{1}\bar{1}0 \rangle$ directions. If, however, the growth is in the $\langle 0001 \rangle$ direction, then at rotation speeds between 60 and 100 rpm ringed and layered domain regions will be broken up and, at 10 to 30 rpm, a spiral domain structure will be formed, which was also confirmed by Garmasch *et al* (1972). The symmetry of the domain structures of crystals grown along the $[000\bar{1}]$ direction has been determined as ∞:m and 2:m, while that of crystals grown along the $[2\bar{1}\bar{1}0]$ and $[\bar{1}2\bar{1}0]$ directions, as 2.m and m.1:m respectively. Adding impurities to the melt lowers, as a rule, the degree of unipolarity of the crystal, which was established by Nassau *et al* (1966), who had added 0.5 wt.% Sc to the melt, and by Parfitt and Robertson (1967), who had introduced Al_2O_3, ZnO, SnO_2 and TiO_2—each in an amount of 0.5 wt.%. An only exception is MgO, which assists in forming a single domain. This effect, also reproduced by Nassau *et al* (1966), was explained by the emergence of an electrically charged layer of Mo^{6+} ions near the growing crystal surface. Addition of Mo to the melt has accordingly been suggested as a method for single-domain formation.

The domain structure of LN crystals grown by other methods has been investigated much less. Domains formed in crystals obtained by the directed crystallization technique have been described by Deshmukh (1972). Apart from the usual, 180 degree domains, which were made apparent only by etching, 35 degree domains have been found in crystals by optical methods. The latter are formed as mechanical twins when crystallization takes place in a confined volume with the twinning plane $\{10\bar{1}2\}$.

The Bridgman–Stockbarger technique makes it possible to grow crystals with the domain structure similar to that of Czochralski-grown crystals.

When LN is grown hydrothermally in the temperature range from 500 to 600 °C, i.e. below the Curie point, single-domain crystals are grown (Niizeki *et al* 1967). The same objective may be achieved by the method of transport reactions (Fushini and Sugii 1974).

Along with macroscopic domains described above, there are also micro-domains in LN crystals. These have been found only in Czochralski-grown crystals by chemical etching on the etched $\{0001\}$ surface of the specimen. Triangular etch pits were mistaken by Nassau *et al* (1966) for emergent dislocations. Ohnishi and Zizuka (1975) have established, however, that the pits appear as a result of etching around disseminations of micro-domains. The latter represent needle-shaped 180° domains stretched along the Z axis; they are 0.1 to 1 μm across and 200 to 400 μm long, their cross sections being polyhedral in shape. Micro-domains are present in grown crystals which have not been subjected to heat treatment; they may also be formed if the crystal surface is mechanically scratched. Such a micro-domain may have a dislocation at its basis.

The observed features of the domain structure of LN along the growth direction [0001] are due to the distribution of impurities, which is in turn caused by the hydrodynamic and thermal flows of the melt in the crucible. The formation of the domain structure of crystals grown in the $\langle 2\bar{1}\bar{1}0 \rangle$ directions takes place because the phase transition does not occur simultaneously in different parts of the crystal and because the growth conditions change with time. The domains may also be formed as a result of the presence of stresses in the para-electric phase during the growth.

4.2 Selective etching of LN crystals

The selective etching technique (SET) allows domains of at least several micrometres in size to be revealed, the domain boundaries being delineated rather clearly. The SET also makes it possible to determine the sign of a domain and to trace the emergence of domain boundaries on the various crystallographic planes, thereby indicating the three-dimensional arrangement of the domains within the specimen. The difference in etching rates and in the morphology of surfaces subjected to selective etching may serve as a basis for choosing the positive direction of the Z axis [0001] in the crystallographic arrangement of a crystal. The crystallographic axes X [$2\bar{1}\bar{1}0$], Y [$\bar{1}2\bar{1}0$], U [$\bar{1}\bar{1}20$], Z [0001] and the crystal physical axes X_1 [$2\bar{1}\bar{1}0$], X_2 [$01\bar{1}0$], X_3 [0001] may also be chosen on the basis of the morphology of the corresponding surfaces and symmetry of the crystal 3m (see figure 1.14).

The optical polarization technique is incapable of revealing the domain structure of lithium metaniobate, because the optical indicatrices in 180° domains have the same orientation; moreover, domains cannot be distinguished by optical extinction. The optical method may, however, reveal domain walls in lithium metaniobate by the presence of birefringence arising from stresses near the domain walls. Figure 4.1(a) shows a many-domain crystal, as seen in cross-polarizers in the direction perpendicular to the polar axis. Figure 4.1(b) shows the same crystal, but with the domain structure disclosed by the SET. The optical polarization technique has a number of disadvantages, as compared with the SET: domains are not clearly determined; their signs are not found;

the interpretation of observed patterns is far from easy, because stressed regions may be due not only to domain walls but also growth dislocations, various inclusions, cracks etc. If domain walls emerge at small angles to the crystal surface, the optical polarization technique only makes it possible to determine the positions of the domain boundaries in rather thin plates where the light flux is not scattered by the domain walls. The application of an electric field improves the determination of domains in the crystal. However, in this case too the interpretation is difficult for a complicated arrangement of domain walls.

(a)

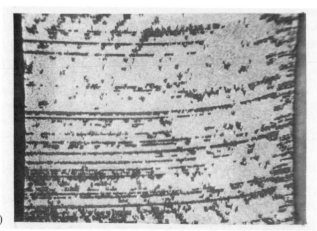

(b)

Figure 4.1. Domain structure of lithium metaniobate disclosed by (a) optical polarization and (b) selective etching techniques. The $(01\bar{1}0)$ surface. Magnified seven times (Evlanova 1978).

Observation of domain walls through an electron microscope enables one to see the configuration of the wall itself, but not the arrangement of domains throughout the crystal bulk (Wicks and Levis 1968).

Ferroelectric domains in LN are equivalent to inversion twins, and may be disclosed by the X-ray topography technique thanks to the difference between the intensities of reflection from the planes (hkl) and ($\bar{h}\bar{k}\bar{l}$) (Wallace 1970). This method allows one to study rather extensive crystal surfaces (of the order of several cm^2), but its resolving power (50×10^{-6} m) is not as good as that of the SET.

The domain structure has also been established by the powder technique in which lycopodium and minium suspended in toluene are used as charged particles precipitated on the crystal surface (Evlanova 1978). The crystal is warmed to 30–50 °C. The domain structure thus revealed turns out to be indistinct, its outlines being smeared and not all domains being disclosed. This seems to be due to the method of treating the crystal surface on which the precipitation occurs. To precipitate powders on a freshly cleaved surface of LN is rather difficult, because the crystal has no perfect cleavage. Powders were precipitated on surfaces obtained by mechanical treatment, as a result of which a deformed layer appeared, thus hindering a clear determination of domains.

Tasson *et al* (1975) precipitated a coloured powder, suspended in oil, on the crystal surface heated to 50–100 °C to reveal the domain structure in LN. The method allows the many-domain specimens to be distinguished from the single-domain ones. Decorating the domain boundaries by pyroelectric charges, arising due to a short-time heating of many-domain specimens to 100–150 °C and their subsequent cooling, has also been used (Belobaev 1976).

The powder and the decoration techniques are greatly inferior to the SET in resolving power and the clarity of determination of domain structure.

There is also the polishing technique, allowing the domain structure to be revealed while examining polished crystal surfaces in reflected light. A relief arises on the surface because of the different hardness of the opposite-sign domains on the planes {0001} and {10$\bar{1}$0}. The resolving power of the polishing technique is considerably lower than that of the SET too.

Thus selective etching seems to be the best method for detecting the growth domain structure in crystals of lithium metaniobate. It was used for the first time by Carruthers *et al* (1971), who indicated the following etchant composition: a mixture of two parts of nitric and one part of hydrofluoric acids, the etching time being 10 min and the boiling temperature of the mixture 110 °C. Nassau *et al* (1966) proposed two more etchants: a KOH melt at 400 °C (etching time 15 s to 1 min) and a mixture of two parts of 30% hydrogen peroxide and one part of sodium hydroxide (etching for 15 min at 50 °C). Experience has shown that the etchant and the whole etching procedure used by Carruthers *et al* (1971), which is commonly referred to as hot etching, gives better results. It has therefore become the chief method of revealing domains in LN crystals.

Hot etching was used by Evlanova (1978) for plates 2 mm thick. The surface was oriented from conoscopic and X-ray back-scattering patterns. The necessity of combining the two methods is due to the fact that some orientations in crystals yield similar X-ray diffraction patterns, for instance $\langle 0001 \rangle$ and $\langle 10\bar{1}1 \rangle$.

The outward appearance of a surface subjected to selective etching depends on its crystallographic orientation. On the (0001) plane, hillocks in the form of pyramids confined by the $\{10\bar{1}2\}$ planes appear (figure 4.2). The process of etching runs in practically the same way on both polished and ground surfaces, the etching rate being 1×10^{-6} m min^{-1} and the etching time from 6 to 10 min. The triangular shape of etch patterns corresponds to the emergence of a three-fold axis on the (0001) plane. When the [0001] direction deviates from the normal to the plane under study, etch pits become asymmetric, extending in the direction [000$\bar{1}$].

The (0001) plane is etched much more slowly than the (000$\bar{1}$) plane, etch patterns not showing after the usual etching time (6 to 10 min). On

Figure 4.2. Etched (000$\bar{1}$) surface of LN. Magnified 350 times (Evlanova 1978).

the (0001) plane, etch patterns appear as pits after 30 to 40 min of etching.

On the $(0\bar{1}10)$ plane, etch patterns have a conical shape (figure 4.3) and if the $[0\bar{1}10]$ direction deviates from the normal to the specimen surface these etch cones are inclined in the $[000\bar{1}]$ direction. The $(01\bar{1}0)$ plane is etched at a slower rate, etch patterns appearing as incompletely faceted rhombuses. The planes $(2\bar{1}\bar{1}0)$ and $(\bar{2}110)$ are also poorly etched and etch patterns are indistinct.

Figure 4.3. Etched $(0\bar{1}10)$ surface of LN, magnified 350 times (Evlanova 1978).

Hot etching is inapplicable for studying the domain structure of large specimens (thicker than 2 mm) and crystal boules. LN crystals are not very plastic and may crack when they are etched or cooled. For such brittle specimens, etchants have been found which are capable of revealing the domain structure at room temperature, e.g. mixtures of hydrofluoric and nitric acids; hydrofluoric acid with an admixture of $KMnO_4$; a mixture of hydrofluoric acid with caustic potash (Evlanova 1978). The domain structure is best revealed by the etchant of the composition $1HF + 4HNO_3$ (the numbers referring to volume parts), with an etching time of about 20 h. The surfaces to be etched should preliminarily be well polished. The difference in the etching rate of two domains on the surface perpendicular to the optical axis leads to formation of a 3×10^{-6} m high jig at the domain boundary after 24 h of etching. In this etchant, the planes $(01\bar{1}0)$ and $(0\bar{1}10)$ are also etched at different rates. They are easily distinguishable when they are obliquely illuminated by reflected light, although some etch patterns do now show themselves.

It has been noted that, while using the SET, on crystal sections perpendicular to the optical axis (Z sections) or perpendicular to the crystal symmetry plane (Y sections), the domain structure is rather clearly revealed and the surfaces of opposite-sign domains are characterized by different etch figures. At the same time, while etching the section cut parallel to the symmetry plane (X section) the domain structure is revealed indistinctly, etching patterns on domains of opposite sign being practically the same.

That an etchant acts differently on different crystal planes can be explained by the crystallography of domain twinning. As is known, the symmetry elements lost in a phase transition serve as elements of domain twinning. In the case of lithium metaniobate, in the phase transition from the symmetry group of the paraphase $\bar{3}\frac{2}{m}$ to the symmetry group of the ferroelectric phase 3m, the symmetry elements $\bar{1}$, $2i,\bar{3}$ ($i = 1, 2, 3$), which serve as twinning elements of ferroelectric domains, get lost. The relative orientation of the crystallographic directions in domains is shown in figure 4.4. From the stereographic projection of the group 3m (figure 4.1) it can be seen that the directions $\langle 000\bar{1}\rangle$ and $\langle 01\bar{1}0\rangle$ are polar (the $\langle 0001\rangle$ direction being the special polar one), while the $\langle 2\bar{1}\bar{1}0\rangle$ directions are non-polar. Polar axes emerge on the Y and Z sections, and etching clearly reveals domains there (figures 4.1 and 4.2). Non-polar directions emerge on the X section, and the $(2\bar{1}\bar{1}0)$ surfaces corresponding to them are etched practically in the same manner.

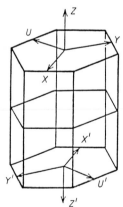

Figure 4.4. Relative position of crystallographic directions in domains of $LiNbO_3$.

The different action of an etchant on the polar (Y, Z) and non-polar (X) sections can be explained by considering the atomic structure of the corresponding planes. Since etching is largely determined by dissolving, it can be associated with a layer-by-layer removal of the atoms. The results of etching will be different if the arrangement of atoms and atomic groups in the direction normal to the plane under study and in the opposite direction will differ. The alternation of atoms and atomic groups lying in the directions $[2\bar{1}\bar{1}0]$ and $[\bar{2}110]$, which are perpendicular to the X section, is the same, resulting in the same etch patterns of the domains on the X section. The directions $[\bar{1}2\bar{1}0]$ and $[1\bar{2}10]$ are normal to the Y section and the $[0001]$ and $[000\bar{1}]$ to the Z section. The arrangement of atoms and atomic groups in each pair of directions is not the same (figures 4.5 and 4.6), leading to different etch patterns of domains in the Y and Z sections.

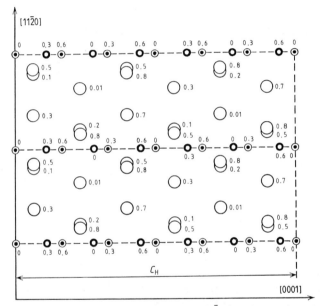

Figure 4.5. Projection of atoms on the $(1\bar{1}00)$ plane in LiNbO$_3$ structure (constructed using the data of Abrahams *et al* 1966a,b,c): Large circles, oxygen; small circles, lithium; dotted circles, niobium.

4.3 Growth domain structure

Crystals of lithium metaniobate grown in the direction of the optical axis $[0001]$ are cylindrical in shape. A boule of a $0°$ crystal consists of the principal (occupying the larger part of the crystal) domain, with the positive direction of the spontaneous polarization vector P_s toward the upper part of the boule, and domains of the opposite polarity. These occur inside the main domain as dome-shaped regions, which often fail to reach the cylindrical surface of the boule, their convex part being

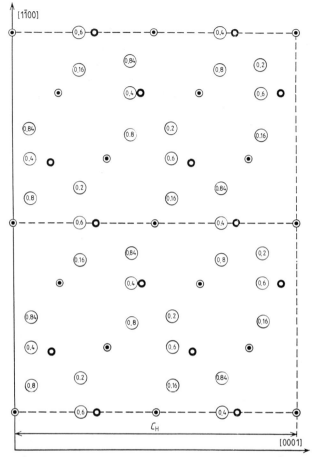

Figure 4.6. Projection of the atoms on the plane (11$\bar{2}$0) in LN structure constructed according to the data of Abrahams *et al* 1966a,b,c). The symbols are the same as those in figure 4.5.

directed towards the lower part of that surface (figure 4.7). On the boule surface, a domain of the same polarity as the inner ones forms an envelope. The size of the envelope domain observed in the direction perpendicular to the optical axis reaches 0.4 mm (Evlanova 1978).

If a crystal growth direction makes an acute angle with the optical axis, then there arises the domain structure shown in figure 4.8. The crystal consists of the main domain which has the same orientation as the 0° crystal. The domed regions of the inner domains do not pass through the whole cross section of the boule, but only pass where their boundaries are perpendicular, or almost perpendicular, to the direction of the optical axis. In those parts of the boule where the boundaries

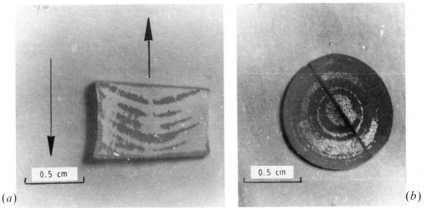

Figure 4.7. Domain structure 0° of the crystal (Evlanova and Rashkovich 1974): (*a*) section parallel to the polar axis, the arrow indicating the direction of the spontaneous polarization vector of the principal domain; (*b*) section perpendicular to the polar axis.

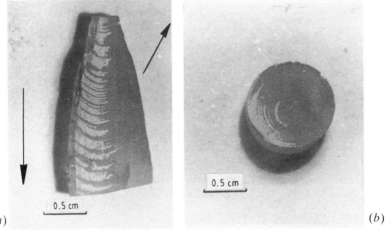

Figure 4.8. Domain structure 57° of the crystal (Evlanova and Rashkovich 1974): (*a*) section parallel to the polar axis, the arrow indicating the P_s direction of the main domain; (*b*) section normal to the growth direction.

would be parallel to the Z axis or would make an acute angle with it, inner domains do not arise, this part of the boule representing the main domain. As in 0° crystals, in boules with the growth direction at an acute angle with the optical axis, the envelope domain may exist.

Crystals grown along the directions $\langle 2\bar{1}\bar{1}0 \rangle$ and $\langle 01\bar{1}0 \rangle$, i.e. 90° crystals, have the same domain structure. Boules of such crystals are elliptic in shape, the optical axis being directed along the minor axis of the cross section. As the crystal is pulled out of the crucible, the ellipticity of the boule cross section increases because of the following growth-rate relation:

$$V_{\langle 2\bar{1}\bar{1}0\rangle} \approx V_{\langle 01\bar{1}0\rangle} > V_{\langle 0001\rangle}$$

where the indices indicate the crystal growth direction. 90° crystals consist of two main domains of opposite polarity and a number of domed inner domains, which are similar to those of 0° crystals. The boundary between the two main domains runs across the plane of the largest axial cross section of the elliptical cylinder of the boule perpendicular to the optical axis (figure 4.9). The orientation of the domains is such that the negative direction of the P_s of each domain is towards the domain boundary (i.e. toward the crystal's interior). Inner dome-like domain regions are oriented relative to the growth direction in the same manner as those of 0° crystals, i.e. with their convex part towards the lower part of the boule. In the case of 90° crystals, however, the inner domains consist of two domains of opposite polarity, and the boundary between them coincides with that between the main domains. In each main domain are parts of the inner domains of the opposite sign. Both main domains may have on the boule surface envelope domains of the opposite polarity, the envelope domains containing no inner domains.

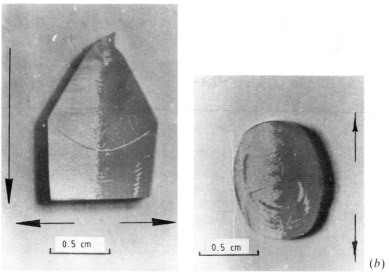

(a)   (b)

Figure 4.9. Domain structure 90° of the crystal (Evlanova and Rashkovich 1974): (a) section parallel to the growth direction; (b) section perpendicular to the growth direction (the arrows show the direction of P_s in main domains).

Thus one can morphologically distinguish three domain types in lithium metaniobate crystals: main domains, inner domains and envelope domains. The types of domain structure formed for different growth directions may be characterized by the number and the orientation of the main domains. The first type of growth domain structure is represented by crystals with a single main domain, in which the positive direction of P_s is toward the seed. The second type includes domains with the positive P_s directed towards the crystal surface.

The main regularities of the growth domain structure are common to crystals obtained by quite different growth regimes. As the pulling rate is varied from 4 to 40 mm h^{-1} and the rotational speed from 4 to 40 rpm, the main features of the domain structure remain unchanged. Nor is it affected by the domain structure of the seed, although one might expect that, in the ferroelectric transition, the 'domain memory' of the seed would restore its domain structure, which will be inherited as the Curie isotherm moves along the crystal. When 0° crystals are grown from single-domain seeds with either positive or negative direction of the Z axis relative to the melt, or when they are grown from many-domain seeds with the domain structure of the first or second type, there will invariably arise the domain structure of the first type. A similar situation also holds for the other directions of crystal growth.

The introduction of impurities does not change either the main regularities of the domain structure or the type of the crystal domain structure. However, impurities do affect the number of the domain boundaries inside the crystal. When Cr, Dy, Er, Nd are added to the melt in amounts from 0.05 to 0.5 wt.%, the number of the inner domains increases so much that it appears impossible to distinguish the main domain. At the same time, crystals grown from a very pure initial material have few inner domains.

4.4 Formation of a stationary domain structure

The experimental studies carried out so far have shown that at temperatures below 1000 °C heating the crystal for up to 10 days leaves its domain structure unchanged. Above that temperature there takes place a gradual change in the initial domain structure and there finally emerges some stable configuration of the domain boundaries which no longer changes as the annealing time is further increased. The time necessary for obtaining this steady state decreases with increasing temperature, but it increases as the specimen thickness increases. It has been found (Evlanova 1978) that for a single-domain crystal specimen 1 mm thick (along the polar axis) it takes 120 h of heating at 1100 °C, 4 h at 1170 °C and only half an hour at 1200 °C to reach the steady state. Although a prolonged pre-heating at about 1000 °C leaves the domain structure unchanged, it reduces the time of reaching the stationary state when the specimen is subsequently heated to higher temperatures. As a rule, specimens are annealed in air. It is known that air is an oxidizing medium for lithium metaniobate at the high temperature. Annealing of the crystals in a vacuum (a reducing medium) has yielded the same results as annealing in the air. The appearance of the steady-state domain structure does not change if one takes various proportions of the principal components in the crystals. Thus, specimens with the ratio Li_2O/Nb_2O_5 ranging from 1.10 to 0.88 and with impurities of Cr, Er, Eu, Nd in the amount of 0.5 wt.% (in the melt), all had one and the same type of stationary structure (Evlanova 1978).

The appearance of the stationary domain structure is also independent of the original arrangement of domains in the crystal. Moreover, the same structure will appear if a single-domain specimen is heat treated.

A typical stationary domain structure is represented by two domains, and the boundary separating them is roughly perpendicular to the ferroelectric axis, often dividing the crystal into two almost equal volumes (figure 4.10). The orientation of the domains is such that the positive direction of the vector P_s is inside the crystal. In the similar type of the growth domain structure, the negative direction of the P_s of each main domain is towards the crystal's interior (see figure 4.9).

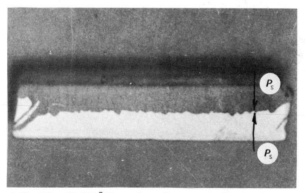

Figure 4.10. The $X_2\{01\bar{1}0\}$ section of a LN specimen with stationary domain structure. Four-fold magnification (Evlanova 1978).

The steady-state structure is attained by both the growth of the already existing domains and the origination and subsequent growth of new domains inside the domains of the opposite sign. The latter phenomenon is always observed in specimens with a small number of domain boundaries. The process of domain structure restructuring starts at the $\{0001\}$ surfaces of the crystal.

In many-domain crystals with a large number of inner domains the domain structure is changed only at the expense of the growth of the already existing domains of the corresponding sign. The process starts from the {0001} surfaces of the crystal. The repolarizing domain first decreases in thickness (in the direction of the ferroelectric axis), falls into isolated regions and then repolarizes completely. The smaller the number of inner domains the more probable is the origination of new domains in the process of change in domain structure. The steady-state domain structure is more rapidly formed in single-domain specimens than in many-domain ones. This seems to be due to the absence of domain walls, which tend to fix the original domain structure, in such crystals. In a single-domain crystal, the domain boundary begins to move away from the (0001) plane, reaching the middle of the specimen in the steady state (figure 4.10).

The domain wall that arises in a crystal represents a jogged surface formed by crystal microfaces, among which the $\{10\bar{1}2\}$ occurs rather often. The larger the crystal size along the polar direction the greater is the deviation of the domain wall from the plane.

If a crystal plate has side surfaces which are not parallel to the direction of the ferroelectric axis, then near such surfaces the domain wall shifts from the middle of the specimen, giving rise to the so-called edge effect (figure 4.11). The shift of the domain wall is observable in a plate in which the direction of the ferroelectric axis deviates from the normal to the plane of the plate. A similar situation is observed in specimens which have the shape of a truncated cone, where the side surfaces are also non-parallel to the polar direction. The shift of the domain wall always occurs in such a way that the negative end of the spontaneous polarization vector P_s emerges to a side surface.

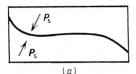 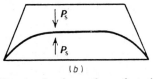

(a)   (b)

Figure 4.11. Scheme of the edge effect in the formation of stationary domains. The arrows indicate the direction of the spontaneous polarization vector: (*a*) specimen as a plate; (*b*) specimen as a truncated cone (Evlanova 1978).

The appearance of envelope domains in the domain structure can be explained as follows. Such domains are best pronounced in the growth domain structure of the second type, the main domains being oriented so that the positive direction of the vector P_s is away from the boule. The orientation of envelope domains is just the opposite. The more slowly the crystal is cooled in the growth installation when the growth is at an end, the larger the size of the envelope domains becomes. They may not arise at all if the crystal is cooled rapidly. In 90° crystals, envelope domains appear on both sides, and they completely cover the main domains. An envelope domain has no inner domains. These properties of envelope domains are due to the partial annealing of the crystal as it is cooled in the growth chamber, thus giving rise to the steady-state domain structure. The domain-structure rearrangement in this process starts from the (0001) surfaces of the specimen; in our case, from the flattened side surfaces of the boule. The size of the resulting domain is proportional to the annealing time. No steady-state domain structure is observed if the annealing time is small. The absence of inner domains in an envelope domain also corresponds to their absence in domains of the stationary domain structure.

In the growth domain structure of the first type (figure 4.12), the envelope domain is less distinctly pronounced. It arises if the polar axis of the crystal is not exactly parallel to the crystal surface and the annealing time in the growth chamber is long enough. The envelope domain is not formed all over the boule's surfaces, but only in that part which makes an acute angle with the direction [0001] of the polar axis. The origination of an envelope domain in the growth domain structure of a 0° crystal can also be explained by the formation of the stationary structure. In contrast to two envelope domains of the growth domain structure, which correspond to the emergence on the boule surface of the positive ends of the polar axis of the main domains, in the 0° crystal a single envelope domain is formed, and this corresponds to the [0001] direction of a single main domain emerging on the boule surface. That envelope domain is situated on one side of the boule surface if the boule is of a regular cylindrical shape, while the non-parallelism of its surface to the polar axis is due to an exact orientation of the seed (figure 4.12(*a*)). However, if the seed is accurately oriented, but the boule's shape is not cylindrical, i.e. the diameter of the 0° crystal is apt to change in the course of growth, then it will be the positive end of the polar axis that will come to the surface in places where the crystal diameter is changed. When annealed, such a crystal will have the envelope domain in corresponding parts of its surface (figure 4.12(*b*)). This envelope domain will show especially distinctly on the boule expansion cone (figure 4.12(*c*)). Crystals grown along this direction, which makes a substantial angle with the polar direction, may also have an envelope domain on the surface—on the side from which the [0001] direction emerges.

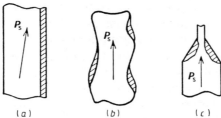

Figure 4.12. Scheme of the arrangement of envelope domains in domain structure of type I: (*a*) growth direction departs from the 0° direction; (*b*) crystal with variable diameter; (*c*) cone of boule expansion. The arrows indicate the direction of the P_s vector of the main domain (Evlanova 1978).

4.5 Repolarization mechanisms of lithium metaniobate crystals

Typical shapes of repolarization pulses observed in LN specimens are shown in figure 4.13 ((*a*), asymmetric; (*b*), symmetric). The shape of a pulse is determined by the character of the repolarization process involved. A symmetric pulse shape characterizes repolarization in specimens of a high optical quality (no micro- or macro-inclusions). An asymmetric pulse, on the other hand, will correspond to low-quality specimens. For some crystals, usually characterized by an asymmetric pulse, it is possible to obtain a symmetric pulse if the higher voltage is applied.

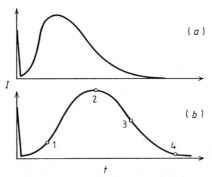

Figure 4.13. The shape of repolarization pulses applied to lithium metaniobate crystals: (*a*), asymmetric; (*b*), symmetric (Evlanova 1978).

A repolarization pulse can be observed when some voltage, which is higher than a threshold value (depending on the specimen temperature), is applied to the specimen. For example, at $T = 130\,°C$ the threshold tension for LN is $80\,kV\,cm^{-2}$, while at $T = 170\,°C$ it is only $60\,kV\,cm^{-1}$. If a tension lower than the threshold one, but close to it, is applied (no repolarization pulse is then observed), then upon etching a chain of hexagonal etch pits will appear on the (0001) surface of the specimen, which are situated along the electrode edge. On the Y section of such a specimen, the etching will bring to light needle-shaped domains penetrating to $(20–30) \times 10^{-6}\,m$ into the specimen. This is a rather stable domain configuration, which will not change as an electric field is applied for a long time (in fact for a period tens of times as long as the repolarization time at the threshold voltage value). On the (0001) surface and in the volume of the specimen, etching fails to reveal any changes in domain structure.

In order to find the correspondence between a repolarization pulse and a real process of displacement of the domain boundaries, as well as to study the repolarization kinetics, Evlanova has performed the following experiment. On single-domain specimens cut out of the same LN crystal (Z section), the repolarization process was interrupted as soon as the repolarization current reached values indicated at points 1–4 (figure 4.13(b)), corresponding to different repolarization times for each specimen. The specimens were then cut at the middle of the electrode along the (0110) plane and the resulting halves were etched. The repolarization times were 7 or 8 min, the electric field intensity being $85\,kV\,cm^{-2}$ and the crystal temperature $130\,°C$.

At point 1, which corresponds to the onset of a repolarization pulse, on the (0001) surface appear hexagonal etch patterns (pits), whose outline is formed by the $\langle 01\bar{1}0 \rangle$ edges (figure 4.14). On the (000$\bar{1}$)

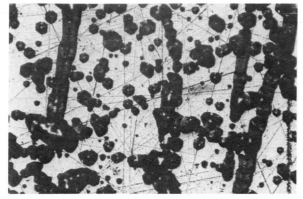

Figure 4.14. Etch figures on the (0001) planc in the initial stage of repolarization. 210-fold magnification (Evlanova 1978).

surface, no new domains are revealed by etching. Figure 4.15 illustrates the Y section of the same specimen. Narrow domains grow from the (0001) surface, new domains originating in the bulk of the specimen. At the (000$\bar{1}$) surface, there are practically no new domains. The domains that have already come to existence are $(50-200) \times 10^{-6}$ m in length and $(2-10) \times 10^{-6}$ m in width.

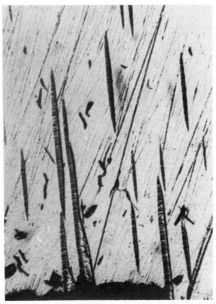

Figure 4.15. Origination of domains in the bulk of a specimen in the initial stage of repolarization (the X_2 section). 210-fold magnification (Evlanova 1978).

Figure 4.16 shows the further growth and merging of domains on the (0001) surface at the maximum repolarization pulse amplitude corresponding to point 2 in figure 4.13(b). The surface in question turns out to be repolarized by 60 to 70 per cent. Figure 4.17 shows the domain structure on the Y section of the same specimen. The domains have grown out from one electrode to the other, joining onto each other in the larger part of the crystal volume. The total area which corresponds to the repolarized volume constitutes 50 per cent.

At point 3 of the above figure the repolarization pulse falls off. In these conditions, the (0001) surface of the specimen is completely repolarized, whereas the (000$\bar{1}$) surface is repolarized only in part. On the Y section of the (000$\bar{1}$) surface there exists a region of non-repolarized domains in the shape of wedges (figure 4.18). Apart from

Figure 4.16. Etch figures on the (0001) plane in the stage of maximum repolarization. 210-fold magnification (Evlanova 1978).

Figure 4.17. Domain structure of the specimen (the X_2 section) corresponding to the maximum value of a repolarization pulse. 210-fold magnification (Evlanova 1978).

this region, the entire crystal volume is repolarized. The repolarization pulse falls off to zero at point 4, the crystal being then repolarized almost completely. At the $(000\bar{1})$ surface, however, there remain some non-repolarized wedge-like parts, the length of these wedges reaching 100×10^{-6} m and their width 5×10^{-6} m. This group of domains usually remains non-repolarized even if the field is applied for a period several times longer than the main repolarization time. The repolarization may be complete in high-quality crystals when the field is impressed for a rather long time.

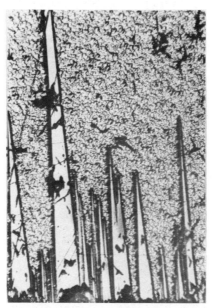

Figure 4.18. Domain structure of the specimen (the X_2 section) at the stage of decreasing repolarization pulse. 210-fold magnification (Evlanova 1978).

The repolarization of a many-domain specimen is peculiar. The specimen is first divided into single-domain and many-domain regions along the boundaries which run in the direction of the field. Single-domain regions are repolarized at normal values of electric-field intensity and temperature, whereas in polydomain regions antiparallel domains fail to originate under these conditions. As the field intensity increases to values approaching the breakdown ones or as the crystal temperature is sufficiently raised, a partial repolarization of the many-domain regions takes place. In those parts of the crystal in which the spontaneous polarization vector is directed opposite to the direction of the field applied, the repolarization occurs by the inter-growth of the needle-shaped domains. The lateral movement of the domain walls is hampered.

The above experiments allow us to estimate the rate at which the domain walls are displaced if the repolarization process is interrupted in the stage when the needle-shaped domains are rather large but have not yet begun to merge. This state of domain structure is reached in the stage corresponding to points 1 and 2 of the repolarization pulse curve (figure 4.13(*b*)). On the Y section of the crystal the widest parts of the new domains will then correspond to those domains which arose at the beginning of period t_s. Estimated in this manner, the rate of the lateral wall movement is $v_g \sim 10^{-5}$ cm s^{-1}, with the electric field intensity $E = 85$ kV cm^{-1} and temperature 130 °C.

The rate of propagation of the needle-shaped domains in the direction of the ferroelectric axis may be estimated also for domains which arose in the initial stage of repolarization. The measurements of the ratio of the length l to the width h of the domains, which have been made on the Y section of isolated domains, yield a value of l/h in the range 18–24.

Taking the above estimate into account, one can find the rate of domain movement along the polar axis: $v_c \approx 2 \times 10^{-4}$ cm s^{-1} at $E = 80$ kV cm^{-1} and $T = 130$ °C.

From the equation

$$2P_s = \int_0^\infty i \, dt = i_{max} t_s f \qquad (4.1)$$

where f is a shape coefficient and t_s is the repolarization time, one can determine the spontaneous polarization of the crystal P_s if the area below the repolarization curve is measured. For nominally pure LN crystals, the value of P_s is 60 ± 5 C cm^{-2}, which is comparable to the results quoted in the literature of 70 C cm^{-2} (Wemple *et al* 1968) and 50 C cm^{-2} (Savage 1966).

The integrated kinetic characteristics of the process of repolarization in LN are given by relations between the electric field intensity E applied to the crystal, the repolarization time t_s, the temperature T and the specimen thickness d.

The dependence of the repolarization rate (which is inversely proportional to t_s) upon electric field intensity is illustrated in figure 4.19 for a specimen 0.45 mm thick at a temperature of 130 °C. This dependence is linear for field intensities above 90 kV cm^{-1}, the deviation from linear behaviour being only observed near the threshold (at this temperature) field intensity ($E_0 = 80$ kV cm^{-1}). A similar dependence is obtained at a temperature of 170 °C.

The dependence of t_s on the inverse temperature $1/T$ at a constant field intensity equal to 85 kV cm^{-1} and a specimen thickness of 0.9 mm is shown in figure 4.20. As can be seen from the figure, in the temperature range from 130 to 180 °C t_s depends linearly on $1/T$.

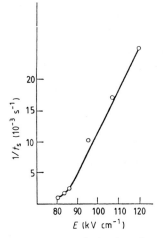

Figure 4.19. The inverse repolarization time as a function of electric field strength at $T = 130$ °C. Specimen thickness = 0.45 mm.

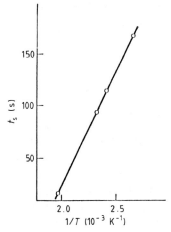

Figure 4.20. The repolarization time as a function of the inverse temperature. The specimen thickness = 0.9 mm, $E = 85\,\text{kV cm}^{-1}$, specimen with admixture of 0.5% Nd.

The repolarization time as a function of specimen thickness is illustrated in figure 4.21 for a field intensity $E = 85\,\text{kV cm}^{-1}$ and temperature 130 °C. One can see from the figure that at a thickness smaller than 0.5 to 0.6 mm the repolarization time depends weakly on the thickness. As the thickness increases, the repolarization time increases very rapidly, i.e. the repolarization rate is (rather strongly) dependent on crystal

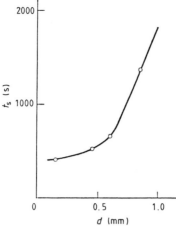

Figure 4.21. The repolarization time as a function of specimen thickness at $E = 85\ \mathrm{kV\,cm^{-1}}$ and $T = 130\ ^\circ\mathrm{C}$ (Evlanova 1978).

thickness, but is relatively independent of the crystal's chemical composition.

One should therefore note the following features of the process of repolarization in lithium metaniobate crystals.

(i) Domains originate mainly on the (0001) surface as wedges and, to a lesser degree, needles in the crystal bulk, domains on the (000$\bar{1}$) surface arising at a later time.

(ii) The growth of domains in the field direction is much more rapid than the lateral movement of the walls.

(iii) In the course of repolarization, nuclei of new domains appear and begin to grow in the crystal volume; domains can move not only by the proper movement of the domain wall, but also by the merger of the individual domains.

(iv) The process of repolarization comes to an end at the (000$\bar{1}$) surface, a certain number of non-repolarized wedge-shaped domains remaining near the surface.

An increase in field intensity leads to a decrease of the non-repolarized crystal volume, but the repolarization is never complete as a rule; the crystal suffers a breakdown by the field.

4.6 The influence of temperature gradients on domain formation in the process of crystal growth and annealing

Ivanova (1979) has investigated the domain formation in Czochralski-grown LN crystals. In the case when crystals grow along the optical axis, the axial temperature gradient, which ensures the growth of a single-domain crystal, is alone in operation. By contrast, for crystals with the growth direction perpendicular to the c axis, the radial gradient is important; therefore in a symmetric thermal field two major domains are inevitably formed.

For lithium metaniobate crystals along a direction making an acute angle with the c axis, the number of major domains and their size vary, depending on the ratio of the components of the axial and radial temperature gradients in the direction of the polar axis of the crystal on the Curie isotherm surface (figure 4.22). As this ratio increases the degree of unipolarity of the specimen also increases, and the crystal will grow single-domain if the projection of the axial gradient upon the c axis is greater than that of the radial gradient all over the surface of the phase transition isotherm. Keeping the specimens grown in the paraelectric phase makes for an increased unipolarity of the crystals, because then their composition becomes homogenized and therefore the emergence of inner domains, which are associated with the instability of the thermal field in the course of growth and with the presence of impurities, is prevented (Evlanova and Rashkovich 1974).

From the study of the effect of growth conditions upon the domain structure of lithium metaniobate crystals carried out by Ivanova, it follows that in order to obtain large single-domain elements out of oblique sections of congruently melted crystals, it is necessary to assure the following conditions.

(i) A flat crystallization front, i.e. a flat isotherm of melting. A flat crystal–melt interface makes for growing crystals of a high optical perfection, with few defects.

A flat front may be attained by reducing the radial gradients in the crystallization region to a minimum. To ensure stable growth, it is desirable to have rather high axial gradients in the growth region, which would make it possible to run the process without interference on the part of the operator and to prevent concentrational overcooling and therefore the streakiness of the crystals, giving rise to inner domains.

(ii) The maximum possible removal of the Curie isotherm from the interface, as well as its flat or slightly convex shape. A prolonged heating of the crystal at temperatures above the Curie point will improve its homogeneity, while a flat Curie isotherm will facilitate the process of crystal polarization (formation of a single domain), because the higher the axial to radial gradient ratio the larger the volume of the crystal that will be polarized. One should note that from the experimental viewpoint these two requirements are contradictory. The trouble is that a flat isotherm can be obtained only in the case when axial gradients are much higher than radial. But if the axial gradient is high, it is impossible to increase the distance between the Curie

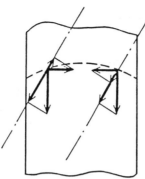

Figure 4.22. Scheme of formation of major domains in the phase transition in lithium metaniobate crystals of 'oblique' direction (Ivanova 1979).

isotherm and the melting isotherm. Lowering the radial gradients in the annealing zone is impeded by the nature of the Czochralski technique itself, which presupposes that the heat of crystallization is released radiatively from the side surface of the growing crystal.

(iii) Constancy of thermal conditions throughout the process of growth. A change in temperature gradients near the growth front leads to growth-rate fluctuations, because the front of crystallization tends to coincide with the isotherm corresponding to the melting temperature. At the same time, growth rate fluctuations lead, even in the case of an almost congruent melting, to variations in composition, thus deteriorating the crystal homogeneity.

Typical results of the study of the domain structure of crystals grown under various conditions are presented in table 4.1, which also lists the values of the axial and radial temperature gradients at different points of the crystal as a ferroelectric phase transition passes through them.

Table 4.1 Temperature gradients in a phase transition and the domain structure of LN crystals (Ivanova 1979).

No	G_{ax} (grad sm^{-1})	G_{rad} (grad sm^{-1})	Domain structure
1	17	10	many-domain
2	17	1	single-domain
3	26	1	single-domain
4	10	7	many-domain
5	10	7	many-domain
6	32	6	many-domain with single-domain parts
7	53	2	the whole crystal is single-domain

4.6.1 Control of domain sizes in shaped LN crystal

The polarization technique for shaped LN crystals is illustrated in figure 4.23 (Red'kin *et al* 1987). The installation for obtaining a single domain was turned on immediately after the crystal was grown to the desirable size. At the end of the process, in order to prevent the dielectric breakdown of the specimen, its neck was reduced to 2 or 3 mm^2 in cross section, whereupon the pulling was stopped. The polarization system was turned off as the die temperature was lowered below the Curie point, $T_c = 100$ °C. The current density was between 1 and 2.5 mA cm^{-2}, the polarity being positive for the seed and negative for the crucible.

It should be noted that in this case the sign-variable electric field can be applied to the crystal, not only in the growth process but also during the cooling, i.e. one can control the domain structure in the part of the crystal which has the temperature above T_c immediately after pulling is stopped. By varying the growth rate, cooling rate and the period during which the sign-variable field is applied, one can obtain domains from several cm to several tens of μm and smaller in size. The shape of the domains is related to the Curie isotherm; therefore it is possible to select optimal thermal conditions for the crystallization process.

4.7 The effect of annealing on the near-surface domain structure

In this section we shall discuss some problems concerned with the change in the near-surface domain structure of LN (Z section), as the crystal is annealed at temperatures between 800 and 1000 °C (Bocharova 1986).

With increasing temperature the probability that some in-homogeneities, due to the emergence of antiparallel domains, will arise in the crystal also increases. The origination rate of the antiparallel domains is proportional to $\exp(-\Delta W/kT)$, where ΔW is the total change in energy in domain formation. The presence of the domains of the opposite sign deteriorates the electro-optical properties of the crystals, also leading to additional light scattering, Barkhausen noises and other undesirable effects (Thaniyavaran *et al* 1985).

The changes that take place on both positive and negative surfaces of LN crystals differ rather greatly. The positive surface is more resistant to etching, shallow dislocational pits appearing on it only after a prolonged boiling in a mixture of acids. Prior to annealing, the original surface of the specimens was single-domain, but after annealing at $T > 800$ °C the etching was able to reveal regions with a higher rate of dissolving (figure 4.24(*a*)). Their size was larger the higher the temperature and the longer the annealing time (figure 4.24(*b*)). Similar pictures were

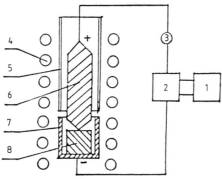

Figure 4.23. Process of polarization of shaped LiNbO$_3$ crystals: (1) power supply; (2) current stabilizer; (3) milliammeter; (4) inductor; (5) platinum shield; (6) crystal; (7) platinum crucible; (8) die (Red'kin *et al* 1987).

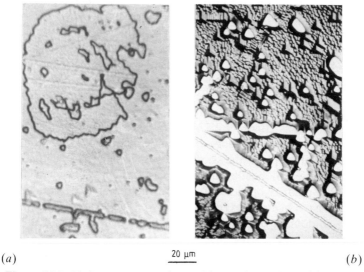

(*a*) 20 µm (*b*)

Figure 4.24. Etch patterns on the positive surface of LN: (*a*) after a 6 h annealing at 850 °C (the boundaries of repolarized regions can be seen); (*b*) after a 6 h annealing at 980 °C. The conditions and time of etching are the same in both cases (Bocharova 1986).

observed by Ohnishi (1977) who conjectured that they should result from the local repolarization. The thickness of the repolarized layer ranges from 1×10^3 to 15×10^3 nm, strongly depending on the conditions of specimen preparation and heat treatment. This explanation is consistent with the discovered existence of the two-domain structure of LN, which arises after a prolonged annealing of the crystals above 1000 °C.

The observed picture of crystal etching reflects the initial stage of repolarization, when antiparallel domains originate and begin to grow near the positive surface (figure 4.23(a)). As regards the repolarization mechanism, one may assume that the origination of flat near-surface domains is initiated by the non-congruent evaporation of lithium niobate. It is known that a prolonged heating of LN plates at a high temperature leads to the depletion of the near-surface region with respect to lithium (Carruthers *et al* 1974). The change in the composition of the near-surface layer will not then exceed 0.01 mole fraction of Li_2O.

To test the above assumption, Bocharova (1986) has performed a series of experiments involving the annealing of LN crystals in vapours of Li_2O. It has been found that in the presence of lithium vapours no repolarization occurs over the entire temperature range under study. This confirms the validity of the assumption about the origination of antiparallel domains on the intrinsic point defects due to the deviation of crystal composition from the stoichiometric one.

A typical picture of etching of the negative surface of the original specimen is shown in figure 4.25(a), where one can see dislocational pits and dark triangular patterns with a basic size of (0.5 to 1.5) $\times 10^3$ nm, their average density being 3.5×10^5 cm^{-2}, or an order of magnitude higher than the density of the individual dislocations. Through a scanning electron microscope, one can clearly see that the dark figures are faceted hillocks (figure 4.25(b)). This is confirmed by the character of the bend of the interference lines and the shadow directions on the replicas separated from the surface of the etched specimens. These formations represent needle-shaped micro-domains stretched along the Z axis with a polarity opposite to that of the die. The pyramidal faces of the hillocks have been indexed as $\{10\bar{2}\}$ (see Niizeki *et al* 1967).

The micro-domain origination in the near-surface layer of the crystals may possibly be due to the mechanical treatment of the specimens and also to a thermal shock. The formation of micro-domain accumulations of a high density (up to 10^9 cm^{-2} in the centre and 10^5 to 10^6 cm^{-2} at the margin) has been observed also after the irradiation of the crystals by laser pulses (Madoyan and Khachaturyan 1983). It should be stressed that the relationship between mechanical stresses and repolarization has been established when studying mechanically introduced twins. The polarity of the twinning band which is formed when LN crystals are strained in the temperature range from 1150 to 1250 °C is opposite to the polarity of the die (Vere 1968). For many-domain crystals, the polarity of the twinning band changes in passing through the domain boundary in such a way that the boundary between two mechanically introduced twins corresponds to the boundary between the domains.

Apart from needle-shaped micro-domains introduced in processing the

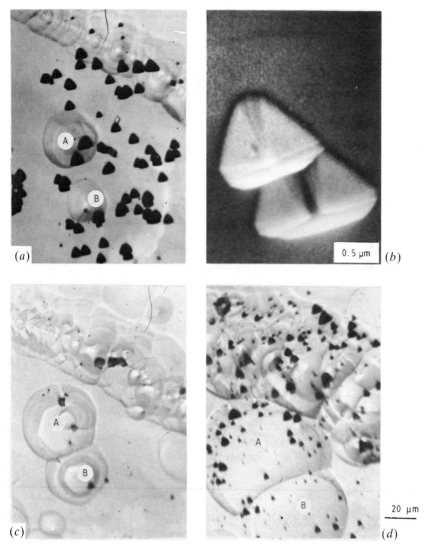

Figure 4.25. Etch patterns on the negative surface of LN: (*a*) initial
specimen, average density of micro-domains = 3.5×10^5 cm^{-2}; (*b*)
micro-domain image obtained in the regime of secondary electrons; (*c*)
morphology after a 5 h annealing at 980 °C and the subsequent etching,
the micro-domain density having been lowered by more than an order
of magnitude; (*d*) micro-domains formed as a result of annealing in the
presence of an admixture of Si, average density = 7×10^5 cm^{-2}
(Bocharova 1986).

LN surface, micro-domains may also be present in the crystal bulk which
have originated because of the growth conditions. For example, the

formation of dagger-shaped micro-domains stretched along the direction of growth is determined by the fluctuations of growth condition and the formation of a cellular crystal structure (Gabrelayan 1978, Xu Run Yuan 1983).

The arrangement of micro-domains cannot uniquely be related to places of emergence of dislocations. Indeed, they are found both near dislocations and in areas free of them. The study of the morphology of the annealed and etched surface has made it possible to establish that the mechanically introduced near-surface micro-domains are stable up to the annealing temperature $T = 950\,°C$. Beginning with that temperature, their reverse repolarization, evidenced by etch patterns, becomes evident. No morphological changes of the surface could be observed immediately after annealing. Figures 4.25(a) and (b) show photographs of the original surface (a) and of the same area after a 5 h annealing at $T = 980\,°C$, followed by etching (b). The letters A and B mark the emergence of near-surface dislocations, which serve as reference points. It can be seen that the domain density has decreased by more than an order of magnitude. The blank experiments have shown that the observed disappearance of the micro-domains results from annealing, rather than the dissolution of LN during etching in acids.

The process of the reverse repolarization of micro-domains on the negative surface occurs non-uniformly, the small micro-domains disappearing more rapidly than the major ones. An increase in annealing time up to 13 or even 15 h at the same annealing temperature leads to a complete disappearance of mechanically introduced micro-domains. The relaxation of elastic stresses near the surface is very likely to occur during that time. The process of repolarization is not influenced much by the presence of Li_2O vapours, because the nature of the process is different from that on the positive side of the crystal.

Impurities affect the positive and negative surfaces of LN crystals differently: the positive surface is stabilized under the influence of the diffusion of silicon, whereas on the negative surface there will arise a new system of micro-domains (figure 4.25(d)). The absence of repolarization on the positive surface can be explained by the fact that silicon slows down the escape of lithium from the crystal by blocking the diffusion channels and compensates for the loss of lithium by occupying the vacant octahedral sites. The emergence of micro-domains on the negative surface seems to be due to stresses which are set up as silicon is incorporated in the lithium niobate lattice, because its octahedral radius, equal to $1.37\,Å$, is twice as long as the ionic radius of Li^+.

Thus in the high-temperature annealing the near-surface domain structure of the Z section of LN will change. A scheme of the formation of repolarized regions on the positive surface and the annealing of micro-domains on the negative surface is presented in figure 4.26.

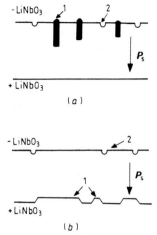

Figure 4.26. Scheme of formation of repolarized regions on the positive surface and micro-domain annealing on the negative surface of the (0001) face of lithium niobate: (*a*) the initial specimen (1, micro-domains, 2, etch pits); (*b*) specimen was etched after a 6 h annealing at 980 °C (1, repolarized regions; 2, etch pits) (Bocharova 1986).

4.8 The regular domain structure in LN crystals

Using external influences upon the crystal near a phase transition, it is possible to change the parameters of the domain structures of ferroelectrics. Exerting influences periodic in time and coordinate allows regular domain structures (RDS) to be formed.

Interest in the RDS of ferroelectrics with a centrally symmetric para-phase is determined by the possibility of applying a sign-variable spatial modulation of the physical properties of a crystal, which are described by an odd-rank tensor. Some problems of using ferroelectrics to excite and receive high-frequency elastic waves, process radio signals and generate harmonics of laser emission have been taken up theoretically by Peurin and Tasson (1976) and Lemanov (1981).

The conditions which are necessary for the RDS to arise may be created both in the course of crystal growth and during the subsequent thermoelectric treatment (by Lin Peng and Bursiu 1982). In the former case, RDS are formed as the crystals are cooled below the Curie point, and the RDS period will be determined by the non-uniform distribution of impurities in the growing crystal. In the latter case, a crystal with homogeneous physical properties is subjected to the influence of external factors: electric field intensity E and temperature gradient near a phase transition.

The change in the symmetry point group in the phase transition $\bar{3}m \to 3m$ makes it possible to form domain structure with a random orientation of domain boundaries relative to the spontaneous polarization vector, because the only element of symmetry lost in the phase transition is the symmetry centre. Such structures can be conveniently described by introducing the vector of spatial periodicity $k = \pi n/d$, where n is the vector normal to the domain boundary surface and d is the domain size.

In an experimental study by Antipov *et al* (1985), the regular domain structure in LN and LT was formed in the process of an after-growth thermoelectric treatment by applying to the crystal some spatially independent vectorial influences: electric field and temperature gradient. The orientation of these vectors relative to the spontaneous polarization of the crystal made it possible to form structures with a given vector K, the domain size having been determined by the period of the sign reversal of the electric field. To produce RDS with $K|P_{s_s}$, the electric field had to be applied to the crystal parallel to the temperature gradient direction, and the domain boundaries formed were normal to the spontaneous polarization vector. The formation of RDS with the opposite direction of P_s in the adjacent domains is possible in LN, because the electric field set up by charged particles is screened due to a high electric conductivity of the crystal near the phase transition. The RDS with $K \perp P_s$ was formed by applying the electric field to the crystal along the Z axis, in the direction normal to that of the temperature gradient. The formation of RDS is decisively affected by the temperature gradient at the Curie point, the current density, and the electric field intensity at the domain boundary.

The regular structures with $K \perp P_s$ and $K\|P_s$ were formed at optimal values of these parameters for each crystal, which made it possible to obtain a minimal visible optical thickness of the domain boundaries. The selective etching of polar faces revealed that the structures obtained represent regions of domains of the opposite sign, whose size is determined by the reswitching time of the external electric influence. No enantiomorphism is present in the class 3m; therefore the sign of the neighbouring domains will reverse along the axes Z and Y, which is confirmed by the observed contrast on the polished surface of a YZ specimen of LN with RDS. The use of optical techniques to observe 180° domains is possible owing to the phase contrast that arises at the domain boundary because of the presence of internal electric fields. The observed thickness of the RDS domain boundaries is determined by the transition region between the domains, due to the refractive index gradient; it is many times larger than the actual thickness of the domain boundaries, measured by the high-resolution transmission electron microscopy technique. The domain structure of LN with $K\|P_s$ has an effective domain boundary thickness of $(40-60) \times 10^{-6}$ m (see figure 4.27(a)). The regular domain structure with $K \perp P_s$ has an effective domain wall thickness of $(2-5) \times 10^{-6}$ m (see figure 4.27(b)).

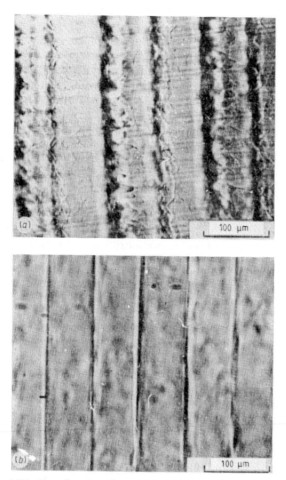

Figure 4.27. Regular domain structure in LN crystal with $K \| P_s$ (a) and $K \perp P_s$ (b) (Antipov *et al* 1985).

As the RDS of LN with $K \| P_s$ is formed near the surface of the specimen in a region $(600 \text{ to } 800) \times 10^{-6}$ m thick, there may take place a change in the direction of the K vector, the angle between the vectors K and P_s reaching 90°.

In some cases the domain boundaries were not flat, which was due to composition variations over the crystal volume, because the domain boundaries repeat the profile of the isotherm corresponding to T_c.

Along with the controlled influences mentioned above, important factors in the process of RDS formation in LN are the electrical conductivity of the crystal, which screens the field of spontaneous polarization, and the concentration of point defects near T_c. The observed bend of the domain boundaries near the surface of LN appears to be due to the surface conductivity.

The development of most devices using RDS ferroelectrics is feasible if RDS is formed with a given **K** direction relative to the spontaneous polarization direction and if it is stable with respect to the influence of controlled external factors. The RDS stability was tested by the cyclic application of a sign-variable electric field and the observation of the light diffraction on the grating which thus arose in the crystal. The change in the intensity of the diffracted light observed after 10^6 pulses of 5×10^5 B m^{-1} were applied to the crystal did not exceed 1% (see chapter 7).

Feng *et al* (1980) have grown LN crystals with periodic domains by utilizing temperature fluctuation during Czochralski growth. Periodic temperature fluctuations may be induced by displacing the rotation axis from the symmetry axis of the temperature field. In order to accentuate the growth striae produced, they doped the melt with 1 wt.% of yttrium. Crystals thus grown have a dopant distribution which varies sinusoidally along the growth (*a*) axis. When cooled below the Curie point, crystals with periodic laminar structure are obtained.

Feisst and Kaide (1985) have reported the growth and application of laminar ferroelectric domain superstructures in chromium-doped LN. The possibility of influencing the domain formation by applying a modulated current through the growth interface was used for Czochralski-grown crystals (Räuber 1978). Those layered crystals represent structures with alternating sign of the effective NLO coefficient. This technique allows the superstructure period and the thickness of the adjacent domains to be adjusted independently in regard to the pulling and rotation parameters. It has previously been shown (Feisst and Räuber 1983) that the distribution coefficient of chromium in LN linearly decreases with current density. Current modulation, therefore, leads to Cr doping striations, and ferroelectric domains that are spatially correlated with these striations are observed. The nature of the pre-polar field, already present in the paraelectric phase, that defines the polarity of the domains when the crystal is cooled below the Curie point, is yet obscure. The results of Ming *et al* (1982) suggest that it is the concentration gradient of charge-uncompensated impurities (e.g. Cr) that initiates the domain formation.

Constant growth rate is needed to grow a spatially periodic domain pattern. Feisst and Kaide (1985) have designed a Czochralski apparatus having a pulling and rotation mechanism with sub-micrometre precision. The platinum crucible was inductively heated. The melt temperature was pyrometrically controlled and was stabilized by adjusting the RF power. No further automatic diameter control was applied. The axial temperature profile above the crucible was adjusted by means of an after-heater to keep the crystal sufficiently conducting and to cool down the crystal after growth. A modulated current was applied between crystal seed and crucible. Square and triangular waveforms, either symmetric around zero or with the positive current domain, were used (positive current means crystal positively biased with respect to the melt). Maximum current densities at the growth interface were $20 \, mA \, cm^{-2}$. $LiNbO_3$ crystals were grown along the polar Y axis from congruent melts with Cr concentrations between 0.01 and 0.2 wt.% Cr_2O_3 in the melt. Domain formation appears to be possible in a wide range of Cr concentrations, current densities and pulling rates. Good results were obtained with 0.05% Cr_2O_3, $I = \pm 10 \, mA \, cm^{-2}$ and $v_p = 5 \, mm \, h^{-1}$. It was difficult, however, to produce domain thicknesses below $8 \times 10^{-6} \, m$, possibly because of the interference between the current induced and the domain formation induced by the change in Peltier temperature.

Shaped LN ribbon crystals were grown by the above technique. The crystals were doped with 1 wt.% $LiTaO_3$ (Red'kin *et al* 1985). The growth direction coincided with the [0001] direction, the $(10\bar{1}0)$ plane being the ribbon plane.

The impurity distribution in the crystals, and consequently their domain structure were controlled by varying the pulling rate rather than by producing periodic temperature fluctuations in the melt under the crystallization front. This is associated with the fact that, due to the thermal inertia of the die, it is very difficult to get a desired frequency of temperature fluctuations. The pulling rate was varied periodically and discontinuously. Under these growth conditions, there occurs a periodic alteration of the shaped crystal cross section, which leads to a change of the effective coefficient of the impurity distribution (Brantov and Tatarchenko 1983). Domains measuring $(200–50) \times 10^{-6} \, m$ were thus obtained. Inasmuch as the shape of the domains reproduced that of the die end surfaces, horizontal, spherical, conical and some other shapes of the domain structure were obtained by varying the die geometry.

5 Electrical Properties of Lithium Metaniobate

5.1 Electric conductivity

The electric conductivity (σ) of lithium niobate has been investigated by various workers, some of whom studied it in the early stage of establishing the properties of this material and some of whom have been engaged in the problem in recent times (Kuz'minov 1975, Raüber 1978). The main findings of these authors are presented in figures 5.1 and 5.2 as well as in table 5.1 (the numbering of the curves of figure 5.1 corresponds to that of the lines in table 5.1). The temperature–conductivity curves can be divided into several segments and in each of them $\sigma(T)$ is fitted by the function

$$\sigma = \sigma_0 \exp\left(-H/kT\right) \tag{5.1}$$

with various values of the pre-exponential term (σ_0) and activation energy (H). An analysis of the literature data has revealed three main temperature intervals, each distinguished by similar H values. On transition from one segment to another, the temperature may vary within 50 to 100 degrees, which seems to be accounted for by various defect structures of samples.

For $T > 700$–800 °C, the activation energy of lithium niobate conductivity (at a normal pressure of oxygen in the environment) for crystals grown from stoichiometric melts is 2.12 to 2.15 eV. For crystals grown with too small or too large (1–2%) amounts of lithium in the melt, the conductivity activation energy is 2.32 and 3.07 eV respectively. Its value may differ (from 1.9 to 2.3 eV) with crystals, depending on their defects.

At temperatures ranging from 100–200 °C to 700–800 °C, the activa-

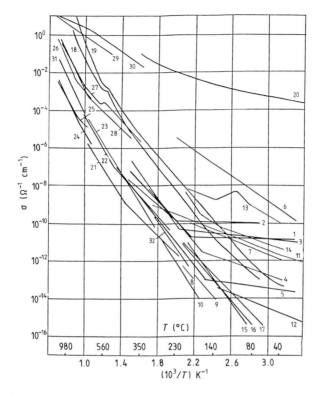

Figure 5.1. Temperature dependences of lithium niobate conductivity (according to Stepanova 1986).

tion energy of lithium niobate conductivity is estimated to be 1–1.5 eV (the *H* values are slightly variable within these limits for non-stoichiometric compositions and in the presence of dopants).

For temperatures under 200 °C, various authors indicate greatly differing values of conductivity activation energy. In some instances, the *H* value remains just the same as at higher temperatures, i.e. 1.0 to 1.28 eV. Other authors report that at $T < 200–150$ °C, the *H* value goes down to 0.34 eV and 0.14 eV for pure crystals. Numerical values over this temperature range are not always available but it is evident from lg $\sigma = f(1/T)$ plots that the slope angles differ by a large factor. Doping lithium niobate with iron in concentrations over 0.3% by weight will increase its conductivity markedly and in this case for $T < 200$ °C the conductivity activation energy proves to be 0.3–0.4 eV.

Conductivity and the energy for its activation are greatly affected by heat treatment and the composition of the environment atmosphere in

Table 5.1. Activation energies of lithium niobate conductivity (according to Stepanova 1986).

No	Crystal composition	Growth atmosphere	Activation energy (eV)(°C)	References
1	Pure	Air	1.4 eV $(200 < T < 400)$	Roitberg et al (1969)
2	0.01% Cr^{3+}	Air	1.4 eV $(200 < T < 400)$	Roitberg et al (1969)
3	0.01% Ni^{2+}	Air	1.4 eV $(200 < T < 400)$	Roitberg et al (1969)
4	Pure	Air	1.1 eV† $(170 < T < 300)$	Lapshin and Rumyantsev (1971)
			0.34 eV $(50 < T < 170)$	
5	Stoichiometric	Air	1.15 eV† $(162 < T < 500)$	Dan'kov et al 1983
			0.14 eV† $(30 < T < 160)$	
6	Stoichiometric	Annealing in inert gas	0.65 eV† $(20 < T < 200)$	Dan'kov et al (1983)
7‡	0.3% Fe	Air	0.37 eV	Barkan et al (1977a,b,c)
8	With Fe	Air	1.1 eV	Ohmori et al (1979)
9‡	Pure	Air	1.1 eV	
10	0.004 wt.% Fe	Air		Barkan et al (1977a,b,c)
11	0.3 wt.% Fe	Air	0.3 eV	Barkan et al (1977a,b,c)
12‡	With Fe	Air	0.41 eV	Levanyuk et al (1980)
13	0.1 wt.% Mn (z-cut)	Air	All activation energies $\leqslant 1$ eV	Kamentsev et al (1983)
14	0.1 wt.% Mn (x-cut)	Air		
15	Pure	Air	1.28 eV	Blistanov et al (1978)
16	0.1% CuO	Air	1.13 eV	Blistanov et al (1978)
17	0.3% Mn	Air	1.1 eV	Blistanov et al (1978)
18	Deficit of Li	Air	1.28 eV $(80 < T < 400)$	Bollman and Gernad (1972)
			2.32 eV $(T > 600)$	

Table 5.1. (*cont.*)

No	Crystal composition	Growth atmosphere	Activation energy (eV)(°C)	References
19	1–2 mol% excess of Li in the melt	Air	1.1 eV ($80 < T < 400$) 3.07 eV ($T > 600$)	Bollman and Gernad (1972)
20	Excess Li	Annealed in vacuum	0.64 eV ($200 < T < 350$) 0.22 eV ($20 < T < 200$)	Bollman and Gernad (1972)
21	0.5% Tm	Annealed in vacuum		Antonov et al (1975)
22	Pure	Annealed in vacuum		Antonov et al (1975)
23	0.5% Nd	Annealed in vacuum		Antonov et al (1975)
24	Stoichiometric	O_2($P = 1$ atm)	2.15 eV ($T > 750$) 1.5 eV ($600 < T < 750$)	Bergmann (1968)
25	Stoichiometric	O_2($P = 1$ atm)	2.12 eV ($T > 700$)	Jorgensen and Barlett (1969)
26	Stoichiometric	Air	0.86–1.3 eV ($400 < T < 700$)	Ivleva and Kuz'minov (1971a,b)
27	Stoichiometric	Air	1.9–2.2 eV ($T > 700$)	
28	Stoichiometric	Air		Jorgensen and Barlett (1969)
29	Stoichiometric	99 CO/1 CO_2	0.72 eV ($T > 400$)	Bergmann (1968)
30	Stoichiometric	99 CO/1 CO_2	1.97 eV ($T > 700$)	Jorgensen and Bartlett (1969)
31	Stoichiometric	O_2 in equilibrium with LiO_2		
32	Congruent	Vacuum	1.25 eV ($200 < T < 300$)	Franke (1984)

†The reference gives a level ionization energy $\Delta E = 2H$.

‡The σ values have been measured indirectly.

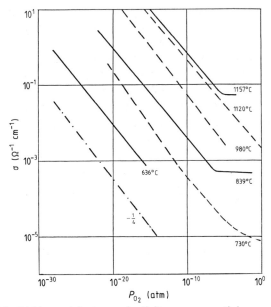

Figure 5.2. Lithium niobate conductivity versus partial oxygen pressure in the environment. The full curves represent the data obtained by Jorgensen and Bartlett (1969); the broken curves show the data of Bergmann (1968).

the heat treatment process. Over a temperature range 400 to 1200 °C, the σ value will increase by several orders of magnitude if measurements are made in an oxygen-free medium; in this case the activation energy is reduced. Preliminary annealing in an inert gas or vacuum exerts a similar effect on lithium niobate conductivity, even at far lower temperatures.

The investigation of conductivity versus environment composition, along with the measurement of electrical transport numbers and thermal EMF, has yielded conclusions regarding the mechanisms of charge transfer in lithium niobate under various conditions.

At temperatures exceeding 700 °C and partial oxygen pressure $P_{O_2} = 1$ atm, the conductivity of lithium niobate depends strongly on the lithium vapour content in the environment; under the conditions of equilibrium with Li_2O, the conductivity will increase (here the activation energy is 1.9 eV) (Jorgensen and Bartlett 1969). With due account of the measured transport numbers, this fact has suggested a conclusion about the ionic nature of lithium niobate conductivity at $P_{O_2} = 1$ atm and $T > 700$ °C, with the current carriers provided by the Li^+ ions. At temperatures exceeding 700 °C, the electrical conductivity of lithium niobate acquires quite a different nature, once the pressure of oxygen in the environment is reduced to $P_{O_2} < 10^{-6}$ atm. The lithium vapour content has no bearing on the conductivity value any longer. On the other hand, the latter is found to depend on the partial pressure of oxygen in the environment. It has been established that at $T > 600\,°C$ and $P_{O_2} < 10^{-6}$ atm the lithium niobate conductivity is proportional to $P_{O_2}^{-1/4}$ (figure 5.2) (Bergmann 1968, Jorgensen and Bartlett 1969). This is attributed to the fact that in a crystal at equilibrium with a low-oxygen atmosphere there are formed oxygen vacancies (V_O). The latter carry an effective positive charge ($+2e$) and trap two electrons, thereby becoming donor centres. Thermal ionization of these centres ensures electron conductivity of the crystal. The experimental dependence $\sigma \sim P_{O_2}^{-1/4}$ is fitted by the model involving singly ionized oxygen vacancies, which is described by the equation

$$O \rightleftarrows V_O^+ + e^- + \tfrac{1}{2}O_2 \uparrow .$$

The electron nature of lithium niobate conductivity at low oxygen pressures and $T > 700\,°C$ is confirmed by the thermoelectrical data (Jorgensen and Bartlett 1969). At low P_{O_2} values the electron conductivity activation energy ($H = 0.72$ eV) is equal to the ionization energy of oxygen vacancies or to half of this energy, depending on the extent of acceptor compensation (Bergmann 1968). Over a pressure range 10^{-6} atm $< P_{O_2} > 1$ atm, the activation energy is 2.15 eV and it is defined by the average of the energy of oxygen vacancies formation and the energy of their ionization.

In a wide range of lower temperatures ($100\,°C < T < 700\,°C$), conductivity is notable for close values of the activation energy, although the mechanisms of charge transfer are quite different. The lithium niobate conductivity at $350\,°C < T < 700\,°C$ with an activation energy of 1 to 1.5 eV is related to the predominant charge transfer by diffusion of impurity ions (Bergmann 1968, Ivleva and Kuz'minov 1971a,b, Ivleva *et al* 1971). The proximity of the activation energy of conductivity at $100\,°C < T < 400\,°C$ (1.1 and 1.28 eV) to that of oxygen diffusion in lithium niobate (1.28 eV), as measured by Jorgensen and Bartlett (1969) for $T \approx 700\,°C$, enabled Bollman and Gernand (1972) to presume that it is the diffusion of oxygen vacancies and interstitial oxygen ions that is responsible for charge transfer. In a majority of works however (Blistanov *et al* 1978, Ohmori *et al* 1979), the conductivity of lithium niobate over a temperature range 100 to 400 °C with an activation energy of 1 to 1.3 eV is considered to be of an electron nature. Electrons are excited at levels in the forbidden band, created by impurities or intrinsic point defects (Ohmori *et al* 1979). Hence the characteristics of lithium niobate conductivity over the above temperature range are completely determined by the system of levels in the forbidden band, i.e by the defect structure of the crystal (see chapter 3).

However, under certain specific conditions at temperatures ranging from 50 to 300 °C, lithium niobate may exhibit ionic conductivity. To this end, the cathode is made of metallic lithium and the Li^+ ions are forced by action of the electric field to diffuse into the lithium niobate, thereby ensuring passage of an ion current. The ions move by the vacancy mechanism (Franke 1984).

To obtain insight into the processes occurring in the operation of lithium niobate based devices and to optimize the parameters of the latter, of particular concern are the conductivity mechanisms coming into play over a temperature range 20 to 100 °C. But at or near room temperature, the value of conductivity of pure unreduced crystals of lithium niobate is very low (10^{-15}–10^{-16} $\Omega^{-1}\,cm^{-1}$) (Krätzig and Kurz 1977). Direct measurement of electric conductivity at the above mentioned temperature is virtually impossible because of the displacement currents induced by the pyro-effect. The values of displacement currents are comparable, even under minute temperature fluctuations (10^{-2}–10^{-3} grad min^{-1}) unavoidably present in thermostatic control, to the bulk conduction currents in this temperature range.

So, lithium niobate conductivity is measured at near-room temperatures using indirect techniques. For instance, variations in an optical inhomogeneity under an applied voltage are determined by the variations of the built-in electric field existing in the crystal, which is established in turn by redistribution of the space charge density on sample inhomogeneities. This makes it possible to estimate the conductivity activation energy from the kinetics of variations in an optical inhomogeneity of lithium niobate induced by the applied voltage. The H values obtained in this way vary from 0.1 to 0.3 eV, depending on the magnitude of the voltage applied to the crystal (Blistanov *et al* 1981).

An alternative approach relies on measurements of optically induced birefringence, since it is governed by the internal electric field set up in the illuminated region of a crystal. The time constant of birefringence growth is defined by the conductance in the illuminated region (σ_Σ). By measuring σ_Σ versus illumination intensity and extrapolating this dependence to zero intensity, one can evaluate dark conductivity for a given temperature (Levanyuk *et al* 1980).

Furthermore, conductivity can be estimated on the basis of the self-erasing time of holograms recorded in lithium niobate after the illumination is discontinued. These methods, however, are largely employed to establish conductivity of iron-doped lithium niobate (Barkan *et al* 1977).

The disadvantage common to all the above described indirect methods used to determine electric conductivity is that they proceed through two steps: first from the variations in birefringence to those of the electric field in a crystal, and second from the latter to conductivity.

As noted earlier, according to various authors, the slopes of the temperature–conductivity curves $\lg \sigma = f(1/T)$ at $T < 100\,°C$ are greatly different (figure 5.1). The directly measured data on conductivity reported in some studies exhibit a great scatter of the values. Hence the available data on lithium niobate conductivity in this temperature range cannot be considered very reliable.

There is no unified standpoint regarding the conductivity mechanism over this temperature range. In particular, Lapshin and Rumyantsev (1971) claim that conductivity is determined by the levels with an ionization energy of 0.68 eV, which are related to the oxygen deficit of the structure. The low H values are attributed to the 'flip-flop' conductivity at deep local levels of oxygen vacancies (Blistanov *et al* 1981). Nevertheless, in any case there is no doubt that the conductivity of lithium niobate at near-room temperatures as well as at $100\,°C < T < 400\,°C$ is determined by a system of energy levels of point defects present in the crystal.

Doping (Barkan *et al* 1977a,b,c, Blistanov *et al* 1978) as a rule contributes to conductivity (table 5.2). The least activation energy and the highest electrical conductivity at low temperatures are noted for Mo-doped crystals. As the concentration of Fe ions increases, the activation energy for the conduction process goes down (Barkan *et al* 1977a,b,c).

Table 5.2. Conductivity (σ_0) defined by formula (5.1) and the activation energy of lithium niobate crystals doped with various elements (according to Blistanov *et al*

Dopant Element	Content wt.%	Conductivity activation energy (H) (eV)	Conductivity (σ_0) ($\Omega^{-1}\,cm^{-1}$)
Nominally pure	—	1.28	89
Mg	0.5	1.41	6.3×10^3
Cu	0.1	1.13	4.0
Mo	0.3	1.00	2.1×10^{-1}
W	0.3	1.28	4.0×10^2
Mn	0.05	1.16	1.12×10

5.2 Dielectric properties

5.2.1 Coordinate system for tensor properties
The coordinate system used to describe the physical tensor properties of

lithium niobate is neither hexagonal nor rhombohedral but rather a Cartesian x, y, z system. The accepted convention for relating the hexagonal axes to the x, y, z principal axes (Weis and Gaylord 1985) is described here. The z axis is chosen to be parallel to the c axis. The x axis is chosen to coincide with any of the equivalent a_H axes. After the x and the z axes are selected, the y axis is chosen such that the system is right-handed. Thus, the y axis must lie in a plane of mirror symmetry. The sense of the z axis is, of course, the same as that of the c axis. The sense of the y axis is determined in a manner similar to that described for the z axis. Upon compression, the $+Y$ face becomes negatively charged. The sense of the x direction, however, cannot be determined in this way because it is perpendicular to a mirror plane. Any charge movement on one side of the plane is 'mirrored' on the opposite side; hence the Y faces do not become charged.

Figure 1.13 illustrates both the standard conventional and a secondary convention for choosing the x and the y principal axes within the hexagonal unit cell in lithium niobate. Figure 5.3 shows the standard convention orientation for the x, y and z principal axes with respect to the crystal boule. When working with tensors, the x, y and z principal axes are often referred to as the x_1, x_2 and x_3 axes respectively.

Suppliers of lithium niobate crystals furnish pieces that are commonly in the form of plates. These slices may be designated x-cut, y-cut or z-cut, corresponding respectively to the x, y or z axes being normal to the large area surfaces. A second letter is often added to the plate orientation description according to the IRE standard. The second letter indicates the direction of the longest dimension of the rectangular plate. Thus, the xy-cut plate is an x-cut plate with its longest dimension parallel to the y axis. Similarly, crystal plates can be xz-cut, yx-cut, yz-cut, zx-cut or zy-cut.

5.2.2 Pyroelectric effect

A pyroelectric solid exhibits a change in spontaneous polarization as a function of temperature. The relationship between the change in temperature ΔT and the change in spontaneous polarization ΔP is linear and can be written as $\Delta P = \gamma \Delta T$, where γ is the pyroelectric tensor. In tensor component form this may be written as $\Delta P_i = \gamma_i \Delta T$. In lithium niobate this effect is due to the movement of the lithium and niobium ions relative to the oxygen layers. Since the lithium and niobium ions move only in a direction parallel to the c axis, the pyroelectric tensor is of the form

$$\gamma_i = \begin{bmatrix} 0 \\ 0 \\ \gamma_3 \end{bmatrix} \tag{5.2}$$

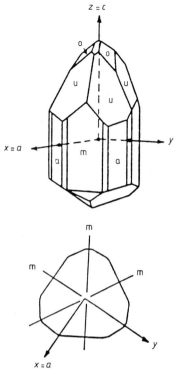

Figure 5.3. Standard orientation of the x, y and z principal axes relative to the crystal boule for lithium niobate.

where $\gamma_3 = -4 \times 10^{-5}\,\mathrm{C\,K^{-1}\,m^{-2}}$ (Savage 1966). Note that the negative value of γ_3 indicates that upon cooling the $+C$ crystal face will become positively charged as previously discussed.

5.2.3 Permittivity

The relationship between electric flux density D and electric field E is linear and can be written as $D = \varepsilon E$, where ε is the second-rank permittivity tensor. In tensor component form this relationship may be expressed as

$$D_i = \sum_j \varepsilon_{ij} E_j \qquad (5.3)$$

where i, $j = x$, y or z. It can be shown through a thermodynamic conservation of energy argument that only the diagonal elements of the permittivity tensor are non-zero for non-gyrotropic materials. Furthermore, since the lithium niobate crystal is symmetrical about the c axis, the permittivity is the same for any electric field direction in the plane perpendicular to the c axis. Thus the permittivity tensor can be represented by the 3 × 3 matrix

$$\varepsilon_{ij} = \begin{bmatrix} \varepsilon_{11} & 0 & 0 \\ 0 & \varepsilon_{11} & 0 \\ 0 & 0 & \varepsilon_{33} \end{bmatrix}. \tag{5.4}$$

Permittivity is often normalized in terms of the permittivity of a vacuum (ε_0). The dimensionless constant $\varepsilon_{ij}/\varepsilon_0$ is called the relative permittivity or the dielectric constant. When measuring permittivity, mechanical constraints imposed on the crystal are important. If the crystal is free (unclamped), the stress in the crystal is zero; if the crystal is rigidly held (clamped), the strain in the crystal is zero. Capacitance measurements at very low frequencies are used to determine the unclamped values of permittivity (ε^T). Capacitance measurements at frequencies well above the mechanical resonance can be used to determine directly the clamped values of permittivity (ε^S). The clamped values can also be determined indirectly from acoustic phase velocity measurements. Measured values of relative permittivity are listed in table 5.3. The differences between the measured values may be due to stoichiometric differences between samples (Turner *et al* 1970). The stoichiometries of the samples are indicated in the table by the value of the Curie temperature T_C.

Table 5.3. Relative permittivity coefficients (according to Weis and Gaylord 1985).

$\varepsilon_{11}^T/\varepsilon_0$	$\varepsilon_{33}^T/\varepsilon_0$	$\varepsilon_{11}^S/\varepsilon_0$	$\varepsilon_{33}^S/\varepsilon_0$	T_C (°C)
85.2	28.7	44.3	27.9	1165
84	30	44	29	—
84.6	28.6	—	—	—
—	—	43.9	23.7	—
84.1	28.1	46.5	27.3	—

The Curie temperature versus Li_2O content in a crystal is plotted in figure 5.4. The results of the T_C measurements cover the range from 47.0 to 49.0 mol.% Li_2O in the initial liquid with the majority of samples in the 48.0 to 49.0 mol.% range. A linear least-squares fit of these data yields

$$T_C = -637.30 + 36.70C$$

or

$$C = 17.37 + 0.02725T_C$$

where T_C is the Curie temperature in °C and C is the Li_2O content in mol.% of the sample. Examination of the data shown in figure 5.4 shows a slight upward curve as one approaches higher Li_2O compositions. A least-squares fit of the data using the form $T_C = a + bC + cC^2$ yields

$$T_C = 9095.2 - 369.05C + 4.228C^2$$

which gives a better fit than the linear model (the curve is not shown in figure 5.4).

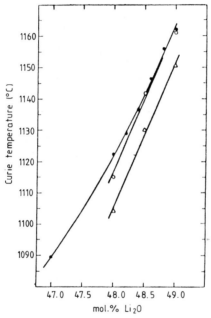

Figure 5.4. Curie temperature versus mol.% Li_2O in lithium niobate: ● O'Bryan *et al* (1985), ○ Geunais *et al* (1981), △ Carruthers *et al* (1971).

5.2.4 Piezoelectric effect

A piezoelectric solid exhibits an induced polarization with applied stress. The relationship between polarization and stress is linear and may be written as $P = d\sigma$, where the vector P is the induced polarization, σ is the second-rank stress tensor and d is the third-rank piezoelectric tensor. In tensor component form, the piezoelectric effect may be expressed

$$P_i = \sum_{j,k} d_{ijk}\sigma_{jk} \tag{5.5}$$

where $i, j, k = x, y, z$. It can be shown through general thermodynamic arguments that

$$\sigma_{jk} = \sigma_{kj}. \tag{5.6}$$

Thus, the d_{ijk} tensor contains only 18 independent elements and can be written as a 3×6 matrix. Customarily the jk subscripts are reduced to a single subscript using the substitutions

$$jk = 11 = 1 \qquad jk = 22 = 2$$

$$jk = 33 = 3 \qquad jk = 23, 32 = 4$$

$$jk = 31, 13 = 5 \qquad jk = 12, 21 = 6. \tag{5.7}$$

Since the lithium niobate crystal possesses a 3m point group symmetry, all tensors describing the physical properties of lithium niobate must have at least that symmetry. They can have more symmetry than the crystal structure but they cannot have less symmetry than the crystal structure. The basic principle was first asserted by Neumann. Neumann's principle cannot be properly applied to the reduced subscript matrix but rather must be applied to the original tensor.

The application of Neumann's principle to the d_{ijk} tensor followed by the use of the reduced-subscript notation gives

$$d_{ijk} = \begin{bmatrix} 0 & 0 & 0 & 0 & d_{15} & -2d_{22} \\ -d_{22} & d_{22} & 0 & d_{15} & 0 & 0 \\ d_{31} & d_{31} & d_{33} & 0 & 0 & 0 \end{bmatrix}. \tag{5.8}$$

Note that $d_{15} = d_{24}$, $d_{22} = -d_{21} = (-d_{16})/2$, and $d_{31} = d_{32}$. Thus the piezoelectric effect in lithium niobate can be described by four independent coefficients d_{15}, d_{22}, d_{31} and d_{33}. Measured values for these quantities are presented in table 5.4. Valuable bibliographic information for the tables presented was obtained by Yamada (1981). Unfortunately, some of the parameter values in the latter work were found to be in error after comparison with the original sources.

Table 5.4. Piezoelectric strain coefficients $(10^{-11}\,\text{C}\,\text{N}^{-1})$ (according to Weis and Gaylord 1985).

d_{15}	d_{22}	d_{31}	d_{33}
6.92	2.08	−0.085	0.60
6.8	2.1	−0.1	0.6
7.4	2.1	−0.087	1.6

The fact that both d_{22} and d_{33} are positive is due to the convention discussed previously for determining the senses of the $y(x_2)$ and $z(x_3)$ axes. The determination of these piezoelectric coefficients is often based on taking differences of ultrasonic wave velocities of comparable magnitudes; thus they are subject to rather large percentage errors (especially d_{33}) (Graham 1976). However, the linear hydrostatic coefficient d_h has been measured to within 1.5% (Graham 1976). For crystals possessing 3m symmetry, $d_h = 2d_{31} + d_{33}$. The magnitude of d_h predicted by values obtained by Werner *et al* (1966), Smith and Welsh (1971) and Yamada *et al* (1967) are 4.0, 4.3 and 14.5 ($\times 10^{-12}\ \mathrm{C\,N^{-1}}$) respectively. However, a direct measurement by hydrostatic loading determined d_h to be $(6.31 \pm 0.014) \times 10^{-12}\ \mathrm{C\,N^{-1}}$ (Graham 1976). Subsequent hydrostatic measurements of d_h for nine samples from five different suppliers (Graham 1977) showed excellent agreement with the value reported by the same author earlier (Graham 1976).

5.2.5 Converse piezoelectric effect
A piezoelectric solid also exhibits a change in shape (strain) with an applied electric field. This is called the converse piezoelectric effect. It can be shown through a thermodynamic argument that the coefficients relating the induced strain to the applied electric field are identical to those relating the induced polarization to the applied stress in the direct piezoelectric effect. In tensor component form, the converse piezoelectric effect may be expressed

$$S_{jk} = \sum_i d_{ijk} E_i \tag{5.9}$$

where S_{jk} is the second-rank strain tensor.

5.2.6 Elasticity
The elasticity of a solid can be described by the strain in the solid that results from an applied stress. Within the material elastic limit, this relationship is linear and is represented by Hooke's law

$$S_{ij} = \sum_{k,l} c_{ijkl} \sigma_{kl} \tag{5.10}$$

where S_{ij} is the second-rank strain tensor and σ_{kl} is the second-rank stress tensor. Thus c_{ijkl}, the elastic stiffness tensor, is a fourth-rank tensor. A reciprocal expression can also be written as

$$\sigma_{ij} = \sum_{k,l} s_{ijkl} S_{kl} \tag{5.11}$$

where s_{ijkl} is the elastic compliance, a fourth-rank tensor. If body torques are ignored it can be shown that $s_{ijkl} = s_{ijlk} = s_{jikl}$ and similarly $c_{ijlk} = c_{ijkl} = c_{jikl}$. Using these relations the reciprocal fourth-rank tensors (c and s) containing 81 elements are reduced to 36 independent elements that can be written as a 6×6 matrix using the index assignments given in (5.7). When this reduction is applied to the elastic compliance tensor the following assignments must also be made:

$s_{ijkl} = s_{mn}$ when m and n are both 1, 2, or 3; $2s_{ijkl} = s_{mn}$ when m or n alone are 4, 5, or 6; and $4s_{ijkl} = s_{mn}$ when m and n are both 4, 5 or 6. The application of Neumann's principle to the elastic compliance tensor followed by the use of the reduced subscript notation gives the 6×6 matrix

$$s_{ijkl} = \begin{bmatrix} s_{11} & s_{12} & s_{13} & s_{14} & 0 & 0 \\ s_{12} & s_{11} & s_{13} & -s_{14} & 0 & 0 \\ s_{13} & s_{13} & s_{33} & 0 & 0 & 0 \\ s_{14} & -s_{14} & 0 & s_{44} & 0 & 0 \\ 0 & 0 & 0 & 0 & s_{44} & 2s_{14} \\ 0 & 0 & 0 & 0 & 2s_{14} & 2(s_{11} - s_{12}) \end{bmatrix} \qquad (5.12)$$

which applies to all crystals in the 3m point group. Note that the coefficient s_{56} was incorrectly labelled in the work by Raüber (1978). Similarly, the elastic stiffness tensor can be represented by the 6×6 matrix

$$c_{ijkl} = \begin{bmatrix} c_{11} & c_{12} & c_{13} & c_{14} & 0 & 0 \\ c_{12} & c_{11} & c_{13} & -c_{14} & 0 & 0 \\ c_{12} & c_{13} & c_{33} & 0 & 0 & 0 \\ c_{14} & -c_{14} & 0 & c_{44} & 0 & 0 \\ 0 & 0 & 0 & 0 & c_{44} & 2c_{14} \\ 0 & 0 & 0 & 0 & 2c_{14} & \frac{1}{2}(c_{11} - c_{12}) \end{bmatrix}. \qquad (5.13)$$

Each of the tensors contains only six independent coefficients. When determining the values of these coefficients, the contribution of the piezoelectric effect must be taken into account. Longitudinal components of elastic waves create corresponding electric fields in the material which contribute additional stiffness terms. Measured values for the elastic compliance and elastic stiffness tensor coefficients and their temperature dependences are given in tables 5.5 and 5.6. The effects of stoichiometry on these coefficients is not well known.

5.3 Thermal diffusion in lithium niobate crystals

While poling the grown crystals of lithium niobate, D'yakov (1982) detected a potential difference at the faces of a crystal boule at temperatures above 1200 K, and that if they were shorted by a conductor, an electric current of an order of 10^{-5} A would flow in an external circuit.

The check experiments have established that the current arising is not caused by thermal EMF or by the pyro-effect. In the experiments, temperature gradients as well as heating and cooling rates were made to

Table 5.5. Elastic compliance coefficients (10^{-12} m^2 N^{-1}) at a constant electric field (according to Weis and Gaylord 1985).

Parameter	s_{11}	s_{12}	s_{13}	s_{14}	s_{33}	s_{44}
s^E	5.831	−1.150	−1.452	−1.000	5.026	17.10
s^E	5.78	−1.01	−1.47	−1.02	5.02	17.0
s^E	5.64	—	—	−0.84	4.94	—
Normalized temperature coefficients ($\times 10^{-4}$ C^{-1})						
s^E	1.66	0.28	1.94	1.33	1.60	2.05
s^E	1.5	—	—	—	1.5	—

Table 5.6. Elastic stiffness coefficients (10^{11} N m^{-2}) at a constant electric field (according to Weis and Gaylord 1985).

Parameter	c_{11}	c_{12}	c_{13}	c_{14}	c_{33}	c_{44}
c^E	2.030	0.573	0.752	0.085	2.424	0.595
c^E	2.03	0.53	0.75	0.09	2.45	0.60
c^E	2.0	0.54	0.6	0.08	2.43	0.60
c^E	2.06 ± 0.03 —	—	—	—	2.36 ± 0.028	—
Normalized temperature coefficients ($\times 10^{-4}$ C^{-1})						
c^E	−1.74	−2.52	−1.59	−2.14	−1.53	−2.04

vary. This has proved that the direction of the current is determined by the position of the 'head' and 'tail' of the crystal domain structure relative to the electrodes, while the current strength depends on the temperature and it is independent of the rate of its variations.

It was shown in the foregoing discussion that at high temperatures and at atmospheric pressure, lithium metaniobate crystals exhibit purely ionic conductivity and the current carriers are lithium ions and oxygen vacancies. So the electric current arising in an external circuit may be attributed to the electrochemical potential gradient of lithium oxide present in the crystal. The value of the current is equal to the difference in the mobilities of positive and negative current carriers diffusing along the crystal in the direction of their lowest concentration.

It follows from the phase diagram of Li_2O–Nb_2O_5 that the melt composition with 49 mol.% of Li_2O is on the right of the congruent melting point (48.6 mol.% of Li_2), and when a crystal is pulled from this melt, the former and the latter are enriched in lithium. It is remarkable that the lithium concentration in the grown crystal will increase from the seed towards the end of the crystal. As this crystal is raised to near-melting temperatures, the lithium ions, being more mobile, start to diffuse towards the seed and the external circuit will experience a current flow.

If the crystal has been grown from a melt with 48 mol.% of Li_2O, which is on the left of the congruent point, the situation will reverse and accordingly the polarity of the current arising in the external circuit will also be reversed.

Calculations have revealed that the current observed in the circuit corresponds to a diffusion flow of Li^+ in the crystal that changes the crystal composition by 0.001 mol.% in 100 h at 1230 °C. In the experiment, no appreciable variations in the current were observed while the temperature was constant. Heating the crystal in a step-like manner, D'yakov was able to establish the temperature dependence of intrinsic conductivity and calculate the activation energy for the conduction process without an external voltage supply. The value obtained is 44.7 kcal mol^{-1}, which is coincident with the lithium ion diffusion energy (Jorgensen and Barlett 1969).

The electrical phenomena accompanying the ion thermal diffusion in lithium metaniobate have been studied using concentration cells made of two melt-grown crystals with different concentrations of lithium oxide.

The temperature of the furnace where experiments were carried out was controlled to within ± 0.5°; the heating and cooling rates were 10 °C min^{-1}. Concurrently, the weight of a sample was recorded up to 0.1 wt.%.

Figure 5.5 plots typical temperature dependences of the current and EMF of lithium niobate-based concentration cells. The curves fall into three characteristic segments.

The first segment corresponds to the temperature range from 600 °C to 950–1000 °C. This range admits only very rough measurements, since the crystal resistance either exceeds or is comparable to the intrinsic resistance of a measuring circuit ($R_{cr} \gtrsim 2$ MΩ). The potential across the electrodes in this range may be caused by the pyroelectric effect. The potential difference between the right and the left lithium niobate half-cells presumably results from the near-electrode local departures from the composition as well as from the effect of the impurities absorbed by the surface. The magnitude of the potential in this range may change and its sign reverse in repetitive investigations of the same cell, but it is not affected by the crystal orientation or by its size.

The second segment is related to temperatures from 1000 to 1150 °C. Here the ionic conductivity in crystals becomes appreciable, depending exponentially on the temperature.

The EMF arises in the cell primarily due to the difference in the

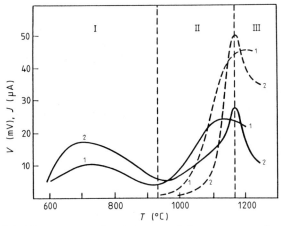

Figure 5.5. Temperature dependences of the current and EMF of a concentration cell formed by a congruent lithium niobate crystal and a crystal grown from a melt with 58% Li_2O. The broken curves represent current variations, while the full curves show voltage variations (1, cooling; 2, secondary heating) (according to D'yakov 1982).

velocities of the lithium and oxygen ions diffusing in the direction of lower Li concentration in the crystal. Another process contributing to the potential to an ever increasing extent with a rising temperature is evaporation of lithium from various-composition crystals at differing rates. On volatilization of Li_2O the crystal acquires an excess positive charge

$$Li_i^+ \xrightarrow{+e} Li \uparrow$$
$$O_x^* \xrightarrow{-e} V_O^{\cdot} + \tfrac{1}{2}O_2 \uparrow .$$

In other words, excess lithium evaporates faster than the rest of the crystal components; it evolves in the form of neutral atoms and gives up its positive charge to the lattice in the form of single-charge oxygen vacancies. Therefore the half-cell containing a high excess of the alkali component will be charged positively, thereby reducing the total EMF of the composite cell.

In experiment, this shows up in a slower growth of the EMF with temperature and in the variations of the activation energy of the conduction process (figures 5.5 and 5.6).

The third segment of the curves plotted in figure 5.5 pertaining to temperatures over 1170 °C is a region where the composition of the surface layer of lithium niobate samples undergoes marked variations. Above the mentioned temperatures, the EMF and the current curves have maxima followed by a sudden fall of the measured quantities. It is remarkable that the temperature–conductivity curves do not display any anomaly in this region. The reduction in the EMF and current is related not only to the increasing effect of Li_2O evaporation on the cell EMF but also to the shunting of the cell by a liquid film formed at the crystal surface as a result of the departure of surface layer composition from the congruent one.

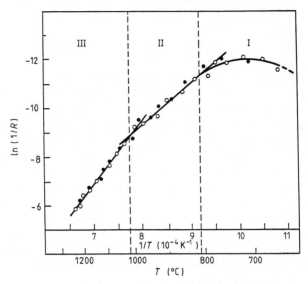

Figure 5.6. Temperature dependence of the conductivity of a concentration cell formed by a congruently grown lithium niobate crystal and a crystal pulled from a melt composition containing 58% Li_2O plotted on the basis of the data of figure 5.5. Open circles refer to heating, full circles to cooling (according to D'yakov 1982).

The method used by D'yakov (1982) to measure the temperature dependence of EMF and current of concentration cells is inferior to the classical EMF measurement technique in terms of capabilities and accuracy. Yet it permits a fairly fast measurement of characteristics of many concentration cells and yields the following information.

(i) On the basis of current direction one can find out which of the two samples contains larger amounts of alkali metal and, knowing the magnitude of the potential and the shape of the EMF and current curves in a high-temperature region, one can roughly estimate a relative content of Li_2O in crystals.

(ii) On the basis of measured EMF and current for various temperatures (figure 5.5) a curve has been plotted that relates the intrinsic resistance of a concentration cell to temperature (figure 5.6); its slope is helpful in establishing the activation energy of the conduction process for various temperatures without resorting to external voltage sources.

The trend of the curve in segment I (figure 5.6) cannot serve, as noted earlier, as a characteristic of the concentration cell. In segments II and III, the curve can be represented by two straight lines whose changing slope reflects the conduction activation energy going from $32 \, kcal \, mol^{-1}$ at 800–1050 °C up to $55 \, kcal \, mol^{-1}$ at temperatures exceeding 1050 °C.

In the above experiments, the region of extrinsic conductivity is fitted by segment I (figures 5.5 and 5.6), while the intrinsic conductivity region falls into two parts differing in the activation energy. The boundary of the two parts lies at 1050 °C; at this temperature the evaporation of Li_2O from the crystal becomes appreciable (Kaminov and Carruthers 1973). This process is an additional source of current carriers in the form of single-charge oxygen vacancies and it may cause changes in the activation energy of lithium niobate conduction at high temperatures.

It is pertinent to mention here that the averaged slope of the curve in segments II–III (figure 5.5) corresponds to an activation energy of $41.5 \, kcal \, mol^{-1}$, which agrees fairly well with the earlier data of $43.7 \, kcal \, mol^{-1}$.

5.4 Relaxation phenomena in lithium niobate crystals

To clarify the mechanism of charge transfer in lithium niobate, exhibiting a low activation energy, Kudasova (1980) employed the method of internal friction. In a piezoelectric crystal, mechanical oscillations give rise to oscillations in the electric field. The relaxation of this field through charge transfer brings about mechanical relaxation. Measurement of mechanical relaxation is superior to direct measurements of dielectric loss, thanks to its high mechanical Q value and low background of mechanical loss.

One of the sources of energy loss in ferroelectrics may be the interaction of domain boundaries with point defects. For instance, a multidomain lithium niobate single crystal has shown two internal friction peaks—at 300 K and 550 K at a frequency of measurement of $v = 5$ Hz, the peaks are absent in single-domain samples (Postnikov *et al* 1971). The internal friction peak observed at 300 K depends on the density of oxygen vacancies. As the latter increases, the peak shifts to a lower temperature region and attains saturation. The activation energy of the process determining the occurrence of this peak is calculated from its frequency shift to be 0.70 ± 0.05 eV. This peak is presumed (Kaverin 1971) to be the result of the interaction of electrically charged oxygen vacancies with fixed 180° domain walls charged due to the piezo-effect. The other peak of internal friction, observed at 550 K, stems from the relaxation process with an activation energy of 1.4 eV and is attributed to the domain boundary interaction with defects whose nature has not been identified. Kaverin (1971) claims that it is charged dislocations that are responsible for the occurrence of this peak, whereas Gridnev *et al* (1978) believe that the reason is lithium vacancies. Vacuum annealing of copper-doped lithium niobate crystals produces at 300 K, along with the internal friction peak, a relaxation peak with an activation energy of 0.1 eV near 55 K. The larger the amount of copper, the higher the peak with its simultaneous shift towards lower temperatures. This suggests the relationship of the peak to the concentration of copper ions near the domain boundaries. The mechanism responsible for the internal friction peak near 130 K, which is observed when lithium niobate has been annealed in hydrogen, is not related to the presence of a domain structure. It does not occur upon vacuum annealing of crystals, while oxygen annealing of hydrogen-reduced lithium niobate samples even reduces it. It can be supposed, therefore, that this peak of internal friction is associated with chemically bound oxygen.

Apart from the above mechanisms basic to internal friction in lithium niobate, ultrasound may attenuate through relaxation processes, specific to piezoelectrics and related to the movement of charged point defects in the electric field accompanying an ultrasonic wave. A kind of relaxation may also come into play, which is characterized by a low activation energy and which involves localized electrons and holes. In this case, in the electric field induced in a piezoelectric by mechanical oscillations, the elastic dipoles prove to be flipped ('flip-flop' mechanism) (Nowick and Berry 1972).

In the study of Kudasova (1980), internal friction was measured by the three-component piezoelectric resonator method in a $1\,\mathrm{N\,m^{-2}}$ vacuum at a frequency $v \approx 10^5$ GHz over a temperature range 290 to 520 K. Mechanical oscillations in the resonator were excited by a square cross section bar of α quartz. A sine wave voltage was applied to the quartz exciter which converted it to mechanical longitudinal oscillations. An ultrasonic wave passed throughout the resonator and was picked up by a quartz receiver in the form of a sine voltage. A long rod of fused silica (a buffer) separated the resonator system from the hot-temperature zone where the sample was mounted. Test samples were cut from lithium niobate single crystals parallel to the growth axis. The length of the sample was chosen so that its own resonance oscillation frequency might coincide with the inherent frequency of the resonator (deviation not more than 0.5%). The sample length (l) was roughly estimated by the formula

$$v_{\mathrm{res}} = 1/2l(\rho/c)^{1/2} \tag{5.14}$$

where c is the elastic stiffness constant of the crystal, ρ is the sample density and v_{res} is the resonance oscillation frequency of the sample itself. A more accurate choice of sample length was made by adjusting the frequency of the resonator–sample system to the inherent frequency of the resonator. During internal friction measurement, by varying the frequency of the applied signal it was possible to obtain resonance of electric and hence of mechanical oscillations. The magnitude of internal friction was calculated by the formula

$$\Delta = k(U_{out}/U_{in}) \tag{5.15}$$

where U_{out} is the output voltage of the receiver, U_{in} is the input voltage of the excitor, k is a coefficient determined by the sample geometry and the properties of the material studied (for the lithium niobate crystal, $k = 4 \times 10^{-4}$). An error of peak temperature measurement lies within $\pm 0.5°$; an absolute error of the activation energy of the process does not exceed 12%.

The temperature–internal friction curves for single- and multidomain lithium niobate samples have a peak (peak A) at $T \approx 370$ K, while the same curves for multidomain samples exhibit a second peak (peak B) which is shifted relative to the first one by about 30° to a high-temperature region (figure 5.7). The temperature–shear modulus curve has an inflexion point characteristic of relaxation processes. The activation energy of ultrasound attenuation calculated from the internal friction measurement data for various frequencies proved to be 0.17 ± 0.02 eV and 0.7 ± 0.04 eV for peaks A and B respectively.

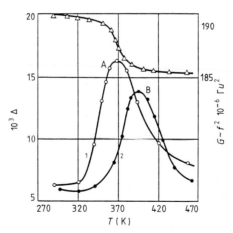

Figure 5.7. Temperature dependence of shear modulus and internal friction for single-domain crystals of lithium niobate. Open circles, $v = 138$ kHz; full circles, $v = 183$ kHz (according to Kudasova 1980).

The presence of peak A in multidomain samples and its absence in single-domain samples suggests its relationship to the interaction of domain walls with point defects. This is also confirmed by the relaxation nature of the peak. Peak A of internal friction has not been found to depend on heating rate, orientation or signal amplitude (over a strain interval from 5×10^{-7} to 10^{-5}). The presence of this peak both in multi- and single-domain samples, as well as its independence of orientation, are evidence that it has nothing to do with the domain structure of the crystal.

An attempt was undertaken to study the effect of crystal heat treatment on the magnitude and position of peak A. Having been annealed in vacuum for 6 h at $T = 770$ K, the sample became dark and its conductivity increased by seven orders of magnitude (from 10^{-13} to $10^{-6}\,\Omega^{-1}\,m^{-1}$). The position of peak A did not change upon annealing but its height increased. Oxygen annealing at $T = 1200$ K for 8 h rendered the sample transparent and the maximum of the temperature–internal friction curve became lower. This behaviour of peak A for a heat-treated crystal suggests that it is related to oxygen vacancies, since annealing of lithium niobate in a reducing medium raises the density of oxygen vacancies, while oxygen annealing lowers it.

The modest activation energy for the internal friction process makes it possible to presume that mechanical relaxation of peak A proceeds by the 'flip-flop' mechanism specific of piezoelectrics and related to electron redistribution in localized energy levels of charged oxygen vacancies found in the electric field of an ultrasonic wave.

Thus the investigation of internal friction of lithium niobate crystals has corroborated the existence of low-activation processes of electric charge transfer over a temperature range 290 to 520 K, and has indicated that there is a relationship between these processes and point defects of crystals, namely electrically charged oxygen vacancies.

5.5 Electric fields in lithium niobate crystals

Lithium niobate is a pyroelectric; therefore variations in the crystal temperature will affect the magnitude of spontaneous polarization ΔP_s. The spontaneous polarization electric charges are compensated by free electric charges through the conduction process, so that the crystal as a whole remains electrically neutral. By using the above data on lithium niobate conductivity, one can show that the rate of relaxation of the pyro-charge electric field due to bulk conductivity at near-room temperatures (say at $T = 80\,°C$ the Maxwellian relaxation time τ is about two hours) that crystal temperature variations as slow as $10^{-1}\,°C\,h^{-1}$ may set up a pronounced electric field due to the pyro-effect. Hence, because of the long τ values and low conductivity, at random temperature fluctuations the crystal is virtually always in a non-equilibrium state which is characterized by a finite value of the intrinsic electric field E_i. The E_i value is defined by the conductivity and spontaneous polarization of the crystal according to the formula

$$\varepsilon\varepsilon_0 E_i = (D - \Delta P_s) \tag{5.16}$$

where D stands for the induction of the electric field of free electric charges and ε is the dielectric constant.

The direction of the internal electric field is opposite to that of spontaneous polarization. For spontaneous polarization $P_s = 70\ \mu C\,cm^{-2}$, the depolarizing internal electric field in the lithium niobate crystal is estimated by Ohmory *et al* (1974) to be about $10^5\ V\,cm^{-1}$. This estimate agrees fairly well with experimental observations made during photoconductivity investigations of lithium niobate (Ohmori *et al* 1975). Internal electric fields in lithium niobate crystals are determined by the heat treatment conditions and dopant. They are $\sim 180\ kV\,cm^{-1}$ and $\sim 85\ kV\,cm^{-1}$ for pure and Fe doped crystals respectively. Vacuum annealing of lithium niobate crystals reduces markedly (down to $\sim 8\ kV\,cm^{-1}$) their value. The same effect is produced, according to the estimates of Blistanov *et al* (1978), in molybdenum-doped lithium niobate.

The existence of an internal electronic field in lithium niobate is also indicated by other experimental facts. For example, there has been discovered an effect of thermally stimulated autoelectronic emission (Malter effect) from the surface perpendicular to the ferroelectrical axis of the lithium niobate crystal, which will be discussed below. The anomalously high energies of emitted electrons (above 300 eV) allow us to hypothesize that they are released sufficiently far from the surface and get accelerated in the intrinsic electric field of the crystal. This latter field induced by the pyro-effect under temperature variations within 25 to $100\,°C$ is estimated for lithium niobate to be $2-1.5 \times 10^7\ V\,cm^{-1}$ (Kortov *et al* 1979). This value seems to be too high, since the adopted model does not take account of the compensation of the pyro-charge field by conduction. The upper temperature limit of the emission ($\approx 170\,°C$) is specified by an increasing crystal conductivity due to the higher density of free carriers excited thermally from local levels.

In lithium niobate crystals, however, the internal electric field E_i coinciding with the direction of the optical axis should not give rise to the transverse electro-optical effect (chapter 6). The cause of appearance of the E_{ix} and E_{iy} field components normal to the optical c axis of the crystal and responsible, in the opinion of Blistanov *et al* (1978), for the residual light current has not yet been definitely recognized. We can suppose that E_{ix} and E_{iy} appear here and there in the crystal as a result of a structural inhomogeneity present therein and causing non-uniformity of the pyro-field.

The structural inhomogeneity of crystals, e.g. a block structure, non-uniform distribution of defects (impurities, dislocations), make the crystal lattice misorient in individual crystal blocks, i.e. it can generally be discribed as misorientation of the crystallophysical axes of individual crystal blocks relative to each other by angles α_i (figure 5.8). In this case, as the temperature of crystals undergoes variations, the pyro-effect may produce a system of pyro-charges whose total field in individual areas of the crystal has the E_{ix} and E_{iy} components. In particular, if a crystal consists of blocks the pyro-effect will establish a pyro-field E_z^m in each individual block m which will be parallel to the z axis of a given block. Taking the coordinate frame of a sample so that the z axis of the sample coincides with its optical axis, one can specify the α_{ij} cosines of the angles between the crystallophysical coordinate frame of the block m and the sample coordinate system. Then, within the frame of the sample coordinates, the internal field components in this block are

$$E_i^m = \alpha_{iz}^m E_z^m. \tag{5.17}$$

In the crystal as a whole we have

$$E_i = \int_0^\alpha \alpha_{ij}^m E_z^m \, d\alpha = \alpha E^c \tag{5.18}$$

where the coefficient α is determined by the size and misorientation of various samples volumes, i.e. it is an averaged characteristic of structural inhomogeneity, which cannot be established by calculation in a general case. The α value has been obtained experimentally and for satisfactory-quality crystals it turns out to be $\approx 10^{-3}$, which is equivalent to a deviation of the crystal optical axis from the system axis of several

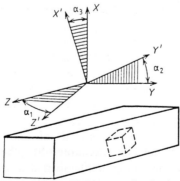

Figure 5.8. Misorientation of the crystallophysical axes of individual crystal blocks relative to the mean direction of the x, y and z axes (according to Kudasova 1980).

minutes. The coefficient α can be regarded as a measure of optical inhomogeneity of a crystal, when no information about the specific causes of the inhomogeneity is available, and it can be used to estimate the electric field component perpendicular to the crystal optical axis. In this case, αE_i is the field to be applied to a crystal normal to its optical axis in order to produce 'bleaching' of the system by means of the electro-optical effect, which will be equivalent to the observed one.

According to Ohmori *et al* (1974), the internal electric fields in lithium niobate are $E_i = 10^6 - 10^7\,\mathrm{V\,m^{-1}}$. The angle of misorientation of the crystallophysical axes of individual crystal parts, which ensures appearance of the internal field components E_{ix}, E_{iy} of the order of $10^4\,\mathrm{V\,m^{-1}}$, must be equal to several minutes. This also confirms the above estimates.

Internal fields in crystals can be studied thanks to the regular domain structure. This structure enables a built-in electric field to be set up in a crystal in a form convenient for measurement, whereas in single-domain samples the field is markedly non-uniform and can be only indirectly estimated to have its averaged value. Despite the different forms of existence of the built-in electric field in a single-domain crystal and in a sample exhibiting a regular domain structure, the nature of the field variations in response to external effects is similar in the two cases, since it is defined by the same physical processes.

As the temperature of a crystal changes, its faces perpendicular to the direction of spontaneous polarization begin to carry pyro-charges. For a crystal with a regular domain structure and equal thickness of domains with different directions of the spontaneous polarization vector, the total charge on each face is zero. In the case of departures of the domain structure from regularity, which may amount to $\pm 5\%$ of the mean value, on the faces perpendicular to the z axis there form uncompensated pyro-charges setting up an electric field E_i in the crystal which is proportional to the value of the pyroelectric field in a single domain E_p:

$$E_i = \eta E_p = \eta\gamma\Delta T / \varepsilon\varepsilon_0 \qquad (5.19)$$

where γ is the pyro-coefficient, ΔT is the change in the temperature and ε and ε_0 are permittivity and dielectric constant respectively. The factor of proportionality η defines the degree of sample unipolarity (Aleksandrovsky 1977):

$$\eta = (S^+ - S^-)/(S^+ + S^-) \qquad (5.20)$$

where S^+ and S^- are the surface areas of different-sign domains in a plane perpendicular to the z axis.

The electric field E_i is nearly uniform, provided the dimension of a crystal along the z axis greatly exceeds the domain thickness. In domains with different directions of the spontaneous polarization vector, this field induces, through the electro-optical effect, variations in the refractive indices, which are the same in magnitude but opposite in sign. In each domain $\Delta n_e = \frac{1}{2} n_e^3 r_{33} E_i$, the difference in the refractive indices of the adjacent domains is $2\Delta n_e$.

If an external electric field E is applied to the crystal faces perpendicular to the z axis, the value Δn_e is defined by the total effect of the built-in and external fields:

$$\Delta n_e = \tfrac{1}{2} n_e^3 r_{33}(E_i + E). \qquad (5.21)$$

Periodic variations in the refractive index n along the y axis of the crystal are a phase diffraction grating for light propagating along the x axis. In the experiment (Stepanova 1986) the light is polarized along the z axis; therefore the phase diffraction grating is formed by variations in the extraordinary refractive index n_e. As Δn_e vanishes, the diffraction efficiency I_1/I_0 tends to zero.

The built-in electric field is measured by compensating it by an external field ($E_1^* = -E_i$). To this end, the external field E^*, applied to a sample is such that the diffraction efficiency is either zero or minimal, if any.

By measuring the dependence of I_1/I_0 upon the external electric field, one can establish the value of E^* (and hence of E_i) corresponding to a minimum value of I_1/I_0 (figure 5.9). In actual crystals, a minimum value of diffraction efficiency is somewhat different from zero. This is attributed to crystal inhomogeneity.

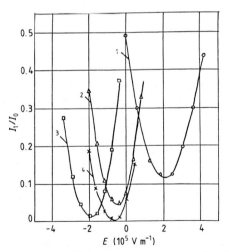

Figure 5.9. Diffraction efficiency versus external electric field: 1, 51 C; 2, 72 C; 3, 83 C; 4, 107 C. The heating rate is $26\,^{\circ}\mathrm{C\,h^{-1}}$ (according to Stepanova 1986).

Obviously, the above method of measuring E_i yields true information about variations in the built-in electric field of a crystal exposed to external factors only in the case where the position of the domain walls in a regular domain structure is preserved, i.e. where the η value remains constant.

In a sample maintained at a constant temperature for a long time, the initial value of the built-in field is usually small. On heating the sample from the initial state up to $T = 60\,°C$, the linear enhancement of E_i is governed by the pyro-effect. (The polarity of E_i is assumed to be positive.) Then, having attained a maximum value at T_{max}, the built-in field decreases, which is accounted for by increasing conduction facilitating the compensation of pyro-charges by free carriers. The shape of the $E_i(T)$ curve plotted using the compensation technique coincides with that of residual light current versus temperature observed by Blistanov *et al* (1978) (see chapter 9).

At temperatures above $150\,°C$, pyro-charges are compensated in the course of conduction processes in a time shorter than the measurement time. As a result, the built-in field can be considered to be zero. Crystal cooling is accompanied by changing ΔP_s to the opposite sign, while the D value is virtually unable to change because of the conduction decreasing at a lower temperature. That is why upon cooling the built-in field at room temperature is negative.

The initial value of $E_i(T_{room})$ in all subsequent measurements taken in equal time intervals is the same and is determined by the field induced in the crystal upon cooling. In the segment corresponding to linear variations of the built-in field, all the values of $E_i(T)$ obtained for various heating rates are virtually the same (figure 5.10), since the strength of the pyro-field is determined solely by temperature variations.

At faster heating rates, the maximum of the $E_i(T)$ curve shifts towards higher temperatures. This is due to the fact that at faster heating the pyroelectric field increases more rapidly, while its compensation by free charges at specific temperatures proceeds at the same rate.

The temperature dependence of the built-in field is important for understanding the behaviour of an optical inhomogeneity exposed to an external electric field at various temperatures (see chapter 9).

The relaxation kinetics of a built-in pyroelectric field under isothermal holding was studied using lithium niobate crystals with a regular domain structure. Samples were raised from $20\,°C$ at a rate of 50 degrees per minute up to a specified temperature lying within 40 to $120\,°C$. At $T = $ const., the built-in pyroelectric field underwent relaxation from the initial value, determined by the temperature history of the sample, and tended to zero. The measured E_i values versus time under isothermal holding are presented in figure 5.11.

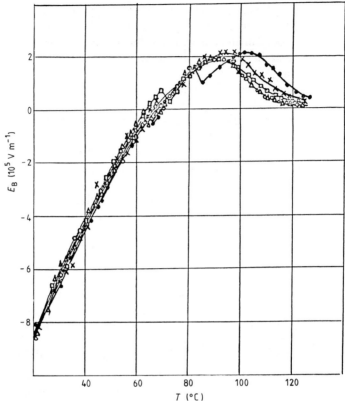

Figure 5.10. Temperature dependence of the built-in field for various rates of heating of a lithium niobate sample: ○ 19 °C h^{-1}; △ 26 °C h^{-1}; □ 31 °C h^{-1}; × 38 °C h^{-1}; ● 50 °C h^{-1} (according to Stepanova 1986).

The pyroelectric field relaxes as a result of compensation of pyrocharges by free carriers. Variations of the pyro-field in time at a constant temperature are fitted by the expression

$$E_i(t) = E_0 \exp(-\sigma t/\varepsilon\varepsilon_0) \tag{5.22}$$

where σ is the conductivity, ε and ε_0 are permittivity and dielectric constant respectively, t is the time and E_0 is the value of the pyro-field at $t = 0$.

5.6 Electret effect and relaxation polarization of lithium niobate

Lithium metaniobate single crystals enjoy great favour in nonlinear

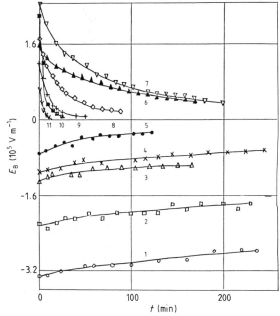

Figure 5.11. Relationship $E_i(t)$ under isothermal conditions: 1, 44 °C; 2, 47 °C; 3, 53 °C; 4, 58 °C; 5, 70 °C; 6, 76 °C; 7, 83 °C; 8, 90 °C; 9, 100 °C; 10, 108 °C; 11, 118 °C (according to Stepanova 1986).

optics. But their electro-optical characteristics are deteriorated by the relaxation polarization processes giving rise to space charge fields. This invokes great interest in the study of slow polarizations and the electret effect in substances of this kind.

Polarizations with long relaxation times τ may result in the development of an electret effect, i.e. the existence of external electric fields. Investigation of the electret effect makes it possible to obtain insight into the relaxation processes with $\tau > 10^6-10^8$ s. The characteristics of the electret effect are the electret difference of potentials P_{el} and the effective surface charge density σ. The values of P_{el} and σ are usually determined by the compensation method. Information about the energy characteristics of the trapping centres of charge carriers responsible for establishment of space charge fields can be derived by measuring the currents of thermally stimulated depolarization. These currents were measured in short circuit conditions in the course of heating at a rate of $0.1 \, \mathrm{K \, s^{-1}}$ (Kuz'minov *et al* 1983). Platinum electrodes were applied by cathode sputtering. A and C disks were cut from single crystals with their cuts perpendicular and parallel to the plane (001) respectively.

The findings of this work characterize the electret effect and relaxation polarization in lithium niobate crystals (Bondarenko *et al* 1985).

Recording was made of the surface charge density $\sigma \neq 0$ for single-domain lithium niobate crystals after they had been stored for a year in an unshorted state at room temperature and normal humidity and had not been exposed to a polarizing field.

On heating up to temperatures close to (but lower than) the Curie point (T_C), the external field disappeared both in the A and C samples of lithium niobate single crystals. Within approximately 9×10^4 s it was re-established only in the C samples. Heating up to such high temperatures (1273 K) must lead to complete vanishing of relaxation polarization and the fact that it did not disappear in the C samples attests to the substantial contribution made by spontaneous polarization in the establishment of an external field of lithium niobate electrets. When heated up to $T > T_C$, the samples went over to a multidomain state and the external field reduced to zero.

The magnitude of the electret effect is related to the orientation of the polarizing field vector relative to the polar axis of a crystal. Polarization in a field of 7×10^5–10^6 V m^{-1} over a temperature range 373 to 423 K for 5.4×10^3 s by means of applied electrodes will develop a surface homo-charge $E \perp [001]$ in all the A samples of studied crystals and either a homo-charge (when $E \Uparrow (001)$) or a space hetero-charge (when $E \| (001)$) converting with time into a homo-charge in the C samples. An examination of the thermally stimulated depolarization currents in the crystals studied has also indicated that they depend considerably on the relative orientation of the vectors of spontaneous polarization and polarizing field. The density of charge q, determined by integrating the areas taken up by the curves plotted for thermally stimulated depolarization currents, is 10 to 15 times higher for C samples as compared to A samples. A comparison of the q and σ values has revealed that the former is several orders of magnitude larger than the latter (say by a factor of 7000 for lithium niobate A samples). We should like to note here that the q value determined in this manner is in units of C m^{-2}. The field induced by the q charge is expected to be compensated by dielectric relaxation in several hours, but as a matter of fact the electret effect is retained for years. For instance, in the case of lithium niobate samples, within 9.7×10^6 s after polarization $\sigma = 8.1 \times 10^{-5}$ C m^{-2}, while the Maxwellian relaxation time is 6×10^2 s.

Prokopalo (1979) has established the important contribution of structural point defects, namely vacancies of oxygen V_O and A ion V_A, in the formation of electrical properties of oxide compounds exhibiting a perovskite-type structure. To check the role of V_O and V_A in the electret effect and relaxation polarization, lithium niobate and lithium tantalate crystals were heat-treated in vacuum to increase their oxygen deficit. In reduced samples, the σ value increased ten-fold and more as compared to unreduced samples (figure 5.12). The q value was also found to rise markedly.

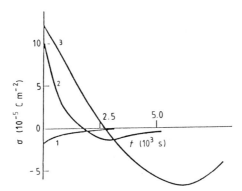

Figure 5.12. Surface density of the σ charge versus storage time of lithium niobate A-cut electrets heat-treated in vacuum: 1, unreduced sample; 2, 873 K, 5 Torr, 1.4×10^4 s; 3, 1073 K, 5 Torr, 1.4×10^4 s (according to Kuz'minov *et al* 1983).

Electret substances are generally notable for two kinds of relationship between the σ value and storage time: (i) the homo-charge undergoes relaxation to zero; (ii) the hetero-charge is converted into the homo-charge, followed by relaxation of the latter (Gubkin 1978). Reduced lithium niobate and lithium tantalate crystals were the first on which sign reversal of the electret charge from the homo- to the hetero-charge has been registered. This fact is indicative of the decisive role of migration polarization in the formation of the electret state in oxygen-octahedral substances. In structures of less close packing (pseudoilmenite), the contribution of migration polarization into the electret effect is larger than in perovskite-structure substances.

To establish the role of charge carrier injection from electrodes, lithium niobate single-crystal A samples were subjected to a polarizing field with the use of insulating layers of flouroplastic. As a result, a hetero-charge was found to form with an initial electret potential difference $P_{el} = 100$ V and $\tau = 2.8 \times 10^5$ s. When the crystals were polarized using applied metallic electrodes a more stable homo-charge was formed with an initial value of $P_{el} = 330$ V and $\tau = 3.2 \times 10^7$ s.

By relying on the experimental data obtained (Kuz'minov *et al* 1983, Bondarenko *et al* 1985), a qualitative model has been put forward describing typical electret phenomena in substances of pseudoilmenite structures. They are as follows.

(i) The major role in the electret effect is played by migration polarization stemming from the movement of vacancies V_O and V_A. Part of the vacancies possess a charge (V_O positive, V_A negative) and form complexes. Exposed to a polarizing field, these complexes are destroyed and the charged vacancies that have shifted to the electrodes contribute to the formation of a hetero-charge. To do this, it suffices that part of the vacancies will shift by a macro-distance (about 1×10^{-3} m) or the whole system of vacancies will shift by a micro-distance (of the order of 10–100 unit cell dimensions).

(ii) The volume polarization in single-domain ferroelectrics which is due to the shifting of vacancies is anisotropic: it is more pronounced along the polar axis and is weaker normal to it. This predetermines anisotropy of the electret effect in the crystals studied.

(iii) The higher density of charged vacancies in the near-electrode layers is a contributing factor in the enhancement of the electric field and provides a stimulus for charge carriers to be injected from the electrodes, thereby producing a homo-charge. The ion vacancies (together with other defects of the crystal lattice) act as trapping centres of charge carriers, thereby reducing their effective charge. The result is a nearly complete mutual compensation of the injected and space charges. This explains the stability of the electret state and the great difference in the densities of σ and q charges (the σ value equal to the difference between the injected and the space charges is about 0.01% of the magnitude of each of them). As a result, to achieve a stable electret state, it is necessary to establish in a single polarization cycle optimal conditions for development of migration polarization as well as for injection of charge carriers from the electrodes. The intensity of the two processes is determined by the temperature and electric field strength (the mobility of vacancies is mainly governed by the polarization temperature, while the carrier injection is primarily specified by the field strength).

(iv) In the course of establishing the thermoelectret state, spontaneous polarization enhances the charge carrier injection from the electrodes. Reduction of the value of permittivity in the near-surface layers, compared to the bulk, of ferroelectrics brings about field redistribution, which promotes penetration of injected charges to a greater depth than is achievable with non-ferroelectrics. This provides more favourable conditions for compensation of migration polarization.

In single-domain crystals, spontaneous polarization charges may directly contribute to the electret external field.

(v) The thermoelectret state that has established itself in lithium niobate is characterized by a higher density of vacancies in the near-electrode layers, whose charge is mostly compensated by that of injected carriers. After the polarizing field has been removed, depolarization in the electret develops by two mechanisms: (*a*) equalization of the V_O and

V_A densities in the whole of the electret volume by diffusion; (*b*) release of the charges localized in the trapping centres, which are compensated due to bulk conduction. These processes are inter-related and the σ value is determined by their combination.

5.7 Thermionic emission of lithium niobate single crystals

In the theoretical work by Chensky (1970) it was shown that spontaneous polarization of ferro-semiconductors below the Curie temperature may affect the emission properties of surfaces lying perpendicular to the crystal polar axis due to the band bending, whose maximum magnitude may be half the width of the forbidden band. If the spontaneous polarization vector is parallel to the emitting surface, the ferroelectrical band bend is zero. The band bend is determined by the magnitude of the polarization vector, the electron density in the conduction zone and the temperature.

A direct proof of internal electric fields existing in a crystal is the thermally stimulated electron emission from the lithium niobate crystal. It was first observed by Rosenblum *et al* (1974).

Electron emission is caused by the electric field induced by the variations in spontaneous polarization P_s of pyro-active crystals including lithium niobate. The air breakdown observed near the lithium niobate crystal is also caused by the pyro-field generated upon crystal heating. The observed phenomena show that thermally generated fields may be strong enough to invite electron emission. The exoelectron emission was observed using photomultipliers and luminescent screens. Experiments were carried out with single- and multidomain nominally pure lithium niobate crystals. Emission was observed from the surface perpendicular to the polar *c* axis. When a single-domain crystal was raised from room temperature to $100\,°C$ at a heating rate of about $20\,°C\,min^{-1}$, an electron emission current of density $10^{-10} - 10^{-9}\,A\,cm^{-2}$ was observed to flow from the C^+ surface of the sample. The emission from a polished surface was weaker than that from a freshly cleaved one. Cooling was accompanied by occasional discharge bursts. Conversely, bursts were observed during heating of the C^- surface of the sample, whereas this surface exhibited a persistent emission during cooling.

Multidomain samples emitted upon heating as well as cooling. The strength of the emission current from each crystal surface depended on the relative size of the surfaces C^- and C^+ of domains on each side. A typical proximity image of the emission during cooling from a multidomain surface is depicted in figure 5.13(*a*). The same surface after etching in HF is seen in figure 5.13(*b*). The image intensity depended

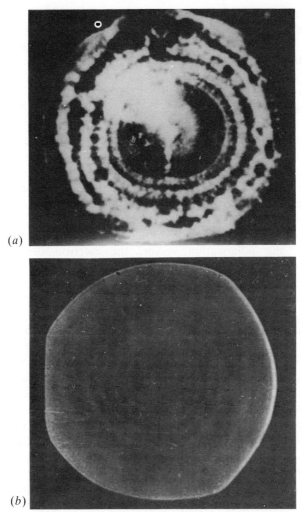

(a)

(b)

Figure 5.13. (*a*) Proximity image of the emission during cooling from a multidomain lithium niobate crystal. (*b*) The same surface after etching in HF. The crystal is 15 mm in diameter (according to Rosenblum *et al* 1974).

on the rate of heating (or cooling) as well as temperature. No emission was observed when the sample was in thermal equilibrium. Emission is accompanied by occasional discharge bursts which make the emission die out. The energy of emitted electrons is dependent upon the sample–detector separation, because the involved surfaces form a capacitor. An appreciable emission is observed for a separation of about

1 cm and a retarding potential of about 5 kV. The observed emission can be explained by relying on the thermally stimulated emission field of a pyro-active crystal. At room temperature, lithium niobate exhibits spontaneous polarization $P_s = 0.7 \, \mathrm{C \, m^{-2}}$. Owing to this polarization, the uncompensated charge of the C surface will have a surface density σ. The polarization charge density is equal to the crystal polarization $|P_s| = |\sigma|$. If a crystal is placed in the vacuum, a potential difference E_0 arises between the heating block on which the crystal is mounted and the detector. In the case of a parallel electrode arrangement the field is

$$E_0 = 4\pi\sigma = 4\pi P_s. \tag{5.23}$$

In actual crystals, the internal field E_i causes compensation of the surface charge due to its finite conductivity σ_c. In vacuum, the charge decays exponentially:

$$\sigma(t) = \sigma(t = 0) \exp(-\sigma_c t / \varepsilon \varepsilon_0) \tag{5.24}$$

where ε is the permittivity. In air, additional compensation stems from the ion sorption on the crystal surface. Polarization effects resulting from temperature variations of a crystal are observed during the time $\varepsilon_0 \varepsilon / \sigma_c$. In the experiments carried out by Rosenblum *et al* (1974) this time was about 30 min.

The change in the polarization ΔP_s under variations of the crystal temperature can be estimated from the expression by Di Domenico and Wemple (1969):

$$\tfrac{3}{2}(P_s/P_0)^2 = [1 + \tfrac{3}{4}(T_c - T)/\Delta T_0]^{1/2} - 1. \tag{5.25}$$

For lithium niobate, $P_0 = 0.71 \, \mathrm{C \, m^{-2}}$ and $\Delta T_0 = 90 \, °\mathrm{C}$ (ΔT_0 being the difference between the actual temperature of a ferromagnetic transition and the Curie–Weiss temperature). As the temperature goes from $T_1 = 25 \, °\mathrm{C}$ to $T_2 = 100 \, °\mathrm{C}$ in a time shorter than the charge relaxation time, the polarization change is $P_s = 0.015 \, \mathrm{C \, m^{-2}}$ and the corresponding strength of the electric field on the crystal surface is $E_0 = 1.35 \times 10^7 \, \mathrm{V \, cm^{-1}}$. This field is sufficiently strong to induce electron emission from the C^+ surface of the lithium niobate crystal. On cooling $P_s > 0$ a positive charge arises at the C^+ surface. In this case, emission comes from the sample holder and the front end of the detector facing the crystal. The emission goes over into a discharge. Fields of the same strength are established between domains of opposite polarity. The emission field and the relaxation mechanisms (i.e. bulk and surface conductivities) reduce the charge build-up. After a certain specific time, a steady-state equilibrium establishes itself which is described by:

$$j = 4\pi(\mathrm{d}P_s/\mathrm{d}t) - j'(T) \tag{5.26}$$

where j is the emission field current density and $j'(T)$ is the relaxation current density. The latter depends heavily on the temperature and tends to extinguish the emission field at elevated temperatures.

Electrical discharges on the positively charged crystal surface are brought about by the ionization of residual gas molecules in the vicinity of the edges of the sample–crystal holder–detector system. The resultant discharge neutralizes a substantial portion of the charge accumulated on the crystal surface. As a result, the electric field is reduced. Prolonged heating renders this process periodic. The electric field strength on lithium niobate crystals is $1.5 \times 10^5 \, \text{V cm}^{-1}$ when the crystal temperature is changed by one degree (Avakian *et al* 1976a). This value is five times that of an electrical breakdown in air. Boikova and Rozenman (1978) examined the spectral dependence of the quantum yield $Y(h\nu)$ of the photoemission from lithium niobate crystals doped by iron ions, $LiNbO_3$:Fe (figure 5.14). Over an excitation energy range 3.9 to 5.2 eV, the quantum yield increases nearly parabolically. The spectral characteristics of the quantum yield in the vicinity of the threshold frequency for the photoelectric effect can be represented as

$$Y = a(h\nu - h\nu_0)^m \tag{5.27}$$

where $h\nu_0$ is the photo-effect threshold frequency, $h\nu$ is the quantum energy and a and m are constants characteristic of a given crystal (say for lithium niobate $m = (2.73 \pm 0.05)$). Equation (5.27) gives the emission properties in the vicinity of the photo-effect threshold frequency $(h\nu - h\nu_0) \sim 1$–3 eV. It is evident from figure 5.14 that $h\nu_0$ for planes perpendicular to P_s is 4.3 and 3.9 eV. The photoelectric work function for a ferromagnetic material is defined by

$$A_1 = E_g + \chi + \Delta\chi_1 \tag{5.28}$$

for a face with a negative direction of P_s and by

$$A_2 = E_g + \chi - \Delta\chi_2 \tag{5.29}$$

for a face with a positive direction of P_s. In these expressions E_g stands for the width of the forbidden band, χ is the affinity for the electron and $\Delta\chi_1$ and $\Delta\chi_2$ are the ferroelectric band bends. It has been established for lithium niobate that $E_g = 3.9$ eV (Ionov 1973), with no account taken of the forbidden band bending $A = 2.2$ eV. Expressions (5.28) and (5.29) imply that $\Delta\chi_1 = 0.15$ eV and $\Delta\chi_2 = 0.25$ eV.

The small values of $\Delta\chi_1$ and $\Delta\chi_2$ are likely to be due to the high densities of surface states N and indicate that spontaneous induction is primarily screened by the surface level charge.

For opposite surfaces perpendicular to the spontaneous polarization vectors, the condition

$$\Delta\chi_1 - \Delta\chi_2 \simeq 2P_0/eN \tag{5.30}$$

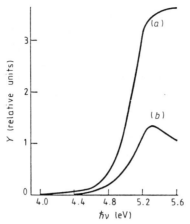

Figure 5.14. Spectral dependence of the quantum yield of the photoemission LiNbO$_3$:Fe single crystals. The curve (*a*) corresponds to the geometry where P_s is directed towards the detector; the curve (*b*) corresponds to the opposite sample orientation (according to Boikova and Rosenman 1978).

is satisfied. It implies that $N = 2 \times 10^{15}$ cm^{-2} eV^{-1}, which corresponds to the high density of surface states observed in ferroelectric materials. Thus, over a small energy interval (about 0.1 eV) there is a high density of energy states forming a narrow surface zone.

Rozenman and Boikova (1979) studied the kinetic behaviour, orientation dependence and the energy of electrons emitted by Fe-doped (0.05%) lithium niobate single crystals under the pyro-effect conditions. Measurements were taken in a 10^{-6} Torr vacuum with the temperature varying linearly from 100 to 200 °C at a rate of 2.5 °C min^{-1}. Just as in the earlier work by Rosenblum *et al* (1974), two types of emission have been observed: stationary (quasi-stationary) and in the form of individual maxima (figures 15(*a*),(*b*)). The former type (figure 5.15(*a*)) is observed to occur from the C^+ surface in the course of heating and from the C^- surface during cooling. The latter emission type (figure 5.15(*b*)) is achieved when the temperature variations reverse, namely from C^+ surface on cooling and from the C^- surface on heating. The temperature range of a stationary emission for the C^+ surface of a crystal is −50 to +170 °C, while for the C^- surface its range is narrower. In the second emission type, discharges take place, as a rule, at temperatures at which the first-type emission current undergoes sudden changes. The energy of emitted electrons is $W \approx 10^3$ eV. A parameter of the pyro-field distribution in a crystal may be the Debye screening length l_D, defined by

$$l_D = (kT\varepsilon_C/16\pi e^2 n_s)^{1/2} \ln(n_s/n_i) \qquad (5.31)$$

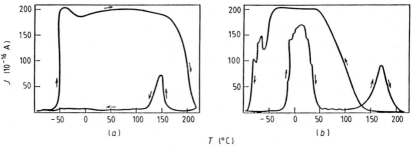

Figure 5.15. Electron emission from the C^+ (*a*) and C^- (*b*) faces of the LiNbO$_3$:Fe single crystal (according to Rosenman and Boikova 1979).

where k is the Boltzmann constant, T is the absolute temperature, ε_C is the permittivity at the Curie point, e is the electron charge, n_s is the electron density in the region of ferroelectric band bending and n_i is the electron density in a quasi-neutral area of a crystal. The n_s value can be calculated from the total ferroelectric band bend defined by formula (5.30):

$$e\Delta\chi = 2kT \ln(n_s/n_i). \tag{5.32}$$

Taking $n_i = 10^9 \text{ cm}^{-3}$, $e\Delta\chi = 0.4 \text{ eV}$ and $\varepsilon_C = 2.6 \times 10^4$, we obtain $n_s = 2.8 \times 10^{12} \text{ cm}^{-3}$ and $l_D = 4 \times 10^{-2}$ cm. With the assumption that a maximum energy of emitted electrons W_e is equal to the potential drop in a layer $l = 2l_D$, the field strength near the C^+ surface of a crystal is

$$E_{in} = W_e/2l_D e = 1.4 \times 10^4 \text{ V cm}^{-1}. \tag{5.33}$$

Obviously, this field ensures stationary emission of electrons from the C^+ surface of a crystal which can be identified as the field emission of electrons (auto-electronic emission). The upper limit of stationary emission, 170 °C, coincides with the annealing temperature of the photo-refractive effect in the lithium niobate crystal (see chapter 8). The drop of the emission current in this region is indicative of relaxation of the field E_i. This may be attributed to the increasing conductivity of the crystal due to the higher density of free carriers thermally excited from local levels. This claim is corroborated by a maximum emission from the C^- surface of a crystal, obtained on heating the latter up to 170 °C (figure 5.15(*b*)). Electrons localized at the donor centres are excited into the conduction band and screen the pyro-field, while some of them leave the crystal. The activation energy for this process is 1.12 eV, which coincides with that of annealing in the case of induced birefringence in these crystals. Evidently, the physical nature of the temperature threshold existing for these effects is similar.

Fridkin *et al* (1978) have noticed the same physical nature of electron exo-emission and photo-refractive effect. The authors examined single-domain single crystals of lithium niobate doped with iron ions. Silver electrodes were applied to the crystal surfaces perpendicular to the ferroelectric axis. Electron exo-emission was measured under the photovoltaic conditions (shorted electrodes) and at photo-EMF (illumination of a crystal with open electrodes). The spectral distribution of exo-emission has been established in a separate experiment and shown in figure 5.14. Concurrently, measurement was made of the photovoltaic current spectral distribution in the same sample. The experimental results are presented in figures 5.16(*a*), (*b*), and 5.17. Figures 5.16(*a*) and (*b*) concern the kinetics of exo-emission from the surfaces (001) and (00$\bar{1}$), respectively. Curves (1) are related to the photo-EMF, while curves (2) correspond to the photovoltaic current conditions.

It is evident from these findings that exo-emission in the direction

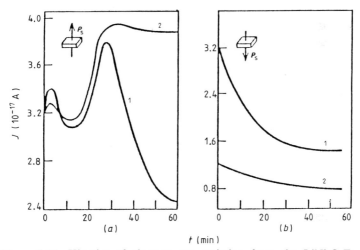

Figure 5.16. Kinetics of electron exo-emission from the LiNbO:Fe crystal: (*a*) in the direction [001]; (*b*) in the direction [00$\bar{1}$]. 1, crystal with open electrodes; 2, crystal with shorted electrodes (according to Fridkin *et al* 1978).

[001] in the photo-EMF regime rapidly decreases with increasing illumination time. On the other hand, exo-emission in the same direction [001] but in the photovoltaic current regime increases and reaches saturation within the relaxation time. In the direction [00$\bar{1}$], exo-emission tends to decrease with time for the two measuring regimes (figure 5.17).

Figures 5.17(*a*), (*b*) show spectral distributions of the quantum yield *Y* of exo-emission for the (001) and (00$\bar{1}$) surfaces respectively. Exo-emission was measured in the photovoltaic current regime. The quantum yield of exo-emission from the two surfaces reaches saturation for a photon energy below 5.3 eV. Figure 5.17 also gives the spectral distribution of the photovoltaic current *J*, which reaches its maximum value at

5 eV. This maximum is also observed for an undoped lithium niobate crystal and may correspond to the internal photo-refractive effect. These results are in accord with the photoconductivity spectral distribution for LiNbO$_3$:Fe (figure 5.14).

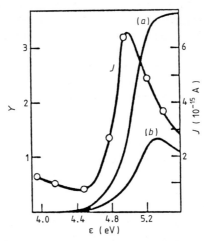

Figure 5.17. Spectral distribution of the quantum yield Y of exo-emission and photovoltaic current J: (a) in the direction [001]; (b) in the direction [00$\bar{1}$] according to Fridkin *et al* 1978).

The results obtained indicate that exo-emission and the photo-refractive effect are inter-related. The magnitudes of exo-emission in the directions [001] and [00$\bar{1}$] are different. Figure 5.17 demonstrates that this difference is the case where electron exo-emission is excited by high-energy photons so that the difference in the photoelectric work functions for the surfaces (001) and (00$\bar{1}$) is negligible. The kinetics of exo-emission in ferroelectric crystals is sensitive to the pyro-effect and spontaneous polarization screening by non-equilibrium carriers. The experimentally observed slow kinetics rules out at least the pyro-effect. The obtained results can be accounted for by the origin of a space charge in a crystal when the photo-EMF is measured. It is common knowledge (Glass *et al* 1974) that the photo-refractive effect in lithium niobate with open electrodes generates a voltage $U \approx 10^4$ V in the direction [001]. In doing so, a negative charge tends to concentrate on the surface (001). This in turn will reduce the exo-emission in the direction [001] as a result of the inverse field establishment. The kinetics of the exo-emission reduction is fitted by curve (1) of figure 5.16(a), which coincides with the kinetics of the photo-refractive effect in lithium niobate and is defined by the Maxwellian relaxation time. The exo-emission kinetics in the shorted electrode regime does not show this tendency (curve (2) of figure 5.16(a)). Production of a positive space charge on the surface ($00\bar{1}$) due to the photo-refractive effect in the open electrode regime will generate an oscillating field, and will accordingly enhance exo-emission (curve (1) compared to curve (2) in figure 5.16(b)). The difference in the spectral distributions of exo-emission from the surfaces (001) and ($00\bar{1}$) may be attributed to the band bend near the surface as well as to the onset of the photo-refractive effect.

5.8 Effective ion charges and spontaneous electric moment of lithium niobate

Of great scientific concern is the nature of chemical bonds in ferroelectric materials. Sophisticated as it is, this issue has been judged only qualitatively. Therefore it is of definite interest to draw on a purely chemical concept of 'element electronegativity'. Electronegativity, as a measure of atom affinity for the electron, permits its utilization for establishing an electric charge asymmetry in a molecule. On transition from molecules to crystals, the coordination number increases, thereby contributing to the ionic nature of bonds. The regularity of this enhancement is not yet understood. We should also like to note that the concept of atom electronegativity is somewhat limited. It makes sense only for a certain specific state of hybridization and cannot allow for the specificity of intermolecular processes.

The chemical calculations involving the use of atomic electronegativity have so far been restricted solely to molecules possessing a structural formula AB_n. It is noteworthy that the larger the n value, the more approximate the results (Batsanov 1962).

Granovsky (1962) employed the atomic electronegativity to calculate the effective charges of the bonds in perovskite-type compounds (ABO_3). The calculating principle relies on the Pauling rule of electrostatic valency (Bokiy 1960). The calculations were performed with due account of the coordination of the A atoms (12) and B atoms (6) and using the values of atomic electronegativity.

Of particular interest is the application of similar calculations to lithium metaniobate, $LiNbO_3$, which is a unique ferroelectric material as far as its physical properties are concerned.

But in the calculation of effective bond charges it seems possible to consider not only the coordination of the Li and Nb atoms but also the interatomic distances known from the neutron-diffraction studies and X-ray crystallographic analyses of lithium metaniobate (see chapter 1). On the basis of the values obtained of effective bond charges one can calculate the magnitude of spontaneous polarization and compare it with the experimentally determined value. This comparison will permit, to a certain extent, a conclusion about the legitimacy of the performed calculations.

At room temperature, lithium niobate crystals are made up by oxygen layers with slight distortions of the hexagonal packing. One third of octahedral vacancies are filled with Nb atoms, another third is filled with Li atoms and the last third remains unoccupied. The structure repeats itself along the triad axis. Octahedra touching at their faces impair the structure stability and change the distances between octahedral atoms (the Pauling rule: Bokiy 1960). Adjacent to the metal atoms are three oxygen atoms, which are graphically shown in figure 1.8. All the octahedra in lithium niobate are distorted; two kinds of metal–oxygen distances are available. The distance of Li–O to the nearest oxygen layer is 2.068 Å, to the next-nearest 2.238 Å. The corresponding distances of Nb–O are 1.889 and 2.112 Å respectively. The two shortest Li–Nb separations are 3.010 and 3.054 Å and these values are close to the interatomic distances in pure metals.

An examination of the interatomic distances has led to the conclusion that there are partially covalent bonds in lithium niobate. In the lithium niobate structure, the Nb atom is displaced from the highest-symmetry position (in the middle of the oxygen layers) by 0.258 Å, while the Li atom is displaced by 0.836 Å. These charge displacements form dipole moments in a ferromagnetic state.

The coordination numbers of the Li and Nb atoms are equal to 6. The values of electronegativity per metal–oxygen bond are $E_{Li} = 0.158$ and $E_{Nb} = 0.284$ (these values are borrowed from the book by Batsanov 1962, table 31). The oxygen atom forms four bonds to the metal atom and six bonds to other oxygen ions. It can be supposed in a first approximation, as is done by Granovsky (1962), that the oxygen electronegativity is independent of the nature of the surrounding atoms. Then the oxygen electronegativity per bond is $E_O = 0.350$. Let us find the number of electrons forming the Li–O valence bond. The contribution of lithium in each bond is one-sixth of the electron charge. The oxygen ion introduces the number of electrons proportional to the electronegativity:

$$n_{Li}^O = -2E_{Li}/(2E_{Li} + 2E_{Nb} + 6E_O) = -0.106. \qquad (5.34)$$

Thus the Li–O covalence bond is formed by the number of electrons defined by

$$N_{Li}^O = n_{Li}^O + \tfrac{1}{6} = +0.061.$$

In a similar way, the Nb–O bond is calculated:

$$n_{Nb}^O = -0.190 \qquad N_{Nb}^O = n_{Nb}^O + \tfrac{5}{6} = +0.644.$$

For the oxygen ion we have

$$n_O^O = -0.235 \qquad N_O^O = -0.470.$$

The following portions of the charges from N_{Li}^O, N_{Nb}^O and N_O^O fall to the share of oxygen:

$$S_{Li}^O = N_{Li}^O E_O/(E_{Li} + E_O) = 0.042.$$

Similarly,

$$S_{Nb}^O = +0.356 \qquad S_O^O = -0.235.$$

The effective charge of oxygen for each bond is defined by:

$$q_O^{Li} = n_{Li}^O - S_{Li}^O = -0.148e$$
$$q_O^{Nb} = -0.546e$$
$$q_O^O = 0$$

where e is the electron charge. The total effective charge of the O ion is

$$q_O = 2q_O^{Li} + 2q_O^{Nb} = -1.388e.$$

The charges of the Li and Nb ions are

$$q_{Li} = 0.888e$$
$$q_{Nb} = 3.376e$$

respectively. This means that the ionicity of the Li–O bond is 0.888 and that of the Nb–O bond is 0.655. Using the magnitudes of the niobium and lithium atoms' displacement from their symmetrical positions as well as the effective charges, one can calculate the spontaneous electric moment according to the formula

$$P_s = N[(q_1 x_1 + q_2 x_2)/d] \qquad (5.35)$$

where N is the number of dipoles per square centimetre of the crystal surface normal to the polarization direction (triad axis), $(q_1 x_1 + q_2 x_2)$ contains the effective charge and displacement of the Li and Nb atoms respectively and d is the distance between the oxygen layers towards the triad axis ($d = 2.310$ Å).

The calculations yield $P_s = 0.535$ N of the electron charge. In each hexagonal cell of lithium niobate, the surface charge is produced by a single Nb ion and a single Li ion. Consequently, the number of dipoles falling within a square centimetre of the surface normal to the polarization direction is equal to the number of cells in the surface layer:

$$N = C_H/V$$

where $C_H = 13.856$ Å and $V = 318.221$ Å^3. Hence we have $N = 4.35 \times 10^{14}$ cells. Substituting this value and expressing electron charges in terms of coulombs, we arrive at the polarization

$$P_s = 37.2\ \mu C\,cm^{-2}.$$

According to the results of Savage (1966), the spontaneous polarization in lithium niobate is $P_s = 50\ \mu C\,cm^{-2}$. For calculations of this kind, such an agreement between the predicted and experimental polarization values is quite satisfactory.

A rough estimate of the effective ion charge in the lithium niobate structure can be obtained as follows. It is known that in the course of its ionization the ionic radius is changed (Batsanov 1962). Figures 5.18 and 5.19 are these plots for the Li and O, Nb and O atoms separated by the distances determined in the lithium niobate crystal by X-ray analysis. It is evident from the plots that the ionicity of the Nb–O bond with interatomic distance of 1.889 Å is 0.88. For the Li–O ions separated by 2.068 Å the ionicity is 0.97. For the other bonds the ionicity is 1.00.

As mentioned earlier, the Nb and Li atoms are surrounded by the adjacent three oxygen atoms. Hence the total ionicities of niobium and lithium are 0.64 and 0.91 respectively. The effective charges are $q_{Li} = 0.91$ and $q_{Nb} = 3.20$ electron charges. The spontaneous polarization calculated according to the above formula is $P_s = 46.8\ \mu C\,cm^{-2}$. The calculated effective charges of lithium, niobium and oxygen in the structure of lithium niobate obtained by the above two methods are close to the results of Gervais (1976).

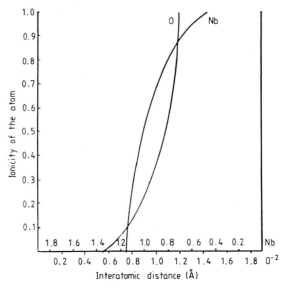

Figure 5.18. Ionicity of the Nb–O bond with an interatomic distance of 1.889 Å.

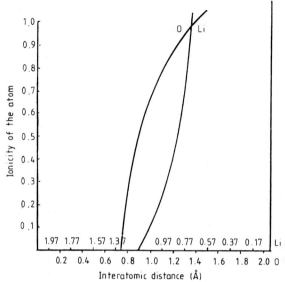

Figure 5.19. Ionicity of the Li–O bond with an interatomic distance of 2.068 Å.

6 Optical and Electro-optical Properties of Lithium Metaniobate Single Crystals

6.1 Optical properties of lithium metaniobate

Lithium metaniobate single crystals are uniaxial negative $(n_o > n_e)$, transparent from about 0.4 to 5 μm (figure 6.1) (Boyd *et al* 1964). The nature of their transmission spectra depends on the conditions of heat treatment and polarization of crystals. Crystals prepared with no direct current maintained through them during growth are clear and colourless. Their transmission spectrum at $\lambda = 0.3$ to 1.2 μm is given in figure 6.1(*a*). In the infrared region there is a narrow absorption band at $\lambda = 2.9$ μm (figure 6.1(*e*)). After annealing in air with concurrent poling (1460 K and current density not more than 1 mA cm^{-2}) crystals acquire a light yellow tint which becomes more intense with an increasing poling current density. The transmission spectrum of such crystals (figure 6.1(*b*)) displays weak absorption bands near $\lambda = 0.5$ μm and $\lambda = 0.8$ μm, while that at $\lambda = 2.9$ μm gets stronger.

The light yellow coloration is also produced in crystals grown after new portions of the starting material have been introduced into the crucible, as well as in crystals grown with an application of a current denser than 1 mA cm^{-2} or crystals cooled too rapidly (at a rate exceeding 100 K h^{-1}) after growth has been terminated. The spectra of such crystals are identical to that of figure 6.1(*c*). The tinted crystals are partially bleached by heating in air at temperatures above 770 K for two or three hours. The bleaching process is accelerated to a large extent by annealing in oxygen. In the transmission spectra of the

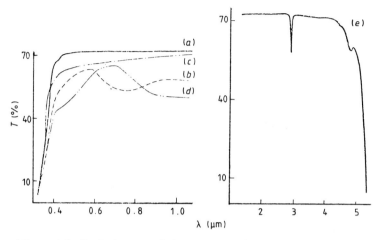

Figure 6.1. Optical transmission spectra for congruently melting lithium niobate single crystals: (*a*) pulled without electric current maintained during the growth process; (*b*) grown with no electric current and subject to poling and concurrent annealing in a separate furnace; (*c*) growth with electric current in the growth process; (*d*) heated in hydrogen; (*e*) optical transmission spectrum of lithium niobate in the infrared region.

resulting crystals, the absorption bands at $\lambda = 0.5$ and $\lambda = 0.8\,\mu$m disappear.

Having been heated in hydrogen at 670–870 K for several minutes, the originally clear lithium niobate crystals become brown and their transmission spectra display a strong absorption band near $\lambda = 2.9\,\mu$m and two newly emerging absorption bands at $\lambda = 0.5$ and near 0.8–1.1 μm (figure 6.1(*d*)). They are bleached again by annealing in an oxygen atmosphere and their spectra retain a single absorption band at $\lambda = 2.9\,\mu$m. Clear colourless lithium niobate single crystals transmit as much as 72% (without corrections for reflection) over a range $\lambda = 0.4$–4 μm.

The nature of the observed optical transmission spectra for lithium niobate single crystals is consistent with the data presented in chapter 3 which relate the absorption near 2.9 μm to the OH^- grouping, 0.5 μm to the Nb^{4+} ions, and 0.8–1.1 μm to the $(2e^-)O$ centres.

The dispersion dependences of the refractive indices n_o and n_e as well as the temperature dependences for several wavelengths $\lambda = 546$, 577 and 579 μm have been established by Agheev *et al* (1968) and are shown in figure 6.2.

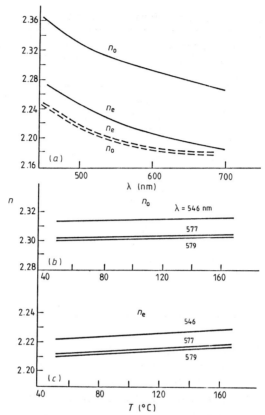

Figure 6.2. Refractive indices of lithium metaniobate (full curve) and lithium tantalate (broken curve) versus wavelength and temperature (*a*); (*b*) $dn_0/dT = (0.8–0.2) \times 10^{-5}$; (*c*) $dn_e/dT = (5 \div 0.2) \times 10^{-5}$ (according to Agheev *et al* 1968).

More comprehensive dispersion dependences of n_o and n_e over a wide frequency range for lithium niobate crystals grown from congruent melt compositions are collected in table 6.1.

For some optical mixing processes, phase matching is achieved normal to the optical axis (non-critical phase matching: see chapter 7). A wide choice of frequencies for optical interaction is determined by the temperature dependence of the birefringence $(n_o - n_e)$ of the lithium niobate single crystal. The temperature dependences of birefringence at wavelengths 632.8 and 1152.3 nm have been investigated by Miller and Savage (1966) as well as by Werner *et al* (1966).

The birefringence of lithium niobate changing with temperature was measured by passing the plane-polarized light from a gas laser operated

Table 6.1. Refractive indices (according to Weis and Gaylord 1985).

λ (nm)	Laser	Stoichiometric ($T = 25\,°C$)		Congruently melting ($T = 24.5\,°C$)	
		n_o	n_e	n_o	n_e
441.6	He–Cd	2.3906	2.2841	2.3875	2.2887
457.9	Ar	2.3756	2.2715	2.3725	2.2760
465.8	Ar	2.3697	2.2664	2.3653	2.2699
472.7	Ar	2.3646	2.2620	3.3597	2.2652
476.5	Ar	2.3618	2.2596	2.3568	2.2627
488.0	Ar	2.3533	2.2523	2.3489	2.2561
496.5	Ar	2.3470	2.2468	2.3434	2.2514
501.7	Ar	2.3435	2.2439	2.3401	2.2486
514.5	Ar	2.3370	2.2387	2.3326	2.2422
530.0	Nd	2.3290	2.2323	2.3247	2.2355
632.8	He–Ne	2.2910	2.2005	2.2866	2.2028
693.4	Ruby	2.2770	2.1886	2.2726	2.1909
840.0	GaAs	2.2554	2.1703	2.2507	2.1719
1060.0	Nd	2.2372	2.1550	2.2323	2.1561
1150.0	He–Ne	2.2320	2.1506	2.2225	2.1519

at 632.8 and 1152.3 nm perpendicular to a crystal plate $\simeq 1$ mm thick with the optical axis oriented parallel to its major faces.

The polarization plane of the incident beam made an angle of 45° with the optical axis of the crystal. The output beam was analysed by a Nicol prism and made to shine on a photoelectric cell. The crystal plate was mounted in a heating system maintained at a constant oxygen pressure of 1.5 atm to prevent oxygen loss from a highly heated crystal.

The plate temperature was controlled by a Pt/Pt–Rh thermocouple placed nearby. The thermocouple EMF was recorded in one of the channels of a two-channel potentiometer.

The relative intensity of the light traversing the plate at a temperature T is given by:

$$I/I_0 = \sin^2\varphi = \sin^2(2\pi l B/\lambda) \tag{6.1}$$

where B is the birefringence, l is the crystal length along the beam propagation and λ is the light wavelength in a vacuum.

The rate of variations in φ with temperature can be cast in the form

$$d\varphi/dT = 2\pi l/\lambda(dB/dT + SB) \tag{6.2}$$

where S is the coefficient of linear expansion perpendicular to the c axis of the crystal.

The intensity of the transmitted radiation was recorded by means of the second channel of the potentiometer. Successive minima of the transmitted light intensity occur at a phase difference $\varphi = \pi$ corresponding to a temperature change from T to $T + \delta T$. The birefringence at a temperature $T + \delta T$ can be expressed by

$$B_{(T + \delta T)} = B_T \pm (\lambda/2l) - B_T S \delta T. \tag{6.3}$$

The fact that the φ value decreases with increasing temperature was verified with the help of a polarizing microscope. The hot plate was put on a stage and the behaviour of an interference pattern in the convergent monochromatic light was observed with a crossed polarizer and analyser. As the plate was cooled down, the hyperbolic interference fringes disappeared and within the field of view new interference fringes were observed to appear. Therefore, the negative sign in equation (6.3) has been chosen correctly.

Birefringence data for room temperature have been calculated by Nelson and Mikulyak (1974). The coefficient of thermal expansion included in equation (6.3) was measured using a wedge and proved to be $S = (21 \pm 1) \times 10^{-6}$ deg^{-1} over a temperature range from 20 to 700 °C.

The calculation results are presented in figure 6.3. The errors of about $\pm 0.5\%$ in the values $B_T - B_{20\,°C}$ are due to the uncertainty in the quantities S and l. In the calculations, allowance was made for the departure of the crystal optical axis from the normal to the plate. Errors in the absolute magnitude of B_T included those in the $B_{20\,°C}$ values. The slope of the curve for 632.8 nm varies within -2.7×10^{-5} K^{-1} at 100 °C to -9.5×10^{-5} K^{-1} at 1000 °C. Miller *et al* (1965) report the value $\mathrm{d}B/\mathrm{d}T = -4.5 \times 10^{-5}$ K^{-1} for the yellow sodium lines, obtained by averaging through a measured temperature range above 300 °C.

For a wavelength of 632.8 nm, the transmitted light intensity showed 361 minima from room temperature to 1205 K (for $\lambda = 1152.3$ nm this number is 178). Successive minima occurred about 1.5° apart.

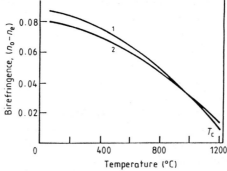

Figure 6.3. Temperature dependence of the birefringence of stoichiometric lithium niobate for $\lambda = 632.8$ nm (curve 1) and $\lambda = 1152.3$ nm (curve 2) (according to Werner *et al* 1966).

Within ± 0.5 °C there was no difference in the diagrams obtained with rising and falling temperature. Above 1205 °C and to a maximum attainable temperature of 1245 °C, no intensity fluctuations have been observed. Below 1205 °C a minimum intensity of the transmitted light was more than ten times smaller than its maximum value. Meanwhile above 1205 °C the transmitted light intensity was more than half the laser beam intensity. These data are evidence that the high-temperature phase of lithium niobate is not cubic.

Hobden and Warner (1966), drawing on the experimental data, derived an analytical expression for the temperature dependence of the refractive indices n_o and n_e, and the birefringence, and also calculated phase-matching angles for various cases.

The refractive indices were measured using a lithium niobate prism with the optical axis parallel to the two major faces. The prism was arranged in a small furnace on a spectrometer stage. The refractive indices were taken at eight temperatures between 19 °C and 347 °C for eight lines of the helium metal vapour lamp at 447.1, 471.3, 492.2, 501.6, 587.6, 667.8, and 707.6 nm.

An analysis of the experimental data has yielded two equations for the temperature dependence giving the refractive indices between 400 nm and 4000 nm:

$$n_o^2 = 4.9130 + \frac{1.173 \times 10^{-5} + 1.65 \times 10^{-2} T^2}{\lambda^2 - (2.12 \times 10^2 + 2.7 \times 10^{-5} T^2)^2} + 2.78 \times 10^{-8} \lambda^2$$

(6.4)

$$n_e^2 = 4.5567 + 2.605 \times 10^{-7} T + \frac{0.97 \times 10^{-5} + 2.70 \times 10^{-2} T^2}{\lambda^2 - (2.01 \times 10^2 + 5.4 \times 10^{-5} T^2)^2}$$
$$- 2.24 \times 10^{-8} \lambda^2$$

(6.5)

where T is the temperature in K and λ is the wavelength in nm.

The standard deviation of 112 experimentally determined values of the refractive indices from those calculated according to formulas (6.4) and (6.5) is 2.2×10^{-4}.

The value of the negative birefringence decreases with increasing temperature and it drops off to zero at 882 °C for $\lambda = 632.8$ nm and at 888 °C for $\lambda = 1152.3$ nm.

The change in $(n_e - n_o)$ with temperature, as predicted by equations (6.4) and (6.5), differs by ± 0.0010 from the experimental data for about 600 °C. Above this temperature, higher order terms come into play.

Equations (6.4) and (6.5) were employed to predict the pre-requisites for phase-matching various optical displacement processes, which were verified experimentally.

As was repeatedly stressed in the foregoing chapters, many physical properties of crystals are governed to a large extent by the melt composition ratios. In the lithium niobate crystal it is the extraordinary refractive index n_e that depends significantly on the melt composition ratio, while the ordinary refractive index n_o remains virtually at a constant level (figure 6.4) (Bergmann *et al* 1968). The composition of melts, and hence the composition of crystals grown from them, may vary throughout the growth process. It is therefore essential to take account of this possibility in creating an electro-optical modulator or deflector. An isomorphous dopant of niobium is tantalum. The starting material may contain more or less tantalum oxide. Sometimes, to reduce the Curie temperature and natural birefringence, mixed $LiNb_{1-y}Ta_yO_3$ crystals are grown. Such crystals have different refractive indices and their dispersions. Table 6.2 is a compilation of the dispersions of the refractive indices n_o and n_e for various contents of tantalum in mixed lithium niobate–tantalum crystals. For practical applications refractive indices for various wavelengths are calculated according to the Sellmeier relation (Di Domenico and Wemple 1969):

$$n^2(\lambda) - 1 = S_0\lambda_0^2/[1 - (\lambda_0/\lambda)^2] \tag{6.6}$$

where λ_0 is the average oscillator position and S_0 is the average oscillator strength. The λ_0 and S_0 values for various tantalum contents

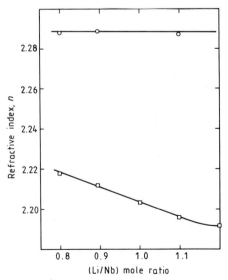

Figure 6.4. Refractive indices n_o (upper curve) and n_e (lower curve) of lithium niobate versus molar ratio Li_2O/Nb_2O_5 in the melt (according to Bergman *et al* 1968).

are listed in table 6.3. The refractive indices n_o and n_e and the birefringences calculated using this relation are given in figures 6.5 and 6.6 respectively.

Table 6.2. Refractive indices n_o and n_e for mixed $LiNb_{1-y}Ta_yO_3$ crystals at 20 ± 0.5 °C (according to Shimura 1977).

λ (Å)	$y = 0.81$		$y = 0.92$		$y = 0.97$		$y = 1.00$	
	n_o	n_e	n_o	n_e	n_o	n_e	n_o	n_e
5893	2.2057	2.1986	2.1984	2.1946	2.1902	2.1933	2.1862	2.1910
6328	2.1954	2.1888	2.1888	2.1853	2.1800	2.1829	2.1766	2.1815
8000	2.1702	2.1638	2.1643	2.1604	2.1561	2.1589	2.1531	2.1579
8500	2.1666	2.1606	2.1598	2.1559	2.1516	2.1545	2.1484	2.1529
9000	2.1615	2.1553	2.1557	2.1519	2.1478	2.1507	2.1446	2.1491
10 600	2.1517	2.1457	2.1460	2.1422	2.1385	2.1413	2.1351	2.1396

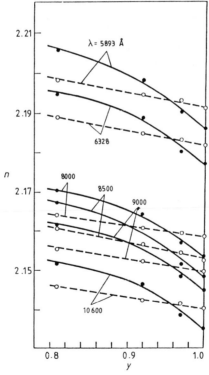

Figure 6.5. Refractive indices n_o (full circles) and n_e (open circles) versus Ta content in mixed $LiNb_{1-y}Ta_yO_3$ crystals for various light wavelengths (according to Shimura 1977).

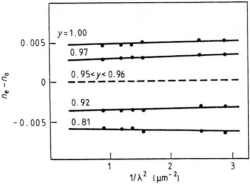

Figure 6.6. Birefringence $n_e - n_o$ in a mixed $LiNb_{1-y}Ta_yO_3$ crystal versus $1/\lambda^2$ for various Ta contents in the crystal (according to Shimura 1977).

The negative birefringence of lithium niobate can be understood qualitatively by considering the effect of the atomic dipoles induced by an applied electric field. When a field is applied, the negatively charged electrons and the positively charged nucleus separate and produce an electric dipole. Once the induced dipoles cause a net reduction of the field at their neighbouring atoms, the dielectric susceptibility χ_e is decreased. This results in a smaller dielectric constant, since $\varepsilon/\varepsilon_0 = 1 + \chi_e$. Likewise, if the induced dipoles produce a net enhancement of the field at their neighbouring atoms, the dielectric susceptibility and the dielectric constant are accordingly increased. The polarizing and depolarizing effects of the induced oxygen dipoles on the nearest-neighbour atoms are the dominant contributors to the dielectric susceptibility in lithium niobate. If the applied electric field is parallel to the optical axis, the induced oxygen dipoles exert a net depolarizing effect upon their nearest oxygen neighbours. But if the applied electric field is perpendicular to the optical axis, the induced oxygen dipoles have a net polarizing effect on their nearest oxygen neighbours. Consequently the

Table 6.3. Sellmeier constants λ_0 and S_0 for calculation of refractive indices of $LiNb_{1-y}Ta_yO_3$ crystals (according to Shimura 1977).

y	S_0 $(10^{-14}$ $m^2)$		λ_0 (μm)	
	n_o	n_e	n_o	n_e
1.00	1.2195	1.2123	0.1687	0.1696
0.97	1.2121	1.2121	0.1695	0.1698
0.92	1.2036	1.2121	0.1709	0.1699
0.81	1.1905	1.2121	0.1724	0.1703

value of χ_e parallel to the optical axis is less than that perpendicular to it. Thus in lithium niobate the principal extraordinary refractive index $n_e = (\varepsilon_{33}/\varepsilon_0)^{1/2}$ is smaller than the ordinary refractive index $n_o = (\varepsilon_{11}/\varepsilon_0)^{1/2}$.

6.2 Electro-optical effect in dielectric crystals

This section is concerned with the linear electro-optical effect (the Pockels effect) based on the first-order nonlinear effect of electromagnetic wave–crystal interaction. This effect was discovered by Pockels (1906). This discovery, prior to the advert of high-power laser sources of light, proved to be possible because this effect, just like the quadratic Kerr effect, shows the influence of a strong uniform constant field $E(0)$ upon the medium properties for the propagation of optical waves. Being a first-order effect, the Pockels effect occurs in crystals with no centre of symmetry. In centrosymmetric media only the quadratic Kerr effect is observable. Under the linear electro-optical effect, the constant field causes the linear susceptibility χ_{ik} to change by a value $\delta\chi_{ik}$ corresponding to the change in the dielectric constant by $(1/\varepsilon_0)\delta\chi_{ik}$ (Schubert and Wilhelmi 1971). The real part of $\delta\chi_{ii}$ proportional to an applied constant field causes variations in the phase linearly increasing along the z coordinate, while the imaginary part determines an exponential decay of the amplitude. That is why a nonlinear medium exhibiting the Pockels effect is equivalent to a medium with a linear susceptibility $\chi_{ii} + \delta\chi_{ii}$ or permittivity $\varepsilon_{ii} + (1/\varepsilon_0)\delta\chi_{ik}$. Since $\delta\chi_{ii}$ depends on the direction of an applied external field, the arising birefringence turns out to be dependent upon the field strength. Just as with the Kerr effect, in the Pockels effect the ordinary and extraordinary refractive indices undergo variations and birefringence is induced. This circumstance is useful in creating systems for controlling light fluxes in phase and amplitude (Sonin and Vasilevskaya 1971, Mustel and Paryghin 1970).

Optical properties of crystals in the visible and near infrared regions are often determined by one-electron transitions in the ultraviolet. Static and low-frequency dielectric properties are dictated by the optical lattice vibrations corresponding to frequencies in the far infrared region. The electro-optical effect is observed under direct interaction of an applied electric field with the coupled electron and lattice modes, thereby changing the optical properties.

The applied field may be constant, microwave or an optical electric one. In some crystals the electro-optical interactions are mainly of an electronic nature; in others they are largely due to the lattice modes. The electro-optical effect may change linearly or quadratically with an applied electric field but in each case it remains a first-order effect.

The simplest electro-optical crystals that have been studied in detail are diatomic ionic crystals like ZnS and CuCl, falling into the hemimorphic hemihedral class of the cubic system. Theoretical calculations of various aspects of the lattice dynamics can be found in the book by Born and Huang (1954). Kelly (1966) has done lengthy calculations of the electro-optical effect in ZnS. The dipole moment of the positive ion is attributed partly to the displacement of ions and partly to the electronic polarization. The electronic polarization is equal to the product of polarizability and effective electric field. The nonlinear polarization terms are determined by the quadrupole terms in the expansion of effective fields. The values of the ion charge and polarizability are phenomenological constants calculated from experimental values of the low-frequency and optical dielectric constants and dispersion frequencies in the infrared region. The measured and predicted values coincide to within several per cent. This makes it possible to estimate the contributions of various terms.

Here, the resemblance of a model to the physical situation in an actual crystal is of critical importance. For instance, to obtain a numerical value of the ion charge Kelly draws in his calculations on the Szigetti relation, while to establish effective polarizabilities he employs the Clausius–Mosotti equation. Thus, it is presupposed that all the assumptions adopted in the derivation of these relations are applicable to the calculation of the electro-optical effect. Plentiful experimental data available on the properties of crystals (piezoelectric, electro-optic and dielectric constants, optical frequency shift, Raman scattering spectra), taken collectively, yield information permitting verification of nuances of various theories.

For non-centrosymmetric crystals including lithium niobate and tantalate, we can apply the theory developed by Bloembergen (1966), which is based on the model of an anharmonic oscillator as well as Miller's rule for establishing nonlinear constants. According to Miller (1964), once free energy is expressed in terms of polarization rather than applied electric fields, the second harmonic generation efficiency and the electro-optic coefficient in all his studied materials differ by a factor of four. This relation, known as Miller's rule, holds good up to 10.6 μm in the infrared region.

Kurtz and Robinson (1967) have elaborated a model of the electro-optical effect in which optical refraction is described by the anharmonic one-electron oscillator equation. The influence of the ionic terms is considered by introducing an additive constant to a local field. In the presence of an external optical field $E(\omega, t)$ and a low-frequency field, $E(0)$, the displacement of an electron is fitted by

$$\ddot{x} + \Gamma\dot{x} + \omega_0^2 x + vx^2 = e/m[E(\omega, t) + \beta E(0)] \qquad (6.7)$$

where Γ is the attenuation constant, ω_0 is the angular resonant frequency lying in the ultraviolet spectral region, v is the anharmonic force constant, e and m are the charge and mass of an electron respectively and β is the parameter specifying a local field $\beta = (K_0 + 2)/3$ (K_0 standing for the low-frequency dielectric susceptibility). The effect of an applied field consists of shifting the resonant frequency $\omega_0 \to \omega_0'$, where

$$(\omega_2')^2 = \omega_0^2 + 2ve\beta E(0)/(m\omega_0^2). \tag{6.8}$$

Since the refractive index n changes as $n \sim [(\omega_0')^2 - (\omega_0^2)]^{-1}$, the expression for a linear electro-optic coefficient turns out to be

$$r = \frac{(n^2 - 1)^2 (K_0 + 2)v}{6\pi n^4 N_0 e \omega_0^2} \tag{6.9}$$

where N_0 is the electron density. The nonlinear term v is found from Miller's constant

$$\delta = mv/(2N_0 e^3). \tag{6.10}$$

The mean δ value through 55 measurements on 22 materials (Kurtz and Robinson 1967) is $\delta = 3 \times 10^{-6}$ cm (stat V)$^{-1}$. The value of v obtained from equation (6.10) is equal to 1.33×10^{39} cm^{-1} s^{-1}. Formula (6.9) was employed to calculate electro-optic coefficients for several inorganic crystals and the agreement with the experimental data, if any, was always better than by an order of magnitude.

In piezoelectric crystals, we can observe oscillation modes active in Raman and infrared scatterings which can be excited by an external electric field. Thus the equations for polarization in the electro-optical effect can be related to similar classical equations of Raman scattering. Kaminov and Johnston (1967) have elaborated, on the basis of this relationship, an experimental technique for measuring the electro-optical effect in lithium niobate and lithium tantalate. A virtue of this method is the possibility of establishing absolute values of the contributions made to the electro-optical effect by the electronic and ionic terms. The method relies on the equations for differential optical polarization under the electro-optical effect and Raman scattering. Under the electro-optical effect, the polarization in tensor notation is

$$P_i = \varepsilon_0 n_i^2 r_{ijk} n_j^2 E_j(\omega) E_k(0) \tag{6.11}$$

where n_i and n_j are the principal values of refractive indices, $E_j(\omega)$ is the strength of an optical frequency field and $E_k(0)$ is the strength of a low-frequency field. The polarization under Raman scattering is expressed in terms of differential optical polarizability α_{ijk} and lattice shifts induced by a modulating field $Q_k^m = \beta_k^m E_k(0)$, where β_k^m stands for factors of proportionality expressed in terms of oscillator strength in the infrared region or (for crystals having a single active mode) in terms of dielectric constant. Thus

$$P_i = (\alpha_{ijk}^m \beta_k^m + \zeta_{ijk})E_j(\omega)E_k(0). \qquad (6.12)$$

The terms ζ_{ijk} describe the change in polarization due to the direct action of a modulating field upon an electron. These two equations yield

$$\varepsilon_0 n_i^2 r_{ijk} n_j^2 = \alpha_{ijk}^m \beta_k^m + \zeta_{ijk}. \qquad (6.13)$$

The terms ζ_{ijk} reflect purely electronic effects and are defined by measuring the second harmonic generation in the optical region. To calculate the electro-optic coefficients r_{ijk}, it is necessary to determine α_{ijk}. For lithium niobate and lithium tantalate these terms are established by measuring the Raman scattering efficiency. The calculated values of the four electro-optic coefficients for lithium niobate and tantalate are in accord with measurements.

The electro-optical effect in lithium niobate and tantalate is attributed by 90% to the ionic or lattice contributions and only by 10% to the electronic ones.

Miller has established that the ratio r_{ijk}/ε_{ij} is independent of the material nature and for a given material it is independent of temperature or frequency. The dielectric constant for lithium niobate along the z axis was measured at 200 MHz and found to be frequency-independent. This suggests that the electro-optical effect does not depend upon the electric field frequency either. The values of the electro-optic coefficients at 60 MHz just coincide with the low-frequency values (Turner 1966). The measurements performed at frequencies up to 6 GHz yielded identical results (Kaminov and Sharpless 1967).

On the basis of the Raman scattering spectra it has been found (Axe and O'Kane 1966) that the infrared transverse optical mode ω_{to} is of the order of 200 cm^{-1}. It is to be expected therefore that in ideal crystals low-frequency electro-optic constants are not variable, at least up to a range of 1 mm: although mention must be made that some data are available which attest to a slight frequency dependence of electro-optic coefficients (Vasilevskaya *et al* 1967). In selecting a sample, an examination of dielectric constants takes far less time than a direct measurement of r_{ij}. Miller's rule may be helpful to judge the effectiveness of particular samples for electro-optical applications.

In the crystals concerned, of major importance is the transverse electro-optical effect. Its magnitude is defined by the half-wave bias voltage $[E.l]_{\lambda/2}$, where E is the electric field strength and l is the optical distance. This product is equal to the voltage required for producing a half-wave delay at $l/d = 1$ (d being the crystal thickness along the applied field). Lithium niobate and lithium tantalate crystals are notable for a substantial transverse effect, so that the field strength requirements can be relaxed with a suitable choice of geometry.

While the numerical values of r_{ij} specify completely the crystal parameters, one should take into account, when comparing different crystals, the nature of a particular device involving the use of a given crystal.

In facilities intended for amplitude modulation as well as for establishing electro-optic coefficients, the De Sénarmont compensator has been widely accepted. Here the polarization plane of an incident wave makes an angle of $\pi/4$ with the major directions of oscillations. A half-wave phase delay generally provides an elliptically polarized wave which converts, while passing through a quarter-wave plate oriented in the most appropriate way, into a plane wave turned by an angle of $\pi/2$. This ensures a 100% modulation.

Wemple *et al* (1968) give linear electro-optic coefficients in terms of crystal polarization rather than strength of an applied external field (as was common practice previously):

$$\Delta(1/n^2)_{ij} = \sum_k f_{ijk}\delta P_k \qquad (6.14)$$

where $(1/n^2)_{ij}$ stands for the components of the tensor defining an ellipsoid of refractive indices. The polarization δP_k induced by a field E_k is given by

$$\delta P_k = \varepsilon_0(\varepsilon_k - 1)E_k \qquad (6.15)$$

where E_k is the electric field along the principal k axis, ε_k is the dielectric constant in the same direction and ε_0 is the permittivity of a vacuum. In the above notation, the electro-optic coefficients are as follows:

$$f_{ijk} = r_{ijk}/[\varepsilon_0(\varepsilon_k - 1)]. \qquad (6.16)$$

Wemple *et al* (1968) presumed that the linear electro-optical effect defined by expression (6.14) is a continuation of the quadratic effect taking place in the paraelectric phase of a ferroelectric. The quadratic effect is fitted by

$$\delta(1/n^2)_{ij} = \sum_{k,l} g_{ijkl}P_k P_l \qquad (6.17)$$

where $P_k = P_s + \delta P_k$ is the total polarization (spontaneous and induced).

The crystals of lithum niobate and lithium tantalate, as well as other crystals having the structure of tetragonal potassium–tungsten bronze, consist of oxygen octahedra, BO_6, making up blocks. In lithium niobate, spontaneous polarization is directed along the trigonal axis of the octahedron $\langle 111 \rangle$ (figure 6.7) and its unit cell is formed by two octahedra turned with respect to one another by about 180°. This cell has a C_{6v} symmetry. Transition from the cubic $\langle 100 \rangle$ axis of the

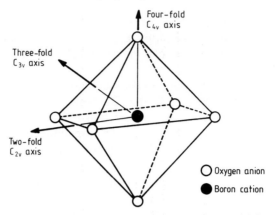

Figure 6.7. Oxygen octahedron BO_6 in the lithium niobate crystal structure.

octahedron to the third-order $\langle 111 \rangle$ axis leads to the following relations between the g and f tensors in a standard notation system (Nye 1964):

$$f_{33} = \tfrac{2}{3}(g_{11} + 2g_{12} + 2g_{44})P_s$$
$$f_{13} = \tfrac{2}{3}(g_{11} + 2g_{12} - g_{44})P_s \qquad (6.18)$$
$$f_{51} = \tfrac{2}{3}(g_{11} - g_{12} + \tfrac{1}{2}g_{44})P_s.$$

When comparing the quadratic electro-optic coefficients of lithium niobate obtained on the basis of the system of equations (6.18) to the g_p values for perovskites in the paraelectric phase, one should allow a correction for the packing density

$$g = g_p / \eta^3 \qquad (6.19)$$

where g stands for the quadratic electro-optic coefficients. Wemple *et al* (1968) have established that the g coefficients are equal, within experimental error, for all oxygen octahedral ferroelectrical materials. The numerical values of g are listed in table 6.4.

Thus the linear electro-optical effect has been related to the fundamental quadratic effect associated with the crystal lattice made of oxygen octahedra.

The linear electro-optical coefficients r_{13} and r_{33} are expressed in terms of quadratic g coefficients of an undeformed octahedron as follows:

$$r_{13} = \tfrac{2}{3}[(g_{11})_p + 2(g_{12})_p - (g_{44})_p]\varepsilon_0(\varepsilon_3 - 1)P_s / \eta^3$$
$$r_{33} = \tfrac{2}{3}[(g_{11})_p + 2(g_{12})_p - 2(g_{44})_p]\varepsilon_0(\varepsilon_3 - 1)P_s / \eta^3. \qquad (6.20)$$

The static birefringence arising from spontaneous polarization is calculated according to the formula

Table 6.4. Dielectric constants ε, quadratic electro-optic coefficients g and relative densities η of some oxygen octahedral ferroelectrics (at room temperature) (according to Wemple *et al* 1968).

Material	η	T_C (°C)	ε_c	ε_a	P_s	$(g_{11} - g_{12})_P$	$(g_{11})_P = \eta^3 g_{11}$	$(g_{12})_P = \eta^3 g_{12}$	$(g_{44})_P = \eta^3 g_{44}$
LiNbO$_3$	1.20	1195	30	84	0.71	0.12	0.16	0.043	0.11
LiTaO$_3$	1.20	610	45	51	0.50	0.14	0.17	0.03	0.12
Ba$_2$NaNb$_5$O$_{15}$	1.03	560	51	242	0.40	0.12	0.17	0.048	–
BaTiO$_3$	1.00	120	160	300	0.25	0.14	–	–	–

$$\Delta n(P_s) = -\tfrac{1}{2}n^3[(g_{44})_p/\eta_3]P_s^2. \tag{6.21}$$

This formula may provide spontaneous polarizations for other oxygen octahedral ferroelectrics. The following values of the quadratic electro-optic coefficients for perovskites in the paraelectric phase (Chen *et al* (1966)) are used: $(g_{11})_p \simeq 0.17 \text{ m}^4\text{C}^2$, $(g_{12})_p \simeq 0.04 \text{ m}^4\text{C}^2$, $(g_{44})_p \simeq 0.12 \text{ m}^4\text{C}^2$, $(g_{11} - g_{12})_p \simeq (0.13 \pm 0.02) \text{ m}^4\text{C}^2$.

To estimate the electro-optical effect, it is convenient to express the half-wave voltage in terms of permittivity ε and spontaneous polarization P_s.

Expression (6.17) implies that

$$\Delta(1/n^2)_i = r_{i3}P_s/[\varepsilon_0(\varepsilon_3 - 1)] \tag{6.22}$$

where Δn_i is the difference between the values of the principal refractive index and the refractive index in the presence of spontaneous polarization P_s. For $n_i = n_e$ from equation (6.22), a differentiation yields

$$n_e^3 r_{33} = -2\varepsilon_0(\varepsilon_3 - 1)(\Delta n_e/P_s) \tag{6.23}$$

where $\Delta n_e \equiv n_e(P_s) - n_e(0)$. A similar expression can be derived for $n_i = n_o$. The difference between these expressions, which is proportional to the half-wave voltage, is cast in the form

$$(n_e^3 r_{33} - n_o^3 r_{13}) = -2\varepsilon_0(\varepsilon_3 - 1)[B(0) - B(P_s)]/P_s \tag{6.24}$$

where B is the birefringence

$$B(P_s) \equiv n_0(P_s) - n_e(P_s) \qquad B(0) \equiv n_0(0) - n_e(0). \tag{6.25}$$

Thus the half-wave voltage is given in the form

$$[El]_{\lambda/2} = \lambda/(n_e^3 r_{33} - n_o^3 r_{13}) = \lambda P_s/\{2\varepsilon_0(\varepsilon_3 - 1)[B(0) - B(P_s)]\}. \tag{6.26}$$

This expression can be used to estimate the change $\delta[El]_{\lambda/2}$ with variations in the chemical composition of the crystal (Turner *et al* 1970). In the latter work it was established that if the Li/Nb melt composition varies within 0.91 to 1.083, the dielectric constant of the grown crystal changes: $\delta\varepsilon_3/\varepsilon_3 = 0.056$. Here the relative variation in spontaneous polarization is $\delta P_s/P_s = +0.019$, and the change in the Curie temperature is $\delta T_C = +56 \text{ K}$.

The change of P_s was estimated using the semi-empirical relation

$$T_C(\text{K}) = 3000P_s^2 \tag{6.27}$$

where the P_s value is expressed in C m^{-2}. The relative crystal density change is $\delta\rho/\rho \simeq -0.002$. The relative change in the refractive index, with no effect of spontaneous polarization, is $\delta n(0)/n(0) = -0.001$ (the value $n(0) \simeq 2.3$ was used). The change in the birefringence is calculated to be $\delta B(P_s) = +0.012$. The low-frequency half-wave voltage was found to be $\delta V_{\lambda/2}/V_{\lambda/2} = 0.02$, while the high-frequency value is $\delta V_{\lambda/2} V_{\lambda/2} \approx + 0.05$.

From these estimates we are led to conclude that, under variations of the Li/Nb melt composition within the above interval, crystals thereby grown hardly change their electro-optical properties, a fact which was verified experimentally (see table 6.5).

Table 6.5. Electro-optic half-wave voltages of lithium niobate for different melt compositions. The measured wavelength was 6328 Å. The uncertainties of $\pm$ 5% refer to reproducibility. The accuracy would be better given by $\pm$ 10% (according to Turner *et al* 1970).

Melt composition $(Li/Nb)_m$	Low frequency (DC) $[E \cdot l]_{\lambda/2}$ (V)	High frequency (76 MHz) $n_e^3 r_{33}$ (mV^{-1})	High frequency (76 MHz) $n_o^3 r_{13}$ (mV^{-1})	High frequency (76 MHz) $[E \cdot l]_{\lambda/2}$ (V)
0.910		3.23×10^{-10}	1.02×10^{-10}	2860 ± 150
0.945	2900 ± 115			
1.000	3000 ± 50	3.28×10^{-10}	1.03×10^{-10}	2810 ± 150
1.083	2930 ± 170	3.24×10^{-10}	1.03×10^{-10}	2860 ± 150

6.3 Phenomenological theory of the electro-optical effect

The modulation of radiation from quantum sources by means of lithium niobate and lithium tantalate crystals involves the linear electro-optical effect.

The phenomenological theory of a linear electro-optical effect explained in the works by Pockels (1906) and Zheludev (1968) applies here to crystals with point group symmetry 3m, including lithium niobate and lithium tantalate.

The refractive index of an anisotropic crystal is represented by an indicatrix described generally by a second-order surface

$$B_{ij} x_i x_j = 1 \qquad (6.28)$$

where B_{ij} stands for the components of the dielectric constant tensor (by definition $B_{ij} = \varepsilon_0 \partial E_i / \partial D_j$).

The B_{ij} values are related to the principal refractive indices by

$$B_{11} = a_0^2 = \frac{1}{(n_1^0)^2} \qquad B_{22} = b_0^2 = \frac{1}{(n_2^0)^2} \qquad B_{33} = c_0^2 = \frac{1}{(n_3^0)^2}.$$

$$(6.29)$$

The principal refractive indices are determined by converting the coordinate system so as to satisfy the conditions $B_{23} = B_{31} = B_{21} = 0$ in equation (6.28).

The conversion is carried out in accordance with the direction cosine matrix

	x	y	z
x'	α_1	α_2	α_3
y'	β_1	β_2	β_3
z'	γ_1	γ_2	γ_3

In an expanded form we have

$$
\begin{aligned}
B_{11} = B_1 &= a^2\alpha_1^2 + b^2\beta_1^2 + c^2\gamma_1^2 \\
B_{22} = B_2 &= a^2\alpha_2^2 + b^2\beta_2^2 + c^2\gamma_2^2 \\
B_{33} = B_3 &= a^2\alpha_3^2 + b^2\beta_3^2 + c^2\gamma_3^2 \\
B_{23} = B_4 &= a^2\alpha_2\alpha_3 + b^2\beta_2\beta_3 + c^2\gamma_2\gamma_3 \\
B_{31} = B_5 &= a^2\alpha_3\alpha_1 + b^2\beta_3\beta_1 + c^2\gamma_3\gamma_1 \\
B_{21} = B_6 &= a^2\alpha_2\alpha_1 + b^2\beta_2\beta_1 + c^2\gamma_2\gamma_1.
\end{aligned}
\tag{6.30}
$$

A slight variation in the refractive index induced by the electric field produces a corresponding variation in the form, dimensions and orientation of the indicatrix. These variations are specified by the increments of the coefficients ΔB_{ij}

$$
\Delta B_{ij} = r_{ijk} E_k
\tag{6.31}
$$

where r_{ijk} stands for the electro-optic coefficients forming a third-rank tensor.

The dimensionless quantities ΔB_{ij} are a measure of the relative distortion of the indicatrix. It is noteworthy that expression (6.31) is similar to that of the converse piezoelectric effect

$$
\varepsilon_{ij} = d_{kij} E_k.
\tag{6.32}
$$

In matrix notation, the equation for the electro-optical effect

$$
\Delta B_{ij} = r_{ij} E_j \ (i = 1, 2, \ldots, 6; \quad j = 1, 2, 3)
\tag{6.33}
$$

differs from that for the converse piezoelectric effect by the succession of indices and absence of factors.

An application of the Curie theorem to a crystal exposed to the electric field reduces markedly the number of independent electro-optic coefficients.

The matrices of electro-optic coefficients for crystals with various point group symmetries are found in the works by Nye (1964) and Zheludev (1966).

The lithium niobate and lithium tantalate crystals fall under a ditrigonal pyramidal class with point symmetry 3m whose elements are depicted in figure 1.13.

The matrix equation for lithium metaniobate and lithium metatantalate in the conventional orientation $m \perp x$ is written as

$$
\begin{vmatrix} \Delta B_1 \\ \Delta B_2 \\ \Delta B_3 \\ \Delta B_4 \\ \Delta B_5 \\ \Delta B_6 \end{vmatrix} = \begin{vmatrix} B_1 - a^2 \\ B_2 - b^2 \\ B_3 - c^2 \\ B_4 \\ B_5 \\ B_6 \end{vmatrix} = \begin{vmatrix} 0 & -r_{22} & r_{13} \\ 0 & r_{22} & r_{13} \\ 0 & 0 & r_{33} \\ 0 & r_{51} & 0 \\ r_{51} & 0 & 0 \\ -r_{22} & 0 & 0 \end{vmatrix} \cdot \begin{vmatrix} E_1 \\ E_2 \\ E_3 \end{vmatrix}.
$$

Here $r_{12} = r_{22}$.

Employing the matrix multiplication rule, we arrive at an equation for the indicatrix in the crystallographic coordinate system of the initial class

$$
(a_0^2 - r_{22}E_2 + r_{13}E_3)\,x^2 + (a_0^2 + r_{22}E_2 + r_{13}E_3)y^2 + (c_0^2 + r_{33}E_3)z^2
$$
$$
+ 2\,r_{51}E_2 zy + 2r_{51}E_1 xz + 2r_{22}E_1 xy = 1. \tag{6.34}
$$

The indicatrix equation takes account of the fact that a crystal not subject to an applied field is uniaxial.

Consider the case where the light propagates along the c axis [0001] and the electric field is applied in the same direction, i.e. $E_1 = E_2 = 0$, $E_3 \neq 0$. In this case, the indicatrix equation is

$$
(a_0^2 + r_{13}E_3)(x^2 + y^2) + (c_0^2 + r_{33}E_3)z^2 = 1. \tag{6.35}
$$

Since $B_4 = B_5 = B_6$, the indicatrix axes coincide with the x, y and z axes. The canonical equation for the indicatrix within the frame of principal coordinates has the form

$$
(a_0^2 + r_{13}E_3)(x'^2 + y'^2) + (c_0^2 + r_{33}E_3)z'^2 = 1. \tag{6.36}
$$

To define changes in the principal refractive indices we will write $B_1 = 1/n_1^2$, whence it follows that $\Delta B_1 = (2/n_1^3)\Delta n_1$.

Within a fair degree of accuracy we can replace n_1 by n_1^o, which yields

$$
\Delta n_1 = -\tfrac{1}{2}(n_1^o)^3 r_{13}E_3 \qquad \Delta n_2 = -\tfrac{1}{2}(n_1^o)^3 r_{13}E_3
$$
$$
\Delta n_3 = -\tfrac{1}{2}(n_1^o)^3 r_{33}E_3. \tag{6.37}
$$

Consequently, within this field configuration, the crystal remains uniaxial. The magnitude of induced retardation of light propagating parallel to the x axis is

$$
\Gamma_{\text{ind}} = (\pi l_1 V_3/\lambda_0 d_3)(n_e^3 r_{33} - n_o^3 r_{13}). \tag{6.38}
$$

For a beam parallel to the y axis we have

$$\Gamma_{\text{ind}} = (\pi l_2 V_3 / \lambda_0 d_3)(n_e^3 r_{23} - n_o^3 r_{13}) \qquad (6.39)$$

where l is the optical distance in a crystal, d is the crystal length along an applied field and V is the potential.

For a field applied along the x axis, $E_1 \neq 0$ and $E_2 = E_3 = 0$ (the crystallographic direction $[2\bar{1}\bar{1}0]$), and multiplying the matrices we obtain

$$\Delta B_1 = 0 \qquad \Delta B_2 = 0 \qquad \Delta B_3 = 0$$

$$\Delta B_4 = 0 \qquad \Delta B_5 = r_{51} E_1 \qquad \Delta B_6 = -r_{22} E_1 \qquad (6.40)$$

$$B_1 = B_2 = a_0^2 \qquad B_3 = c_0^2.$$

The equation for the indicatrix is written as

$$a_0^2(x^2 + y^2) + c_0^2 z^2 + 2 r_{51} E_1 xz - 2 r_{22} E_1 xy = 1. \qquad (6.41)$$

The magnitude of induced retardation of light propagating along the z axis and for the field applied in any direction in the xy plane is defined by

$$\Gamma_{\text{ind}} = (2\pi l_3 V_1 / \lambda_0 d)(n_o^3 r_{22}). \qquad (6.42)$$

The orientation of the major axes of the elliptical cross section normal to the z axis is governed by the direction of an applied field.

Designating the angle between the x axis and the field E by θ, we have the angle rotation of the major axes system Ψ defined by

$$\tan(2\Psi) = \cot\theta = E_1 / E_2 \qquad \Psi = 45° - \theta/2 \qquad (6.43)$$

where E_1 and E_2 are the electric field components within the system of crystal axes.

Once the electric field is applied along the y axis (the crystallographic direction $[\bar{1}2\bar{1}0]$), we obtain the following increments ΔB:

$$\Delta B_1 = r_{22} E_2 \qquad \Delta B_2 = r_{22} E_2 \qquad \Delta B_3 = 0$$

$$\Delta B_4 = r_{51} E_2 \qquad \Delta B_5 = 0 \qquad \Delta B_6 = 0. \qquad (6.44)$$

The indicatrix equation has the form

$$(a_0^2 - r_{22} E_2)x^2 + (a_0^2 + r_{22} E_2)y^2 + c_0^2 z^2 + 2 r_{15} E_2 yz = 1. \qquad (6.45)$$

In the latter two cases a uniaxial crystal goes over to a biaxial one.

The canonical equation of the indicatrix for this geometry takes the form

$$\left(a_0^2 + r_{22}E_2 + \frac{r_{51}^2 E_2^2}{a_0^2 - c_0^2 + r_{22}E_2}\right)y'^2 + (a_0^2 - r_{22}E_2)x'^2$$

$$+ \left(a_0^2 - \frac{r_{51}^2 E_2^2}{a_0^2 - c_0^2 + r_{22}E_2}\right)z'^2 = 1 \qquad (6.46)$$

and the ellipsoid of the principal refractive indices will turn about the c axis by an angle

$$\tan 2\psi = 2r_{51}E_2/(a_o^2 - c_0^2 + r_{22}E_2). \tag{6.47}$$

Even assuming that $r_{51} \gg r_{22}$, the ψ angle will be only several minutes for a field strength of 10^4 V cm^{-1}.

Disregarding the second-order infinitesimal quantities r_{51}^2, the induced retardation is defined by

$$\Gamma = \pi l_1 V_2 n_o^2 r_{22}/(\lambda_0 d_2) \qquad \text{light } x \text{ axis}$$

$$\Gamma = \pi V_2 n_o^3 r_{22}/\lambda_0 \qquad \text{light } y \text{ axis} \tag{6.48}$$

$$\Gamma = 2\pi l_3 V_2 n_o^2 r_{22}/(\lambda_0 d_2) \qquad \text{light } z \text{ axis}.$$

The induced birefringence in lithium niobate single crystals is presented in table 6.6. We should like to note that these expressions do not allow for the term r_{51}^2, since it has manifested itself as a second-order infinitesimal effect.

6.4 Establishment of electro-optic coefficients

6.4.1 Evaluation of the total linear electro-optical effect

The earliest studies of electro-optical properties of a lithium meta-niobate multidomain crystal were carried out under static conditions at a frequency of 21 MHz (Peterson *et al* 1964). The crystal orientation was determined by Laue back-reflection photography; the choice of axes was consistent with the standard orientation. No birefringence was observed in the absence of field. The electric field was applied perpendicular to the z axis. In the case the induced birefringence is fitted by equation (6.42), while the rotation of the ellipse major axes is described by expression (6.43).

The multidomain nature of the crystal has predetermined the presence of a great many areas exhibiting birefringence in the plane perpendicular to the z axis; when an electric field was applied these areas became unstable. This circumstance imposed a limitation on the accuracy of the electro-optic coefficient $r_{22}(r_{12})$, which was estimated approximately to lie within 10^{-9}–$10^{-8} \text{ cm V}^{-1}$ (under static conditions). At 21 MHz modulation occurred at an elevated temperature of the sample. The signal increased in proportion with temperature, a maximum value being attained at 75 °C. Further heating was due to the dielectric loss; an equilibrium temperature established itself at 80 °C.

In the work by Lenzo *et al* (1966a), thought was given to lithium metaniobate single-domain crystals. The electro-optic coefficients were established using a polarizing conoscope. The light was transmitted

Table 6.6. Induced birefringence in lithium niobate crystals.

Field	Light		
	$\|x$	$\|y$	$\|z$
$E\|x$	$\Delta n = n_o - n_e$ $\Psi = 0°$	$\Delta n = n_o - n_e + \frac{1}{2}(r_{51}E)^2 n_o^2 n_e^2(n_o^3 + n_e^3)/(n_o^2 - n_e^2)$ $\tan 2\Psi = -2r_{51}En_o^2 n_e^2/(n_o^2 - n_e^2)$	$\Delta n = n_o^3 r_{22}E$ $\Psi = 45°$
$E\|y$	$\Delta n = n_o - n_e + \frac{1}{2}(r_{51}E)^2 \dfrac{n_o^2 n_e^2(n_e^3 + n_o^3)}{n_o^2 - n_e^2}$ $\tan 2\Psi = -\dfrac{2r_{51}E}{(1/n_e^2) - (1/n_o^2)} - r_{22}E$	$\Delta n = n_o - n_e + \frac{1}{2}r_{22}n_o^3 E$ $\Psi = 0°$	$\Delta n = n_o^3 r_{22}E$ $\Psi = 0°$
$E\|z$	$\Delta n = n_o - n_e + \frac{1}{2}(r_{13}n_o^3 - r_{33}n_e^3)E$ $\Psi = 0°$		$\Delta n = 0$ $\Psi = 0°$

parallel to the z axis and interference figures were observed with static electric fields applied parallel to the x and y axes. Single-domain crystals did not contain inclusions with birefringence, which are inherent in multidomain crystals. But the crystals were slightly biaxial under no-field conditions, which was attributed to the residual thermal stress causing the interference effects to change with field reversal.

Electro-optic coefficients were measured using a system consisting of a 6328 Å He–Ne laser (radiation source), a chopper, a De Sénarmont compensator, a photodetector and a vacuum-tube amplifier. Measurements were taken of the angle ψ by which the analyser deviated from zero to a given light beam direction as a function of the magnitude and direction of the field applied to a crystal. The resultant delay $\Gamma = 2\psi$ was used to calculate electro-optic coefficients for particular cases of table 6.7. This table also contains the values r_{22} and $(0.9r_{33} - r_{13})$ measured for various directions of an applied field and values of the voltage $V_{\lambda/2}$ required for producing a delay by $\lambda/2$ in a sample having the geometry $l/d = 1$.

Table 6.7. Electro-optic coefficients r_{ij} (10^{-7} cm (stat V)$^{-1}$) and half-wave voltages $V_{\lambda/2}$ (V) of lithium niobate crystals for $\lambda_0 = 0.633$ μm (according to Lenzo *et al* 1966a,b).

Light propagation direction	Electric field orientation		
	$E\|x$ axis	$E\|y$ axis	$E\|z$ axis
Parallel to x axis		$r_{22} = 2.2$ $V_{\lambda/2} = 7230$	$0.9r_{33} - r_{13} = 5.4$ $V_{\lambda/2} = 2940$
Parallel to y axis		$r_{22} = 1.8$ $V_{\lambda/2} = 9000$	$0.9r_{33} - r_{13} = 5.1$ $V_{\lambda/2} = 3160$
Parallel to z axis	$r_{22} = 4$ $V_{\lambda/2} = 4000$	$r_{22} = 1.9$ $V_{\lambda/2} = 4250$	

To achieve electro-optical modulation in a radio-frequency range, samples of about 1 mm^2 in cross section and 10 mm in length were used. Signals were recorded by means of a photodiode with a superheterodyne amplifier. Modulation was accomplished at 200 MHz. The values r_{ij} at this frequency have not been determined. It was found that at lower frequencies the electro-optical effect was contributed to by a second-order effect (piezoelectric).

The total array of electro-optic coefficients determining the behaviour of lithium metaniobate was established by Turner (1966). The wavelength of a light beam was $\lambda_0 = 0.633$ μm, and the modulating field frequency was made to vary between 50 and 86 MHz. Measurements were taken with heterodyne techniques.

Phase modulation was related to one of the four electro-optic coefficients by an appropriate choice of the direction of a modulating field and beam polarization relative to the crystal axes. The electro-optic coefficients were calculated according to the above formulas. Table 6.6 lists the values of the electro-optic coefficients which are independent of the modulation frequency. The strength of the modulating field was changed from 85 to 1000 V cm^{-1}. Over this range the modulation is linear.

It is remarkable that the dielectric constant of the lithium niobate single crystals at room temperature over a frequency range 20 to 250 MHz remains the same: $\varepsilon = 28$. Precisely this value has been established at a frequency of 20 kHz (Zakharova and Kuz'minov 1969).

The signs of the electro-optic coefficients were obtained by two methods. Firstly, the light was transmitted through two crystals oriented in a manner so that Δn_{33} of the first crystal was added to Δn_{13} of the second one. The modulating fields in the two crystals had the same direction relative to the polar axes. Within the second method, a transmission band was measured with one crystal used as an amplitude modulator (Kaminov and Turner 1966). The signs of the coefficients r_{33} and r_{13} coincided with those obtained for lithium tantalate.

To determine the coefficient r_{22}, two crystals were used with the fields directed along the x and y axes respectively. The amplitude of the used fields ran to 1000 V cm^{-1} for various frequencies of the modulating field. For field strength above 50 V cm^{-1}, the frequency dependence was less than 5% of the r_{22} value. Yet in crystals containing multidomain inclusions the frequency dependence of the r_{22} value is found to be strong.

Sharp resonances occurring at individual frequencies and repeating themselves in about 2 MHz enhance the modulation. To effect modulation at resonant frequencies, the laser beam traversing a single crystal should be variable in space. The resonances are particularly pronounced if the modulating field is directed along the y axis. This effect may be explained by the resonance of higher-order harmonics of the fundamental sound resonance. Once the field is directed along the z axis, no frequency dependence is observed.

The electric field normal to the z axis gives rise to birefringence due to the coefficient r_{42}. This is a first-order change with respect to the field and it depends solely on r_{42}, provided the modulating field is directed along the x axis, and the optical polarization makes an angle $\pi/4$ with the z axis and lies in the plane determined by the electric field and the z axis. The refractive index for such a beam is

$$n'_e - \tfrac{1}{2} n'^3_e r_{42} E \qquad (6.49)$$

where $n'_e = 2^{1/2} n_e n_o / (n_o^2 + n_e^2)^{1/2}$.

The coefficient r_{42} given in table 6.8 has been established within a lesser degree of accuracy than the other coefficients. This is accounted for by the short optical distance, the impossibility of setting up an appreciable electric field along the x axis and, finally, by a certain dependence of the measured quantity on the modulation frequency.

Agheev *et al* (1968) determined the electro-optic coefficients of lithium niobate using elements in the form of rectangular parallel-epipeds, with their edges parallel to the coordinate axes x, y and z. The accuracies of the element orientation with respect to the z axis and the plane of symmetry were 30' and 2 to 3° respectively. The $+z$ and $+y$ directions were identified from the micro-hardness of lithium niobate elements. A preliminary investigation has revealed that the micro-hardness of the opposite faces of a cube perpendicular to the z and y axes is different. The high-micro-hardness cube faces are related to emergence of $-z$ and $-y$ axes. The faces parallel to the plane of symmetry displayed no difference in micro-hardness.

To establish the electro-optic coefficient r_{22} the electric field was applied along the x axis and the light was made to propagate along the z axis. The crystal was placed between crossed polaroids. The incident light was polarized parallel either to the x or to the y axis. With an applied electric field, the crystal transmitted light whose intensity varied according to the law

$$I = I_0 \sin^2 (\pi \Gamma_z / \lambda) \qquad (6.50)$$

where I_0 is the light intensity for parallel polaroids, λ is the wavelength, and Γ is the difference in the paths of ordinary and extraordinary rays that arises when the field is on, which is defined by

$$\Gamma_z = 2n_o^3 r_{22} l_z V / d. \qquad (6.51)$$

The light transmission is maximal when the path difference is $\Gamma = \lambda/2$. For $\lambda_0 = 6328$ Å it is found that $r_{22} = 3.5 \times 10^{-10}$ cm V^{-1} and its value increases with decreasing wavelength, which is typical of electro-optical materials. As the temperature rises to 500 °C, the electro-optic coefficient remains constant.

To establish the coefficient r_{51}, the field was applied along the x axis and the light was made to propagate along the y axis. Unlike the previous case, with no field applied, the crystal exhibits structure birefringence.

In this system the light intensity varies as

$$I = I_0 \sin \frac{\pi d_y}{\lambda} \left[n_0 - n_e + r_{51} \left(\frac{V_x}{d_x} \right)^2 \frac{n_e^2 n_o^2 (n_e^3 + n_o^3)}{n_o^2 - n_e^2} \right]. \qquad (6.52)$$

The maximum amplitude corresponds to a path difference of $\tfrac{1}{4}\lambda$. As the voltage increases further, we observe the appearance of a double-

Table 6.8. Electro-optic coefficients of lithuim niobate.

Coefficient	Numerical value (10^{-10} cm V^{-1})	Frequency	Structure	λ (μm)	Reference
r_{22}^t	6.4	(Static field)	Single domain	0.63	Iwasaki et al (1967)
	6.8	—	—	—	Zook et al (1967)
	3.5	—	—	—	Agheev et al (1968)
	7.0	—	—	—	Solov'yeva (1969)
	3.0	—	—	—	Miller and Savage (1966)
	5.15	—	—	—	Zakharova and Kuz'minov (1969)
	5.25	—	—	3.39	Mustel et al (1968)
	5.3	—	—	1.15	Smakula and Claspy (1967)
	6.7	100 kHz	—	0.63	Smakula and Claspy (1967)
	6.7	1 kHz	—	—	Lenzo et al (1966)
	3.0	—	—	—	Belobaev (1973)
	3.3	60 Hz	—	—	Lenzo et al (1966a)
r_{22}^s	4.0	—	Single domain	0.63	Iwasaki et al (1967)
	3.4	50–86 MHz	—	—	Turner (1966)
	3.5	100 MHz	—	—	Solovyeva (1969)
	3.5	300 MHz	—	3.39	Mustel et al (1968)
r_{13}^t	10	(Static field)	—	0.63	Zook et al (1967)
r_{33}^t	32.2	—	—	—	Zook et al (1967)
$r_{33}^t - 1.1 r_{13}^t$	21	—	—	—	Zook et al (1967)
	15	—	—	—	Agheev et al (1968)
	14.9	—	—	3.39	Mustel et al (1968)
	17.4	60 Hz	Multidomain	0.63	Bernal et al (1966)

Table 6.8. (*cont.*)

Coefficient	Numerical value (10^{-10} cm V^{-1})	Frequency	Structure	λ (μm)	Reference
$0.9r_{33}^t - r_{13}^t$	18.0	1 kHz	Single domain	0.63	Lenzo et al (1966a, b)
	18.0	100 kHz	—	—	Smakula and Claspy (1967)
	17.0	(Static field)	—	1.05	Smakula and Claspy (1967)
	16.0	—	—	3.39	Smakula and Claspy (1967)
r_{13}^s	8.6	50–86 MHz	—	0.63	Turner (1966)
	8.6	100 MHz	—	—	Solovyeva (1969)
r_{33}^s	30.8	50–86 MHz	Single domain	0.63	Turner (1966)
	28.0	100 MHz	—	—	Solovyeva (1969)
$r_{33}^s - 1.1r_{13}^s$	24.0	300 MHz	—	3.39	Mustel et al (1968)
r_{51}^t	32.6	(Static field)	—	0.63	Zook et al (1967)
	32.0	—	—	—	Bernal et al (1966)
	60.0	—	—	—	Agheev et al (1968)
	26.0	—	—	—	Zakharova and Kuz'minov (1969)
r_{51}^s	26.0	50–86 MHz	—	—	Turner (1966)
	28.0	100 MHz	—	—	Solovyeva (1969)
r_{42}^s	28.0	250 MHz	—	0.63	Avakayn et al (1976a, b)

frequency component in the oscillogram. The numerical value of r_{51} calculated according to formula (6.52) is $r_{51} = (6 \pm 0.2) \times 10^{-8}$ cm V^{-1}.

To determine the difference of the coefficients $(r_{33} - 1.1r_{13})$, the electric field was applied along the z axis and the light propagated either along the x or the y axis. The variations in the path difference due to the electro-optical effect are defined by

$$\Delta\Gamma_x = \frac{1}{2}\frac{V_z}{d_z}(n_e^3 r_{33} - n_o^2 r_{13})l_x. \tag{6.53}$$

The numerical value of $(r_{33} - 1.1r_{13})$ calculated by equation (6.53) is 1.5×10^{-9} cm V^{-1}.

Of the greatest practical concern is the case where the field is directed along the x axis and the light propagates along the z axis. Here, owing to the transverse Pockels effect as well as to the large refractive index, fairly low control voltages (tens of volts for the 20:1 length-to-thickness ratio and 50% modulation) are achievable.

Iwasaki *et al* (1967) examined the temperature dependence and dispersion of the electro-optic coefficient r_{22}^t. Figure 6.8 represents the temperature dependence of r_{22}^t for four different wavelengths. The measurements were carried out on three samples and the values obtained were mutually consistent. The constant r_{22}^t exhibits a dispersion with decreasing wavelength of the light and rising temperature. Kurtz and Robinson (1967) have recently calculated the optical frequency dependence of an electro-optic constant on the basis of a bounded oscillator model under a nonlinear effect. It implies that the electro-optic constant should be proportional to $(n^2 - 1)^2/n^4$. The calculation may apply to a clamped electro-optic constant but Iwasuki *et al* tentatively compared $(n_o^2 - 1)^2 n_o^4$ to r_{22}^t at room temperature. Figure 6.9 shows the relation between r_{22}^t and $(n_o^2 - 1)^2/n_o^4$ for four different wavelengths, where one can see that the linearity is fairly good. These

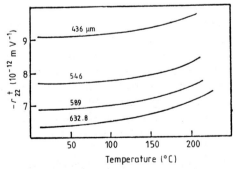

Figure 6.8. Temperature dependence of r_{22}^t (according to Iwasaki *et al* 1967).

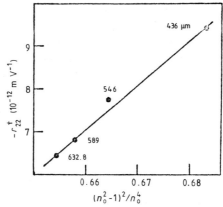

Figure 6.9. Dispersion of the electro-optic coefficient r'_{22} of lithium niobate (according to Iwasaki *et al* 1967).

values of electro-optic coefficients measured at a constant and variable voltage are collected in table 6.8.

6.4.2 Photoelastic effect

A change in the refractive index of a material as a result of the strain developed therein is known as the photoelastic effect. The anisotropic photoelastic relationship between the strain and the refractive index is usually written as

$$\Delta(1/n^2)_{ij} = \sum_{k,l} p_{ijkl}S_{kl} \tag{6.54}$$

where $\Delta(1/n^2)_{ij}$ is the second-order tensor describing the change in the dielectric constant of the material, S_{kl} is the second-rank strain tensor and p_{ijkl} is the fourth-rank photoelastic (strain-optic) tensor. However Nelson and Lax (1971) have shown that the independent elastic variable is not strain but is instead the displacement gradient. Thus, rotation (of the volume elements within an acoustic wavelength) as well as strain contributes to the change in the refractive index of a material caused by acoustic shear waves. Therefore, the previously assumed symmetry of the photoelastic coefficients (corresponding to the acoustic shear waves) upon the interchange of the two elastic indices is incorrect for anisotropic materials. For lithium niobate, the photoelastic coefficients p_{44} and p_{55}, which are numerically equal, consist of symmetric and antisymmetric components (with respect to the elastic indices) (Nelson and Lax 1970). All other photoelastic coefficients contain only symmetric components.

Since the available data do not allow resolution of the components of p_{44}, the interchange of elastic indices will be used here for ease of notation. In addition, as discussed previously, it can be shown that

$\Delta(1/n^2)_{ij} = \Delta(1/n^2)_{ji}$ (Nye 1964). Using these two relations one can reduce the fourth-rank tensor containing 81 elements to 36 independent elements that may be written as a 6×6 matrix. The application of the Neumann principle to the p tensor followed by the use of the reduced-subscript notation gives the 6×6 matrix

$$p_{ijkl} = \begin{bmatrix} p_{11} & p_{12} & p_{13} & p_{14} & 0 & 0 \\ p_{12} & p_{11} & p_{13} & -p_{14} & 0 & 0 \\ p_{31} & p_{31} & p_{33} & 0 & 0 & 0 \\ p_{41} & -p_{41} & 0 & p_{44} & 0 & 0 \\ 0 & 0 & 0 & 0 & p_{44} & p_{41} \\ 0 & 0 & 0 & 0 & p_{14} & \frac{1}{2}(p_{11} - p_{12}) \end{bmatrix}. \quad (6.55)$$

Note that this tensor contains only eight independent coefficients p_{11}, p_{12}, p_{13}, p_{14}, p_{31}, p_{33}, p_{41} and p_{44}. When determining the values of these coefficients one should also take account of the contribution made by the secondary (indirect) photoelastic effect. The latter manifests itself just like the primary effect but it is virtually a coupling of two other effects. In this case, an applied strain causes an electric field due to the piezoelectric effect. This electric field makes the refractive index of a crystal change due to the linear electro-optical effect (Avakyants *et al* 1976). Thus, to the observer who has applied the strain and is measuring the change in the refractive index this secondary contribution is inseparable from the primary photoelastic effect. This indirect photoelastic effect cannot be represented as an ordinary tensor dependent on the acoustic wave direction. In lithium niobate, the secondary effect is significant in the determination of the values of p_{13} and p_{33}. The relationships between the measured coefficients and those describing the primary effects are (Marlescu and Hauret 1973): $p_{13}(\text{pri}) = p_{13}(\text{meas}) - 0.043$, and $p_{33}(\text{pri}) = p_{33}(\text{meas}) - 0.154$. These relationships are consistent with the coefficient values given in tables 5.4 and 6.8. Measured values for the photoelastic coefficient are given in table 6.9.

Table 6.9. Photoelastic (strain) coefficients (dimensionless) at a constant electric field (according to the Weis and Gaylord 1985).

p_{11}	p_{33}	p_{44}	p_{12}	p_{13}	p_{14}	p_{31}	p_{41}
0.036	0.066	—	0.072	0.135	—	0.178	0.155
0.025	0.068	—	0.079	0.132	0.1	0.168	0.158
0.034	0.060	0.30	0.072	0.139	0.066	0.178	0.154
—	0.069	0.152	0.088	0.126	0.080	0.176	0.134
0.045	0.076	0.019	0.096	0.149	0.055	0.138	—
−0.026	+0.071	+0.146	+0.090	+0.133	−0.075	+0.179	−0.151
−0.02	+0.07	+0.12	+0.08	+0.13	−0.08	+0.17	−0.15

To compare the usefulness of acousto-optical materials for device applications, various figures of merit have been developed (Chang 1976). Four of these are

$$M_1 = n^7 p^2 / \rho v, \qquad M_2 = n^6 p^2 / \rho v^3$$
$$M_3 = n^7 p^2 / \rho v^2, \qquad M_4 = n^8 p^2 v / \rho \qquad (6.56)$$

where n is the average refractive index, p is the applicable photoelastic coefficient, ρ is the bulk mass density and v is the bulk acoustic wave velocity being $(\rho s)^{-1/2}$, where s is the applicable elastic compliance coefficient. M_1 is proportional to the product of the bandwidth and diffraction efficiency associated with a bulk acoustic wave for a given level of acoustic power if the height of the acoustic beam is constant. It is appropriately used to select materials for devices, such as modulators requiring an optimum efficiency and bandwidth. M_2 is proportional to the diffraction efficiency associated with a bulk acoustic wave for a given level of acoustic power and it is appropriately used to select materials for narrow-band devices. M_3 is proportional to the diffraction efficiency associated with a bulk acoustic wave for a given level of acoustic power if the acoustic beam height can be made as small as the optical beam size in the wave interaction region. M_4 is proportional to the product of the square of the bandwidth and the diffraction efficiency associated with a bulk acoustic wave for a given acoustic power density. It is appropriately used to select materials for wide-band applications where power density is the limiting factor. For a longitudinal bulk acoustic wave propagating in the x direction, and $n = 2.20$, $\rho = 4.7 \times 10^3 \, \text{kg m}^{-3}$, $v = 6.57 \times 10^3 \, \text{m s}^{-1}$ and an optical beam extraordinarily polarized at an angle of 35° relative to the y axis (measured in the y–z plane), the following typical vaules of the figures of merit have been reported for lithium niobate:

$$M_1 = 6.65 \times 10^{-7} \, \text{m}^2 \, \text{s kg}^{-1}$$
$$M_2 = 6.99 \times 10^{-15} \, \text{s}^3 \, \text{kg}^{-1}$$
$$M_3 = 1.01 \times 10^{-16} \, \text{m s}^2 \, \text{kg}$$
$$M_4 = 62.9 \, \text{m}^4 \, \text{s}^{-1} \, \text{kg}^{-1}.$$

6.4.3 Separation of the secondary electro-optical effect
A piezoelectric crystal exposed to the electric field also experiences, along with the primary electro-optical effect, lying in the direct influence of the electric field upon the optical indicatrix, a secondary electro-optical effect induced by the variations in the optical indicatrix as a result of the piezoelectric strain of a crystal. For centrosymmetric crystals, the secondary electro-optical effect is caused by the electrostriction strain of a crystal.

Pockels was the first to show (Pockels 1906) that the piezoelectric effect in crystals is fitted by a linear equation relating the component of the second-order tensor of polarization constant increments to those of the second-rank tensor of mechanical strain and stresses

$$\Delta a_{ij} = \pi_{ijkl} X_{kl} \qquad \Delta a_{ij} = p_{ijkl} x_{kl} \qquad (6.57)$$

where X_{kl} stand for the stresses, x_{kl} represents the strains, and π_{ijkl} and p_{ijkl} are the fourth-rank tensors of piezo-optic and elasto-optic constants. These tensors are similar by their symmetrical properties to the tensor of the quadratic electro-optical effect.

To separate the primary electro-optical effect from the total is rather problematic, since the piezoelectric strain of a crystal due to the inverse piezoelectric effect and by virtue of the symmetry of the corresponding tensors results in the same deformation of the optical indicatrix as that produced by the electric field. Two methods are available for separating the primary effect. One of them (employed by Pockels) lies in comparing the variations in the polarization constants, produced by the electric field and by the stress separately. The other method (Carpenter 1950) involves a comparison of the variations in the polarization constant produced by a static (low-frequency) electric field when a sample is mechanically free (denoted by the superscript t in table 6.8) and by a high-frequency field when no piezoelectric strains occur, since they are not able to follow the changing field direction (denoted by the superscript s in table 6.8). Thus the electro-optic coefficient of a mechanically clamped crystal r^s characterizes the primary electro-optical effect. If a crystal is mechanically free it experiences, upon application of an electric field, a piezoelectric strain and the inverse piezoelectric effect gives rise to shearing strain therein. This piezoelectric strain will promote further variations in the optical properties of the crystal, according to the shape of the matrix of elasto-optical coefficients p_{ik}.

In contrast to the earlier work by Lenzo *et al* (1966a), Japanese researchers Iwasaki *et al* (1967) have established that r^t_{22} carries a negative sign. The contribution of stresses into r^t_{22} proved to be substantial and the temperature dependence of r^t_{22} turned out to be stronger than was expected from the temperature dependence of the dielectric constant and spontaneous polarization.

The free and the clamped electro-optic coefficients are related by

$$r^t_{ij} = r^s_{ij} + p_{ki} d_{jk} \qquad (6.58)$$

where p_{ki} and d_{jk} are the elasto-optic and the piezoelectric coefficients respectively.

For lithium niobate belonging to the point group symmetry 3m, expression (6.58) is cast in the form

$$r^t_{22} = r^s_{22} - (p_{12} - p_{11})d_{22} + p_{12}d_{24}. \qquad (6.59)$$

The elasto-optic and the piezoelectric constants at room temperature have been presented earlier (see tables 6.9 and 5.4):

$$p_{12} = 0.063 \qquad p_{11} = 0.032 \qquad -p_{42} = p_{41} = 0.136$$

$$d_{24} = 7.4[1 + (T - 20) \times 2.8 \times 10^{-4}] \times 10^{-11} \ \mathrm{C\,N^{-1}}$$

$$d_{22} = 2.1[1 + (T - 20) \times 2.4 \times 10^{-4}] \times 10^{-11} \ \mathrm{C\,N^{-1}}$$

where T is the temperature (°C).

When these constants and $r_{22}^{\mathrm{t}} = 6.4 \times 10^{-12} \ \mathrm{m\,V^{-1}}$ ($\lambda = 0.6328 \ \mu\mathrm{m}$) are substituted into equation (6.59), the contribution of stresses into r_{22}^{t} is $10.4 \times 10^{-12} \ \mathrm{m\,V^{-1}}$ and $r_{22}^{\mathrm{s}} = 4 \times 10^{-12} \ \mathrm{m\,V^{-1}}$, which is comparable with the value $3.4 \times 10^{-12} \ \mathrm{m\,V^{-1}}$ reported by Turner (1966).

The primary electro-optical effect constitutes a substantial proportion of the total at low frequencies. For lithium niobate electro-optic coefficients it is: r_{22}, 50%; r_{33}, 90–95%; r_{13}, 20%; r_{51}, 15–20% (Bernal *et al* 1966).

The dispersion dependence of electro-optic coefficients has been studied by Vasilevskaya *et al* (1967) and Iwasaki *et al* (1967). The works by Smakula and Clapsy (1967) and Sonin and Lomonova (1967) report the results on the temperature dependence of various electro-optic coefficients. The authors claim that it is more expedient to work with elevated temperatures, since the electro-optic activity of lithium niobate crystals increases with their heating. The differences in the magnitude of electro-optic coefficients are attributed to different optical qualities of studied crystals. The scatter of values of electro-optic coefficients has been accounted for by various melt compositions from which crystals are grown.

6.4.4. Measurement of the half-wave voltage by the dynamic method

We shall now consider the method used to measure the half-wave voltage $V_{\lambda/2}$ under the dynamic conditions. This method is very simple and has been widely accepted in practical investigations of electro-optical crystals. The block diagram of the measuring system is shown in figure 6.10. The AC source is provided by a high-voltage transformer. The voltage in its primary winding is controlled by an auto-transformer. The voltage is applied to electrodes perpendicular to the ferroelectric c axis. A collimated light beam of 0.5 mm in diameter is transmitted in a transverse direction. The light source may be a He–Ne laser with a wavelength $\lambda = 0.628 \ \mu\mathrm{m}$. The radiation is polarized by polarizing filters. The polarized beam is focused onto a crystal by a lens of 110 mm focal length. The crystal is mounted in a crystal holder permitting its movement in the vertical and horizontal. The AC voltage applied to the crystal is measured by a digital voltmeter. The modulated light beam is recorded with a photomultiplier energized to 1400–1600 V by a high-voltage stabilized source.

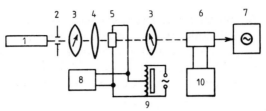

Figure 6.10. Block diagram of the system for measuring the electro-optical effect: 1, laser; 2, aperture; 3, polaroids; 4, lens; 5, lithium niobate crystal; 6, photomultiplier; 7, oscilloscope; 8, digital voltmeter; 9, voltage regulator; 10, high-voltage stabilized supply.

The applied voltage is swept on the oscilloscope screen. As soon as a half-wave voltage is achieved, the adjacently swept maxima and minima are equalized (figure 6.11).

The half-wave voltage referred to units of electrode length and gap are calculated according to the formula

$$V_{\lambda/2} = 1.41(l/d)U \qquad (6.60)$$

where l is the sample length along the light beam, d is the gap and U is

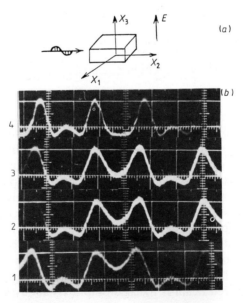

Figure 6.11. (*a*) Oscillograph sweeps for dynamic determination of $V_{\lambda/2}$ of the lithium niobate crystal. The voltage applied to the crystal is: 1, 2800 V; 2, 2900 V; 3, 3010 V; 4, 3660 V. (*b*) Experimental set-up. $V_{\lambda/2} = 3010$ V.

the voltage indicated by the measuring instrument. The factor 1.41 is related to the fact that the voltmeter indicates an RMS voltage rather than the peak value. The value of $V_{\lambda/2}$ obtained in this way is plotted versus the distance from the sample edge. The points of a straight line along the axis are evidence that the crystal is entirely single-domain.

6.5 Specific features of lithium niobate crystal applications in electro-optical devices

The electro-optical properties of lithium niobate crystals, and in particular the values of electro-optic coefficients, enable the laser beam to be controlled by both the longitudinal and transverse control fields. With transverse control electric fields applied, the values of the voltage are smaller by a factor of l/d (l being the crystal length along light propagation and d being the crystal thickness along the applied field). Of practical importance is the case where the electric field is directed along the x axis, while the light propagates along the z optic axis. Thanks to the transversality of the Pockels effect and to large refractive indices, the control voltage may have very low values in this case. For other field and light directions, crystals always exhibit structure birefringence. Nevertheless, they admit application in systems not requiring a high modulation.

In lithium niobate crystals, with the field E directed along the z axis and the light polarized along the x and z axes, intensity modulation arises owing to the interference of ordinary and extraordinary rays. The half-wave voltage becomes equal to

$$V_{\lambda/2} = \frac{\lambda}{n_e^3 r_{33} - n_o^3 r_{13}} \frac{d}{l}. \tag{6.61}$$

Since in lithium niobate the signs of r_{33} and r_{13} are the same and $\Delta n_e > \Delta n_o$, the voltage required to change the phase of a z polarized wave by π is small compared to that required to change the difference between a y and an x polarized wave by the same amount π. Thus, phase modulation is more efficient than intensity modulation.

Relation (6.20) yields

$$r_{33} - r_{13} \approx 2\varepsilon_0(\varepsilon - 1)P_s g_p/\eta^3 \tag{6.62}$$

where $g_p/\eta^3 = g_{11} - g_{12} \approx g_{44} \approx 0.13/\eta^3$ m^{4}C^{-2} and g_{11}, g_{12}, g_{44} are nonzero elements of the g tensor. Using relations (6.61) and (6.62) we obtain

$$V_{\lambda/2} = \left(\frac{\lambda\eta^3}{2n_o^3\varepsilon_0 g_p}\right) \frac{1}{\varepsilon P_s} \tag{6.63}$$

assuming that $\varepsilon \gg 1$ and $n_o \approx n_e$. Equation (6.63) also applies to the paraelectric phase of the crystal, provided P_s is replaced by displacement polarization induced by the field.

Modulators are often analysed using the parameter corresponding to the energy

$$U_\pi = kV_{\lambda/2}^2. \qquad (6.64)$$

This parameter is evaluated according to relation (6.63):

$$U_\pi = \left(\frac{\lambda\eta^3}{2n_o^3\varepsilon_0 g_p}\right)^2 \left(\frac{1}{\varepsilon P_s^2}\right). \qquad (6.65)$$

It is remarkable that expressions (6.63) and (6.65) relate $V_{\lambda/2}$ and U_π for all oxygen octahedral ferroelectrical materials at any temperature only to two parameters of a material, namely ε and P_s. The remaining cofactors included in these expressions, except for η^3, are roughly independent of material or temperature. For specified ε and P_s values, the value of $V_{\lambda/2}$ of lithium niobate type crystals is 1.7 times higher than that with other crystals, which is due to the large value $\eta \approx 1.2$. Expression (6.63) also implies a restriction imposed on ferroelectrics applicability to modulators: at small $V_{\lambda/2}$ values we have, as a rule, large ε values. Yet the $V_{\lambda/2}$ and the U_π values are smaller than those of non-ferroelectrics. Therefore, ferroelectrics are important electro-optical materials. The inverse relationship between ε and $V_{\lambda/2}$ was verified experimentally using, as an example, ferroelectrics of a tetragonal bronze structure (Kuz'minov 1982). It is shown experimentally that at room temperature the product $\varepsilon V_{\lambda/2}$ remains about the same for a wide range of ferroelectrics. However, as the temperature approaches the Curie point T_C (either from the lower or the higher temperature side) $V_{\lambda/2} \sim (1/\varepsilon P_s)$ and $U_\pi \sim (1/\varepsilon P_s^2)$ decrease monotonically. Meanwhile, a number of factors render the operation of modulators at near-Curie temperatures highly undesirable. The static birefringence $\Delta(n_e - n_o)$ fitted by formula (6.21) changes markedly with temperature due to its P_s^2 dependences, while dP_s/dT increases as $(T - T_C)$ tends to zero. Chen (1970) considers a scheme minimizing the adverse effect of the fluctuations in $\Delta(n_e - n_o)$ on the modulation characteristic, on condition that $d\Delta(n_e - n_o)/dT$ is a space-homogeneous quantity. The space-inhomogeneous static birefringence arising from the internal crystal heating by an applied frequency electric field is difficult to compensate. So temperature sensitivity of static birefringence is a significant factor in the choice of a modulator operating temperature. In the practically interesting case $T < T_C$, the temperature dependence of $V_{\lambda/2}$ is not as strong as to cause any problems (Chen 1970).

In the vicinity of T_C, ferroelectrics tend to depolarize if no electric field preventing this process is applied. It is evident from expressions

(6.63) and (6.65) that completely depolarized ferroelectric materials do not exhibit any linear electro-optical effect. Hence the choice of $(T - T_C)$ should be aimed at compensating these opposing factors.

Figure 6.12 gives the temperature dependences of the half-wave voltage $V_{\lambda/2}$ referred to the value of the latter at room temperature for measured electro-optic coefficients r_{13}, r_{22} and r_{33} of lithium niobate. Zook *et al* (1967) point out that above 150 °C the behaviour of the temperature dependence of r_{22} is different from that of other electro-optic coefficients.

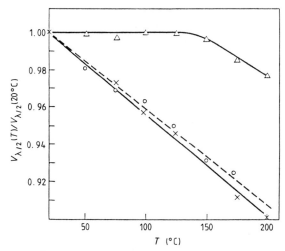

Figure 6.12. Temperature dependences of the half-wave voltage $V_{\lambda/2}$ normalized to room temperature for the $r_{13}(\bigcirc)$, $r_{22}(\triangle)$ and $r_{33}(\text{X})$ electro-optic coefficients of lithium niobate (according to Zook *et al* 1967).

The main parameters of the lithium niobate crystal as an operating substance of a modulator are given in table 6.10. This crystal is distinguished by the following features (Ninomiya and Motoki 1972).

(i) It is notable for a small value of $V_{\lambda/2}$, especially if the modulating field is applied along the z axis, while the optical wave propagates either along the x or the y axis. The $V_{\lambda/2}$ value measured with a 0.6328 μm helium–neon laser is around 2800 V and it is almost identical to that for lithium tantalate.

(ii) Lithium niobate permits creation of a modulator free of structure birefringence, provided the modulating field is applied along the y axis, while the light is made to propagate along the z axis. In this case $V_{\lambda/2} \simeq 7200$ V at high frequencies and $V_{\lambda/2} \simeq 4200$ V at low frequencies. These $V_{\lambda/2}$ values correspond to those for modulators based on the KDP crystal cut at an angle of 45° to the z axis. If the light intensity is high, one must not neglect the crystal heating by absorbed radiation. Crystal heating reduces the transmission coefficient ratio of a modulator. In a modulator with the lithium niobate crystal orientation suggested by Ninomiya and Motoki (1972), the crystal parameters are hardly affected by heating and it withstands relatively high levels of power. In practice, the lithium niobate crystal heated up to 175 °C has a 500:1 transmission-to-noise ratio for a 100 mW argon laser radiation focused by a 360 mm lens. In these conditions the density of transmitted energy is under 100 W mm^{-2}. The lithium tantalate crystal cannot be used in this orientation because of the small r_{22} coefficient.

(iii) The lithium niobate crystal has a relatively low dielectric constant: the ε_{33}^s value for the lithium niobate crystal is twice as low as that for the lithium tantalate crystal. Hence, with the same geometric dimensions of an element, the efficiency of a lithium niobate modulator is twice as high as that on lithium tantalate. In a lithium niobate based modulator with lumped constants, the effect of lead-out inductances is less prominent than in a lithium tantalate modulator. Moreover, in the former modulator, the velocities of light and of a modulating electromagnetic wave are matched more readily.

(iv) The lithium niobate crystal is more liable to optical damage than the lithium tantalate crystal (Chen 1969). This drawback is overcome by operating the modulator at an elevated temperature, as described by Ninomiya and Motoki (1972).

(v) The lithium niobate crystal exhibits high piezoelectric coefficients. The piezoelectric coupling constant of lithium niobate is not smaller than that of lithium tantalate. Therefore provision should be made for damping piezoelectric resonance. This problem is common to all electro-optical light modulators, except for the one based on the ADP crystal cut at an angle of 45° to the x axis.

The phenomenon of optical damage is discussed in chapter 8. The cause of optical damage, i.e. change of the extraordinary refractive index, is the build-up of space charge of retrapped electrons which have been released from the traps by photons. The phenomenon of optical damage that has been given the name of photorefraction disappears upon heating the crystal above 175 °C.

As the beam traverses a crystal exhibiting strong optical damage, the transmission ratio becomes smaller. Figure 6.13 is a plot of this ratio versus temperature for an x-cut lithium niobate crystal. A conclusion can be drawn that x-cut lithium niobate crystals heated up to 180–200 °C are applicable to low-voltage modulators controlling the beam of a helium–neon laser with powers up to 45 mW.

Several workers (Mustel and Paryghin 1970, Mikaelyan *et al* 1970,

Table 6.10. Characteristics of the lithium niobate crystal required for calculation of electro-optical devices.

Characteristic	Numerical value	Reference
Electro-optic coefficients (mV^{-1})	$r_{33} = 30.8 \times 10^{-12}$ $r_{13} = 8.6 \times 10^{-12}$ $r_{22} = 3.4 \times 10^{-12}$ $r_{51} = 2.8 \times 10^{-12}$	Kurtz and Robinson (1967)
Dielectric constant	$\varepsilon_{33}^s = 44$ $\varepsilon_{33}^s = 29$	Kurtz and Robinson (1967)
Refractive indicies at $\lambda =$ 632.8 nm	$n_o = 2.2967$ $n_e = 2.2082$	Hobden and Warner (1966)
Thermal expansion coefficients for 0–500 °C (K^{-1})	$\alpha_a = 1.54 \times 10^{-5}$ $\alpha_c = 0.75 \times 10^{-5}$ $\beta_a = 5.3 \times 10^{-9}$ $\beta_c = 7.7 \times 10^{-9}$	Kim and Smith (1969)
Curie temperature (°C)	1165	
Piezoelectric interaction coefficients (%)	x quasi-transverse 68 y quasi-transverse 60 z quasi-transverse 0 y quasi-longitudinal 30 z quasi-longitudinal 18	Chen (1969)
Velocity of sound (for the longitudinal mode directed towards z $(m\,s^{-1})$	7330.59	Chen (1969)
Density $(g\,cm^{-3})$	4.64	Nassau *et al* (1966)
Path difference temperature coefficient $\dfrac{1}{l}\dfrac{\partial l(n_e - n_o)}{\partial T}$ (K^{-1}) (l = crystal length, T = temperature)	4.3×10^{-3}	

1971, Lee and Zook 1968) have announced that they have created discrete systems for a ray deflection, based on lithium niobate crystals. The latter are also used for amplitude and phase modulation (Kaminov and Turner 1966, Kaminov and Sharpless 1967). The device for discrete ray deflection consists, as a rule, of an electro-optical cell (lithium niobate crystal) and birefringence elements, i.e. uniaxial highly aniso-tropic crystals (Iceland spar plates cut at an angle of 51° to the optic axis) capable of displacing the extraordinary ray from its original position and preserving the direction of the ordinary ray.

The direction of the light beam polarization plane is changed by an

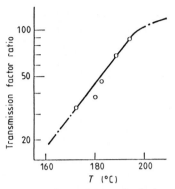

Figure 6.13. Temperature dependence of relative transmission of the lithium niobate crystal for $\lambda = 0.6328\ \mu m$, $W = 100\ \text{W cm}^{-2}$, $l = 25$ mm (according to Ninomiya and Motoki 1972).

electric field applied to the cell. Using m binary elements, we can make the light beam deviate into one of the 2^m possible positions. To avoid light beam mixing, the thickness of successive birefringent crystals should decrease as $l_1 : l_2 \ldots l_m = 1 : 2 \ldots 2^{m-1}$. In particular, consideration has been given to a five-stage 32-point deflection system and a ten-stage system at whose output the ray may be brought to 1024 positions arranged in the form of a 32×32 matrix.

Application of crystals to multi-stage laser-beam control systems imposes specified requirements for their properties. Here of critical importance are optical homogeneity and high stability of parameters under the external electric field.

Figure 6.14 presents the lines of equal half-wave voltage, $V_{\lambda/2}$, for

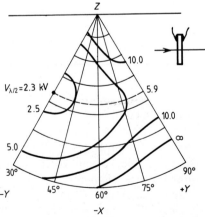

Figure 6.14. Lines of equal half-wave voltage for oblique cuts of lithium niobate crystals in the standard coordinate system (according to Hulme 1971).

oblique cuts of lithium niobate single crystals, allowing us to optimize cuts in devices based on the electro-optical effect (Hulme 1971).

6.5.1 Light modulation in regular domain structure crystals

Light diffraction by a regular domain structure was achieved for the first time in barium–sodium niobate crystals, this phenomenon being of an electro-optical nature (Aleksandrovsky 1977, Aleksandrovsky and Nageav 1983). In the absence of an internal electric field, the phase diffraction grating is contributed to not only by the refractive index variations due to growth layers but also to those due to the pyroelectric field that has not been able to relax (Aleksandrovsky and Nagaev 1983).

The influence of the external electric field on the light intensity distribution in the diffraction pattern permits a claim that the observed diffraction by a regular domain structure is the result of periodic variations in the crystal refractive indices due to the electro-optical effect. With no external field applied, periodic variations in the refractive indices may be caused by the internal pyroelectric field of a crystal. This claim is supported by the fact that when a crystal is raised above 150 °C no light diffraction by a regular domain structure is observed. The diffraction grating disappearance upon heating to 150 °C must not be related to the disappearance of the domain structure itself, since at this temperature, which is so far from the Curie point, no unpoling takes place. On the other hand, the conductivity of lithium niobate at $T \geqslant 150$ °C increases so much that it can rapidly compensate the change in the pyroelectric charge. As a result, there is no pyro-field in the crystal, nor the phase diffraction grating. The diffraction efficiency is defined as the ratio of intensities of the first diffraction maximum (I_1) and of the central beam (I_0), and it is determined by the change of the refractive index

$$I_m/I_0 \simeq (k_0 l \Delta N_{(m)}/2) \qquad (6.66)$$

where $k_0 = 2\pi/\lambda_0$, and λ_0 is the fundamental wavelength, m is the order of an intensity maximum, l is the optical length of a crystal and $\Delta N_{(m)}$ stands for the coefficients of expansion of the function describing periodic variations in the birefringence into a Fourier series.

In the work by Blistanov *et al* (1986) the regular domain structure was oriented so that the y axis was orthogonal to the planes of domain boundaries. The high uniformity of a built-in domain structure ensured a high optical homogeneity of the lithium niobate crystal in the absence of an external electric field. In annealed lithium niobate crystals, owing to the effect of domain boundaries, only 0.01% of the incident light intensity was diffracted.

On application of an external electric field to the lithium niobate crystal with a regular domain structure, a phase diffraction grating is formed because of the different sign of the electro-optic coefficient r_{33} in adjacent domains, the grating period being equal to the double domain thickness. The equations fitting light diffraction by this grating are similar to those of acousto-optics (Gulyaev *et al* 1978).

Investigations were carried out with crystals of about 5 mm in length (along the z axis) and 2 mm in width (along the x axis). The electric field was applied along the z axis; the light at $\lambda = 0.63\ \mu$m was polarized along the z axis and propagated along the crystal x axis (figure 6.15). The coefficient used was $r_{33} = 0.9 \times 10^{-8}$ stat V. The nature of diffraction is defined by the parameter $Q = \lambda l/\Lambda^2 n$, where l is the diffraction grating depth, Λ is its separation. In the studied samples Q was 0.09 to 0.1, which corresponded to the Raman–Nath regime with a great number of diffraction orders. In this case, a maximum diffraction efficiency is observed at normal incidence of light. The experimentally measured light intensities at zero, first and second diffraction maxima versus control voltage are plotted in figure 6.16. A maximum diffraction efficiency of 98% was reached at a voltage $U = 5.3$ kV. Limitations on the maximum diffraction efficiency are presumably imposed by the domain size fluctuations across the optical beam. For instance, normally distributed fluctuations with a relative dispersion of a mean size domain of about 0.05 yield precisely the observed value of maximum diffraction efficiency.

Utilization of the lithium niobate crystal with a regular domain structure for the purpose of radiation modulation ensures high-speed control over the light intensity. The major factor limiting the speed is the presence of piezoelectric resonances determining the transient time

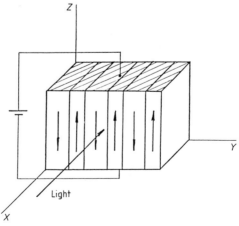

Figure 6.15. Orientation of a studied sample of lithium niobate having a regular domain structure (according to Blistanov *et al* 1986).

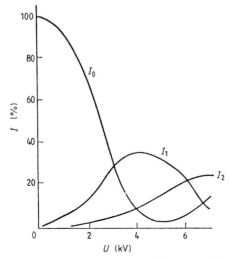

Figure 6.16. Light intensity at the zero (I_0), first (I_1) and second (I_2) diffraction maxima versus control voltage (according to Blistanov *et al* 1986).

in a modulating medium. As is shown by Gurevich *et al* (1973), in a regular domain structure, piezoelectric resonances occur at sound wavelengths that are multiples of the domain structure period. Hence, in lithium niobate with a regular domain structure, the piezoelectric resonance frequencies are shifted towards the high-frequency region. In a crystal with a regular domain structure of about 40 μm, resonances have not been observed up to frequencies of the order of 100 MHz.

The electro-optical modulators based on this effect are superior to the existing ones in terms of simplicity and high speed of operation (for 40 μm regular domain structure, the switching time does not exceed 10 ns). We should like to stress here that in modulators based on regular domain structures the dimensions of the light window are not directly related to the modulator speed, which is the case with other designs of electro-optical modulators.

7 Nonlinear Optical Properties of Lithium Niobate

7.1 Elements of nonlinear optics

The electro-optical phenomena were considered under the assumption that electronic polarizability of a crystal is independent of the strength of the light beam field. As the laser light propagates, the strength of the light wave field becomes comparable with the internal fields in crystals. The latter circumstance is responsible for nonlinear interaction of the light wave field with a medium, thereby violating the principle of superposition and creating the conditions for higher harmonic generation, for sum and difference frequencies. Nonlinear effects may also cause a constant polarization of a crystal exposed to a strong field of an optical frequency. All these effects can be described in terms of phenomenological relations derived by differentiating the expression for free energy with respect to the field (Smolensky *et al* 1971):

$$P_i^{2\omega} = X_{ijk}^{2\omega} E_j^{\omega} E_k^{\omega} \tag{7.1}$$

for the second harmonic generation,

$$P_i^0 = X_{ijk} E_j^{\omega} E_k^{\omega} \tag{7.2}$$

for constant polarization, and

$$P_i^{\omega} = X_{ijk}^{\omega} E_j^0 E_k^{\omega} \tag{7.3}$$

for a linear electro-optical effect, where P_i stands for the polarization components, E_j and E_k are the electric field components and X_{ijk} represents the components of the nonlinear susceptibility coefficient which is a third-rank tensor. The superscripts denote the frequency of an electric field. All three effects are only possible, for symmetry reasons, in non-centrosymmetric crystals. The tensor X_{ijk} may have 18 independent components. In the case of second harmonic generation, Kleinman (1962) has shown that for crystals with negligible dispersion and optical loss the permutation relations

$$X_{ijk}^{2\omega} = X_{kij}^{2\omega} = X_{jki}^{2\omega} \tag{7.4}$$

hold good and, as a result, the number of independent tensor terms reduces to 10.

The conditions of Kleinman are valid for all crystals studied. The nonlinear susceptibility coefficients X_{ijk} differ by more than two orders of magnitude with crystals. Miller (1963) has introduced another coefficient (δ) giving crystal nonlinearity and has obtained an expression relating the nonlinear susceptibility coefficients to the corresponding linear susceptibilities χ. For second harmonic generation, this relation has the form

$$X_{ijk}^{2\omega} = \chi_{ii}^{2\omega} \chi_{jj}^{2\omega} \chi_{kk}^{2\omega} \delta_{ijk}^{2\omega}. \tag{7.5}$$

The values of the coefficient δ_{ijk} are hardly different for various nonlinear effects taking place in the same crystal, or for various crystals. This is evidence that nonlinear effects are intimately linked to crystal polarization. The coefficient δ is a normalized nonlinear susceptibility and represents a measure of nonlinearity of electron processes occuring in a crystal. Within the model of an anharmonic oscillator, δ values are proportional to the anharmonic force constant. The nonlinear optic coefficients responsible for harmonic generation are related to Miller's coefficients δ by Di Domenico and Wemple (1968):

$$d_{ijk}^{\alpha\beta\gamma} = \varepsilon_0 \chi_{ii}^{\alpha} \chi_{jj}^{\beta} \chi_{kk}^{\gamma} \delta_{ijk} \tag{7.6}$$

where χ_{ii}^{α} is the linear susceptibility at a frequency ω_α along the ith principal axis of a crystal ($P_i^{\alpha} = \varepsilon_0 \chi_{ii}^{\alpha} E_i^{\alpha}$) and ε_0 is the permittivity of vacuum. The tensor δ_{ijk} gives optical nonlinearity only due to electron processes and since it is independent of the field its values are the same for electro-optical (EO) and nonlinear optical (NLO) effects. Di Domenico and Wemple (1968) have represented Miller's linear coefficient as follows:

$$\delta_{ijk} = -f_{ijk}^{\text{s}}/(2\varepsilon_0 N_{ii}) \tag{7.7}$$

where $N_{ii} = (1 - 1/n_{ii}^2)^2$ and f_{ijk}^{s} is the clamped electro-optic tensor expressed via polarization applied to a given crystal

$$f_{ijk}^{\text{s}} = r_{ijk}^{\text{s}}/[\varepsilon_0(\varepsilon_k - 1)] \tag{7.8}$$

where r_{ijk}^{s} is the tensor of clamped electro-optic coefficients. Combining expressions (7.7) and (7.8) and substituting them into equation (7.6), we can establish (NLO) coefficients for second harmonic generation (SHG). To do this, we set $\omega_\beta = \omega_\gamma = \omega$ and $\omega_\alpha = 2\omega$:

$$d_{ijk}^{2\omega} = -\tfrac{1}{2}\varepsilon_0^2 n_{ii}^4 (\chi_{ii}^{2\omega} \chi_{kk}^{\omega} / \chi_{ii}^{\omega}) f_{ijk}^{\,s}. \tag{7.9}$$

It is evident from expression (7.9) that the nonlinear optical coefficients δ and d are directly related to the clamped electro-optic tensor $f_{ijk}^{\,s}$.

For perovskite-type compounds ABO_3, the coefficients d_{ijk} obey permutation symmetry, i.e. Kleinman conditions within an accuracy better than 5%.

The lithium niobate crystal which is under consideration has an axial C_{6v} symmetry and the spontaneous polarization P_s is directed along the triad axis. As already shown in the work by Wemple *et al* (1968) and mentioned in the previous section, in oxide compounds of the same type ABO_3, the linear electro-optical effect is equivalent to the quadratic effect displaced due to the spontaneous polarization. Then, for C_{6v} symmetry, the electro-optic coefficients $f_{ijk}^{\,s}$ can be expressed through coefficients of the quadratic electro-optical effect (Zook *et al* 1967):

$$f_{13}^{\,s} = \tfrac{2}{3}(g_{11}^{\,s} + 2g_{12}^{\,s} - g_{44}^{\,s})P_s$$

$$f_{51}^{\,s} = \tfrac{2}{3}(g_{11}^{\,s} - g_{12}^{\,s} + \tfrac{1}{2}g_{44}^{\,s})P_s \tag{7.10}$$

$$f_{33}^{\,s} = \tfrac{2}{3}(g_{11}^{\,s} + 2g_{12}^{\,s} + 2g_{44}^{\,s})P_s.$$

Disregarding the birefringence effects, we obtain expressions for the coefficients δ:

$$\delta_{15} = (g_{11}^{\,s} + 2g_{12}^{\,s} - g_{44}^{\,s})P_s/(3\varepsilon_0 N)$$

$$\delta_{33} = (g_{11}^{\,s} + 2g_{12}^{\,s} + 2g_{44}^{\,s})P_s/(3\varepsilon_0 N). \tag{7.11}$$

For C_{6v} symmetry we have $\delta_{31} = \delta_{15}$.

The calculated NLO coefficients can be compared to experimental values only for few materials due to the lack of relatively good single crystals. Table 7.1 lists, for the sake of comparison, nonlinear optical coefficients of several compounds, including lithium niobate. The values of the coefficient are given relative to that of quartz standard δ_{11}^Q. An absolute value has been established within a high degree of accuracy for d_{36} of $NH_4H_2PO_4$ (ADP) and it proved to be $d_{36} = (5.0 \pm 0.6) \times 10^{-14}$ units of MKS $\delta_{11}^Q \approx 0.25 \pm 0.05 \times 10^{10}$ MKS units. The theoretical δ values have been calculated in accordance with the above formulae.

Harmonic generation requires that the phase-matching conditions should be fulfilled. These conditions are equivalent to the conservation of momentum. Once waves of three frequencies ω_1, ω_2 and ω_3 are interacting in a medium (say $\omega_1 + \omega_2 = \omega_3$), harmonics will be generated, provided the phase relations

$$\mathbf{k}_1^{(1)} + \mathbf{k}_2^{(2)} = \mathbf{k}_3^{(3)} \tag{7.12}$$

Table 7.1. Nonlinear optical coefficients δ_{ij} of some oxygen octahedral ferroelectrics relative to the δ_{11}^O value for quartz (at room temperature) (according to Di Domenico and Wemple 1968).

Crystal	P_s (°C m^{-2})	$\delta_{33}/\delta_{11}^O$	$\delta_{15}/\delta_{11}^O$	Experiment ($\times 10^{10}$)		Theory ($\times 10^{10}$)	
				δ_{33}	δ_{15}	δ_{33}	δ_{15}
K$_3$Li$_2$Nb$_5$O$_{15}$	0.24	2.0 ± 0.2	1.2 ± 0.2	0.5 ± 0.15	0.3 ± 0.12	0.71	0.17
BaTiO$_3$	0.25	0.55 ± 0.11	0.79 ± 0.16	0.14 ± 0.06	0.2 ± 0.08		
		0.49 ± 0.05	1.1 ± 0.1	0.12 ± 0.01	0.28 ± 0.02	0.74	0.17
Ba$_2$NaNb$_5$O$_{15}$	0.4	2.4 ± 0.2	1.4 ± 0.14	0.6 ± 0.18	0.35 ± 0.1	1.2	0.28
LiTaO$_3$	0.5	3 ± 0.3	0.2 ± 0.04	0.8 ± 0.23	0.05 ± 0.02	0.82	0.22
LiNbO$_3$	0.71	4.2 ± 0.6	0.7 ± 0.1	1.1 ± 0.3	0.18 ± 0.06	1.15	0.31
		4.7 ± 1.2	0.56 ± 0.06	1.2 ± 0.5	0.14 ± 0.04		

(k being the wave vector) between the propagating waves are preserved. For second harmonic generation, the phase-matching condition is

$$2k_1 = k_2 \qquad (7.13)$$

or $\Delta k = 0$, where $\Delta k = 2k_1 - k_2$. In a limited medium, energy may be stored also at $\Delta k \neq 0$, but the value of Δk should not be too high. The length of samples still admitting energy storage in second harmonic generation is defined by

$$l\Delta k \approx \pi/2. \qquad (7.14)$$

The phase-matching condition is satisfied if the refractive indices of the initial frequency wave and of the second harmonic, for a specified direction of wave propagation, are equal.

The phase-matching conditions can be met for a combination of waves of different polarizations, say,

$$K_1^o + K_1^o = K_2^e. \qquad (7.15)$$

For uniaxial negative crystals including lithium niobate, the phase-matching condition is

$$K_1^o + K_1^e = K_2^e. \qquad (7.16)$$

The angle θ_m between the direction in which the phase-matching condition is met and the optic axis of a crystal is called the phase-matching angle.

A very important quantity determining nonlinear properties of a crystal is the mismatch gradient $d\Delta k/d\theta$, which defines the rate of variations in the magnitude of mismatch upon deviation from the phase-matching angle θ_m.

It is pertinent to mention the possibility of the so-called non-critical second harmonic generation at a phase-matching angle $\theta_m = \pi/2$. In some crystals, lithium niobate being among them, this becomes feasible due to the temperature dependence of ordinary and extraordinary refractive indices. In this case, the velocities of the fundamental and the SHG waves turn out to be the same, no birefringence is present, and the coherence length of crystal over which energy is stored during second-harmonic generation is dictated only by the laser beam divergence. Similar phase-matching conditions are observed with various variants of parametric amplification (for instance, amplification of waves at frequencies ω_1 and ω_2 by means of high-frequency pumping $\omega_p = \omega_1 + \omega_2$). The phase-matching condition for three waves complying with the frequency relation $\omega_1 + \omega_2 = \omega_3$ is as follows (Armstrong *et al* 1962):

$$\Delta k = k_1 + k_2 - k_3 = 0. \qquad (7.17)$$

In the case of three parallel wave vectors, the expression

$$\Delta k = |k_1 + k_2 - k_3| = 2\pi[n_1(\theta)/\lambda_1 + n_2(\theta)/\lambda_2 - n_3(\theta)/\lambda_3] \quad (7.18)$$

should tend to zero. The refractive indices $n_i(\theta)$ for an ordinary and an extraordinary ray in a birefringent crystal can be cast in the form

$$n_i(\theta) = n_{i,o}\{1 + F[(n_{i,o}/n_{i,e})^2 - 1]\sin^2\theta\}^{-1/2} \quad (7.19)$$

where $F = 0$ for an ordinary ray and $F = 1$ for an extraordinary ray, θ is the angle of deviation of the vector k relative to the optic axis. The condition $\Delta k = 0$ is satisfied if one or two of the three waves are extraordinary.

In the vicinity of θ_m, expression (7.18) can be approximated by linear terms of the Taylor series

$$\Delta k = (\mathrm{d}\Delta k/\mathrm{d}\theta)_{\theta_m}(\theta - \theta_m). \quad (7.20)$$

It is remarkable that only extraordinary rays yield an expression for mismatch gradient

$$(\mathrm{d}\Delta k/\mathrm{d}\theta)_{\theta_m} = (\mathrm{d}k_1/\mathrm{d}\theta)_{\theta_m} + (\mathrm{d}k_2/\mathrm{d}\theta)_{\theta_m} - (\mathrm{d}k_3/\mathrm{d}\theta)_{\theta_m}. \quad (7.21)$$

For more than one extraordinary ray, the terms of expression (7.21) are partially cancelled, thereby reducing the mismatch gradient. The latter quantity is of prime significance in the calculation of the conversion ratio in optical mixing of divergent beams. The process of optical mixing of exactly phase-matched waves has been treated by Armstrong *et al* (1962).

According to Kleinman (1962), the conversion ratio for optical mixing of divergent beams in thick crystals is inversely related to the mismatch gradient.

Weber *et al* (1966) have calculated various possible phase-matching angles for second harmonic generation and for mixing of two beams having parallel wave vectors for several directions of polarization, as well as mismatch gradients. Their calculation results are presented in table 7.2.

7.2 Methods used to establish nonlinear coefficients

Lithium niobate is noted for exhibiting large amounts of second harmonic generation. By virtue of the substantial negative birefringence of lithium niobate ($\Delta n = -0.08$) in the visible and near infrared spectral regions, the fundamental and the secondary waves can be phase-matched. The nonlinear coefficients d_{31} of lithium niobate are about 11 times more than d_{36} in KDP crystals. The rest of the crystals, except for barium sodium niobate, have coefficients equal to or smaller than d_{36} of KDP. Second harmonic generation is proportional to d^2l^2, where l is the

Table 7.2. Phase-matching direction and mismatch gradients for LiNbO$_3$ (according to Weber *et al* 1966).

	First ray			Second ray			Third ray			Kind of ray			Phase-matching angle θ_m	Mismatch gradient $d\Delta k/d\theta$ (10^{-7}Å^{-1} deg^{-1})
λ (Å)	n_o	n_e	λ (Å)	n_o	n_e	λ (Å)	n_o	n_e	1	2	3		
11 523	2.229	2.148	11 523	2.229	2.148	5761	2.305	2.215	o	o	e	66° 25'	11.59
11 523	2.229	2.148	10 600	2.233	2.154	5521	2.313	2.223	o	o	e	72° 17'	9.74
10 600	2.233	2.154	10 600	2.233	2.154	5300	2.235	2.232	o	o	e	84° 08'	4.69

crystal length. Hence, the power of second harmonic generation in lithium niobate (in the absence of saturation) may be 120 times higher than that with KDP crystals.

Let us transform formula (7.1) to apply it to lithium niobate crystals. For point group symmetry 3m, the second-order polarization is given by (Franken *et al* 1961)

$$P_x = 2d_{15}E_zE_x - 2d_{22}E_xE_y$$
$$P_y = -2d_{22}E_x^2 + 2d_{22}E_y^2 + 2d_{15}E_yE_z \qquad (7.22)$$
$$P_z = d_{31}E_x^2 + d_{31}E_y^2 + d_{33}E_z^2$$

where E_i stands for the components of the electric field strength of a primary wave. Once the primary wave, having an ordinary ray polarization, propagates in the plane xz at an angle of θ_m to the crystal optic axis and a wave possessing an extraordinary ray polarization is generated, the two waves may happen to be phased-matched. The θ_m angle is calculated from the known refractive indices. The polarization transverse to the propagation direction is

$$P_{2t}^e = P_{2z}^e \sin \theta_m = d_{31}(E_{1y}^o)^2 \sin \theta_m. \qquad (7.23)$$

Here the subscripts 1 and 2 refer to the primary and the secondary waves respectively, while the superscripts o and e denote ordinary and extraordinary polarizations of the ray respectively. If propagation occurs in the yz plane, the extraordinary polarization transverse to the propagation is

$$P_{2t}^e = (d_{31} \sin \theta_m + d_{22} \cos \theta_m)(E_{1x}^o)^2. \qquad (7.24)$$

The phase-matching conditions for the combination of d_{31} and d_{22} are just the same as for d_{31} alone. If a crystal exhibits phase matching at $\theta = -\theta_m$ rather than at $+\theta_m$, the second term in equation (7.24) is subtracted. The ratio of the SHG powers in the two cases is

$$R = [(d_{31} \sin \theta_m + d_{22} \cos \theta_m)/(d_{31} \sin \theta_m - d_{22} \cos \theta_m)]^2. \qquad (7.25)$$

Irrespective of the relative signs of d_{31} and d_{22}, with a suitable choice of either a plus or a minus sign of θ_m, we shall be able to employ the two coefficients simultaneously. The refractive indices of lithium niobate measured in the wavelength range from 0.4 to 4 μm are shown in table 6.1.

For a gas laser at $\lambda = 1.152$ μm, the phase-matching conditions are attainable both for a single coefficient d_{31} and for a combination of d_{31} and d_{22} at $\pm \theta_m$. The θ_m value calculated from the refractive indices is 65.3°, while the experimentally measured one is $(68 \pm 1)°$. The discrepancy may be attributed to the uncertainies in the quantities determining the phase-matching condition.

The positive pole of a crystal $+C$ is obtained from the positive charge arising across the electrode upon the lithium niobate crystal cooling due to the pyroelectric effect. If spontaneous polarization increases with decreasing temperature, which is the case with most ferroelectric materials, P_s has a $+C$ direction.

The $+Y$ direction is chosen so that the sum of indices $(-h + k + l)$ might be divisible by three at a standard crystal orientation. The SHG output power at $+\theta_m$ is higher than at $-\theta_m$. It is presupposed that the change in sign of θ_m will not affect the signs of d_{22} or d_{31}.

Expressions (7.23) to (7.25) have yielded the values $d_{31} = 10.6 \pm 1.0$ and the ratio $R = 2.19$. Measurements taken under pulsed lasing indicated that $d_{22} \cos \theta_m < d_{31} \sin \theta_m$; the calculated value $d_{22} = 5.1 \pm 2.0$. All the measurements have been carried out relative to d_{36} of the KDP crystal taken for unity. An absolute value of d_{36} of KDP is $(3 \pm 1) \times 10^{-9}$ cm (stat V)$^{-1}$.

Under second harmonic generation in a nonlinear crystal, the Poynting vector of the primary wave deviates generally from that of the secondary wave by an angle

$$\rho \approx (B/n) \sin^2 (2\theta_m) \qquad (7.26)$$

where $B = n_2^o - n_2^e$ and n is the mean refractive index. Miller *et al* (1965) were the first to demonstrate experimentally SHG and frequency mixing in lithium niobate crystals at $\theta_m = 90°$ and $\rho = 0$, in which case no birefringence effects are present and the laser beam divergence is determined solely by its optical path l in a crystal. For $\rho = 0°$, the primary and the secondary waves propagate normally to the optic axis and the velocity-matching condition is fulfilled subject to the following equality:

$$n_1^o - n_2^e = (n_2^o - n_2^e) - (n_2^o - n_1^o) = B - D = 0. \qquad (7.27)$$

An examination of the temperature dependence of dispersion (D) and birefringence (B) has revealed that velocity-matching of the primary and secondary waves under SHG at $\rho = 0$ is possible at certain specific laser wavelengths and with a suitable choice of crystal temperature. For the yellow line of sodium, $dB/dT = -4.5 \times 10^{-5} \,°C^{-1}$ averaged over 300 °C. The value $dD/dT = 1.21 \times 10^{-5} \,°C^{-1}$ has been established from the temperature dependence of coherence length for the nonlinear coefficient d_{22} measured with a YAG:Nd laser. Consequently, for a wavelength close to the present measurements, $d(B - D)/dT = -5.7 \times 10^{-5} \,°C^{-1}$. To achieve SHG, it is essential to specify the temperature at which phase-matching conditions are met with $\rho = 0°$.

For a gas laser at $\lambda = 1.153 \,\mu m$ and 23 °C, $(n_1^o - n_2^e) \approx 0.11$ and phase-matching is attained at $T_m \approx 220$ °C. The experimentally determined value of T_m is 193 °C for a 1.153 μm He–Ne laser line and 0 °C for a 1.058 μm CaWO$_4$:Nd laser. Figure 7.1 plots the temperature

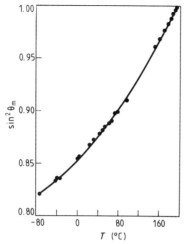

Figure 7.1. Temperature dependence of $\sin^2\theta \simeq (n_2^o - n_1^o)/(n_2^o - n_2^e)$. θ_m is the phase-matching angle for second harmonic generation by a lithium niobate crystal, $\lambda_1 = 1.15\ \mu m$ (according to Miller *et al* 1965).

dependence of $\sin^2\theta_m$ for the lithium niobate crystal over a temperature range from $-80\ ^\circ C$ to $200\ ^\circ C$ at $\lambda = 1.153\ \mu m$.

The strong temperature dependence of $(B - D)$ observed in experiments enables light parametric oscillators to be created by varying the temperature of a lithium niobate crystal.

The second harmonic generation power P_2 is related to the angle of deviation $(90^\circ - \theta_m)$ for the fundamental wave λ_1 as follows (Miller *et al* 1965):

$$P_2 \propto l^2 \frac{\sin\left(2\pi Bl/\lambda_1\right)[(\varphi/n_1^o)^2 - \gamma^2]}{\{(2\pi Bl/\lambda_1)[(\varphi/n_1^o)^2 - \gamma^2]\}^2} \tag{7.28}$$

where γ is the deviation of the phase angle θ_m from 90° and φ is the deviation of the angle made by a laser beam with an optic axis of a nonlinear crystal from 90°.

The full width at half-maximum, as calculated for $\gamma = 0^\circ$, is 5.7°, which is consistent with the experimental data obtained by Miller *et al* (1965). It is noteworthy that the phase-matched half-width observed in KDP is 0.1°. The full width at half-temperature for $T = T_m$ is $0.5\ ^\circ C$ for a gas laser and $2\ ^\circ C$ for a YAG:Nd laser. The authors of the above mentioned work have also pointed out that if a sample was made to turn relative to the laser beam about the z optic axis or about the y axis, the generated harmonics were found to oscillate because of the resonance of the primary wave in the cavity formed by the opposite sides of the sample, the oscillation depth running to 25%.

The nonlinear coefficient d_{33} and the coherence length of a crystal

$$l_c = \lambda_1/[4(n_2^e - n_1^e)] \tag{7.29}$$

are obtainable from the temperature dependence of the second harmonic intensity $I_2(T)$ (Kaminov 1965).

We will denote the laser wavelength λ_1, and the refractive indices of the fundamental and second harmonic extraordinary rays n_1^e and n_2^e respectively. Then the second harmonic intensity is related to the above quantities by

$$I_2(T) = \frac{8\pi}{c}\{64\pi d_{33}I_1 l_c/[\lambda_1(n_1^e + 1)^2(n_2^e + 1)]\}^2$$

$$\times [(1 - a)^2 + 4a\sin^2(\pi l/2l_c)] \tag{7.30}$$

where I_1 and I_2 represent the intensities of the primary and the secondary radiations outside the crystal respectively and c is the speed of light.

The term $a \approx \exp(\alpha_2 l)$ (α_2 being the absorption coefficient for light at $\lambda_1/2$) gives the effect of second harmonic absorption which becomes important at elevated temperatures. The oscillations in $I_2(T)$ observed as the temperature is changed enable one to obtain $l_c(T)$ and $d_{33}(T)$. The temperature dependences of the latter two quantities are plotted in figure 7.2. With a 1.06 μm neodymium glass laser, it was possible to extend to the phase transition temperature.

The experimental data give the Curie temperature to be $T_c = 1195 \pm 15\,°C$. However, temperatures outside these limits have been observed with non-stoichiometric crystals and crystals containing impurities. The SHG data indicate unambiguously that the ferroelectric phase transition is not near 585 °C, as was announced by Ismailadze *et al* (1968) and Smolensky *et al* (1966). Since the coefficient d_{33} tends to zero as the temperature approaches the Curie point, it is concluded to be a second-order transition, in agreement with the heat capacity data. The paraelectric phase is uniaxial with the same optic axis as the ferroelectric phase and exhibits no SHG, indicating that it is probably centrosymmetric. The optical properties of the paraelectric phase are consistent with the space groups proposed on the basis of X-ray data, i.e. R$\bar{3}$ or R$\bar{3}$c. These data do not allow one to distinguish between these groups (Abrahams *et al* 1966a,b,c). However, the general theory of ferroelectricity (Aizy 1965) suggests that it is R$\bar{3}$c space group for lithium niobate. The data on $d_{33}(T)$, $B(T)$ and $n(T)$ show that the unique nonlinear properties of lithium niobate can be utilized through thermal tuning of its indices in interesting frequency regions.

Bebchuk *et al* (1968) have endeavoured to determine the phase-matching angle and the second harmonic yield as functions of the primary radiation power.

The block diagram of the experimental arrangement is shown in figure 7.3. A Q-switched neodymium laser was used and its output pulses were

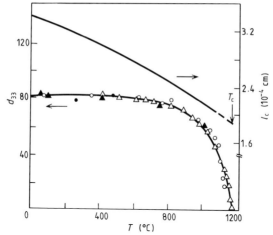

Figure 7.2. Temperature dependences of the coherence length $l_c = \lambda_1/4(n_{o2} - n_{o1})$ cm and of the nonlinear optical coefficient d_{33} for a lithium niobate crystal (the d_{33} value has been measured relative to d_{36} of KDP) (according to Miller and Savage 1966).

amplified in a single-stage amplifier. The power emitted in a pulse was $P_1 = 120$ MW, the beam cross-sectional area was $S = 0.8$ cm^2, its horizontal divergence was $\varphi_x = 4 \times 10^{-3}$ while the vertical divergence was $\varphi_y = 7 \times 10^{-3}$. More than half of the laser output power was contained in a component with a horizontally polarized electric field E_1. To study harmonic generation in an unfocused beam, the scheme of figure 7.3(a) was useful where the beam was additionally attenuated by a neutral glass filter.

Harmonic generation in a focused laser beam was investigated in the arrangement of figure 7.3(b), with a focusing lens put in place of the aperture.

While studying harmonic generation, the authors measured the phase-matching angle and width, and the conversion efficiency in both an unfocused and a focused laser beam. They employed both single- and multidomain lithium niobate crystals, which were made in the form of prisms cut from crystals such that their z axis was normal to the end face and the x and y axes were normal to the side faces. The $+z$ direction was obtained from the rate of etching at the end faces (the negative end is liable to faster etching). The $+y$ direction was identified from the form of an etch pit (the y axis points to the angle of a triangular etch pit).

Measurements were taken for four phase-matching directions: two in the plane xz ($\pm \theta_{mx}$ phase-matching) and two in the plane yz ($\pm\theta_{my}$ phase-matching). The angle was measured from the $+z$ direction:

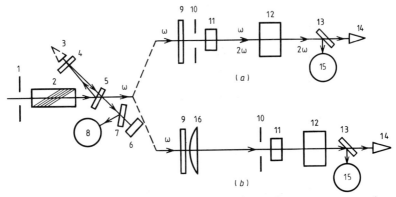

Figure 7.3. Schematic diagram of the experimental arrangement used to study the conversion efficiency in an unfocused (*a*) and a focused (*b*) beam (according to Bebchuk *et al* 1968): 1, 10, apertures; 2, pile of glass plates; 3, reference calorimeter; 4, 5, 7, 13, deflecting plates; 6, phase-matching sensor; 8, 15, photomultiplier; 9, neutral filter; 11, crystal studied; 12, CuSO$_4$; 14, measuring calorimeter; 16, lens.

$+\theta_m < 90°$, $-\theta_m > 90°$. In the experiment, they determined the relationship of the harmonic signal amplitude to the angle of crystal turning relative to the laser beam direction. The obtained angular distributions of the harmonic radiation plotted in figure 7.4 provided the phase-matching width (at half-maximum) and the phase-matching angle

$$\theta_m = 90° - \tfrac{1}{2}(\Delta\theta_1 + \Delta\theta_2)$$

where $\Delta\theta_1$ and $\Delta\theta_2$ are the phase-matching angles measured from the normal to the crystal face from which the harmonic is emitted in the xz and yz planes, respectively. The error of θ_m is within $\pm 20'$ and depends upon the beam-divergence to phase-matching width ratio.

Measurements have shown that in the crystals studied the θ_m value lies within an interval 75 to 87°, $\theta_{mx} = \theta_{my}$ and the multidomain structure is not found to have any bearing on the magnitude of the phase-matching angle. The latter is calculated according to the formula

$$\theta_m = \sinh^{-1}\{n_2^e/[n_1^o(n_1^{o2} - n_2^{o2})/(n_2^{e2} - n_2^{o2})]\}^{1/2} \qquad (7.31)$$

to be $\theta_m = 80°$. This value was calculated on the basis of the refractive indices cited in table 6.1, which are $n_1^o = 2.2336$, $n_2^o = 2.3225$, $n_2^e = 2.3212$ for $\lambda_1 = 1.06\ \mu m$ and $\lambda_2 = 0.53\ \mu m$. It is noteworthy that the theoretical value of θ_m presented in table 7.2 is $\theta_m = 84°08'$.

The variations of the θ_m value with different crystals and the discrepancy between the calculated and experimental values are probably attributable to the slightly differing refractive indices as a result of various impurity contents in crystals.

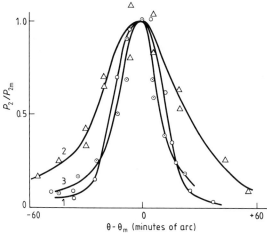

Figure 7.4. Angular distribution of the second harmonic in a single-domain crystal (curves 1, 3) and multidomain crystal (curve 2) of lithium niobate (according to Bebchuk *et al* 1968).

In single-domain crystals, the phase-matching width was measured to be $\delta\theta_m = 20\text{--}40'$, while in multidomain crystals this value is $\delta\theta_m = 1\text{--}1.5°$. Theoretically, the phase-matching width was found by the formula

$$\delta\theta_m = 5.6(\gamma l)^{-1} \tag{7.32}$$

where for lithium niobate the phase-matching parameter $\gamma = 3.4 \times 10^3$ $(\text{cm rad})^{-1}$, where l is the crystal length in the phase-matched direction.

For $l = 0.5$ cm, the calculated value is $\delta\theta_m = 3.3 \times 10^{-3}$ ($\approx 10'$), which is smaller than the experimentally determined value by a factor of 2 to 4. This may be caused by the crystal facets in which the directions of axes differ by $10'$ to $50'$.

The conversion ratio is $\eta = P_2^e / P_1^o$, where P_2^e is the extraordinary harmonic power and P_1^o is the power of polarized radiation incident onto a lithium niobate crystal as an ordinary wave. Over an energy range $0\text{--}0.1$ MW, the conversion ratio increases linearly with the primary radiation power P_1^o. At higher energies, the conversion ratio is smaller because of harmonic saturation at high radiation intensities. For instance, the power $P_1^o = 5$ MW is related to $\eta = 4 \times 10^{-2}$, whereas at P_1^o we have $\eta = 0.16$. In the latter case, lithium niobate crystals become yellow and burnt through; their surface is damaged and shows minute fractures.

The conversion ratio in a focused beam was measured using the version of figure 7.3(*b*). A lithium niobate crystal was placed in a convergent beam at a distance of 500 mm from the lens in order to preclude its burning through.

The measurement results on η are given in figure 7.5 presenting, for the sake of comparison, data for KDP and ADP crystals. As a glance at the figure shows, the conversion ratio goes up rapidly in a single-domain crystal and runs to a value $\eta = 0.25$ at a primary radiation power of 5 MW. It is evident from these results that ratios of at least 0.25 are attainable in focused beams. The values of the nonlinear optical coefficients responsible for SHG are tabulated in table 7.3. Bjokkholm (1968) has found out that the nonlinear optical coefficients d_{31} and d_{22} in lithium niobate crystals have different signs. This finding is of prime significance for optimizing nonlinear interaction of light waves in lithium niobate, which is determined by the relative signs of d_{ij}. Once a lithium niobate crystal is cut in such a manner that the interacting waves are phase-matched in the plane yz, the effective nonlinear optical coefficient has the form

$$d_{\text{ef}} = d_{31} \sin (\theta_{\text{m}} + \rho) \pm d_{22} \cos (\theta_{\text{m}} + \rho) \qquad (7.33)$$

where θ_{m} is the phase-matching angle and ρ is the angle of double reflection (Boyd *et al* 1965). The plus sign corresponds to the case where the beam propagation direction lies in the first quadrant containing the $+z$ and the $+y$ axes or in the third quadrant (positive

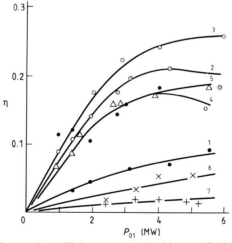

Figure 7.5. Conversion efficiency versus incident radiation power in LiNbO$_3$ crystals exposed to a lens-focused laser beam at a low level of input power ($P_{01} < 6$ MW). Lithium niobate crystals; 1, $l = 0.12$ cm, $+\theta_{\text{my}}$; 2, $l = 0.53$ cm, $+\theta_{\text{my}}$; 3, $l = 0.43$ cm, $+\theta_{\text{my}}$; 4, $l = 0.73$ cm, $+\theta_{\text{my}}$; 5, $l = 0.57$ cm, $+\theta_{\text{my}}$; 6, single-domain KDP, $l = 0.6$ cm, ooe, $+\theta_{\text{m}}$; 7, $l =$ multidomain ADP, $l = 0.5$ cm, ooe (according to Bebchuk *et al* 1968).

quadrants). For the second and fourth quadrants (negative quadrants) the sign is minus. The relative signs of d_{31} and d_{22} are determined by the direction choice of the z and y axes.

Table 7.3. Relative and absolute values of nonlinear optical coefficients of lithium niobate (according to Räuber 1978).

λ (μm)	d_{33}	$d_{31} = d_{15}$	d_{22}	Reference
Relative to d_{36}(KDP)				
1.058	107 ± 20	11.9 ± 1.7	6.3 ± 0.6	Boyd et al (1964)
1.058	83 ± 21	11.9 ± 1.7	6.3 ± 0.6	Ismailzade et al (1968)
1.15	—	10.9 ± 1.7	5.4 ± 2.2	
1.06	7.2 ± 7	11.6 ± 1.2	5.6 ± 0.6	Miller et al (1971)
1.06	54.5			Miller and Nordland (1970)
Relative to d_{33} (LiIO$_3$)				
2.12		4.24 ± 0.43		Choy and Byer (1976)
1.328		4.66 ± 0.56	0.87 ± 0.06	Choy and Byer (1976)
Absolute values (10^{-12} m V^{-1})				
0.488			6.18 ± 0.68	Choy and Byer (1976)
1.06	3.44	5.95		
1.15	33.4	5.77		
1.318	31.8 ± 6.4	5.54 ± 0.61		
2.12	29.1 ± 5.2			

In practice a standard piezoelectric method is employed: the electrodes corresponding to the $+z$ and $+y$ directions will carry a negative charge, with compression applied along these axes. For the lithium niobate crystal, the value of d_{ef} is found to be $1.13 \pm 8\%$. This is indicative of the opposite signs of the coefficients d_{31} and d_{22}.

To exemplify the correct consideration of the signs of the nonlinear optical coefficients d_{ij}, we will take a lithium-niobate light parametric oscillator pumped by a 0.6943 μm ruby laser (Giordmaine and Miller 1965). The ratio $d_{22}/d_{31} \simeq 0.53$ suggests that the d_{ef}^2 value in the negative quadrant is larger by a factor of 6.8 than in the positive one. On the other hand, the oscillation threshold for a phase-matching angle in the positive quadrant is higher by a factor of 6.8 than that in the negative quadrant. The importance of the signs with d_{ij} was also pointed out by Byer et al (1974).

Here it should be stressed that while making nonlinear optical elements for light parametric oscillators it is essential to fix correctly on the crystallographic direction of a seed crystal. The direction of an applied field in crystal polarization is of no importance.

7.3 Relationship of birefringence and phase-matching temperature to lithium niobate crystal composition

Several works report that the Curie temperature, birefringence and phase-matching temperature of lithium niobate are determined by the melt composition from which crystals are grown.

Bergmann *et al* (1968) changed the mole ratio $R = Li/Nb$ in the melt from 0.8 to 1.2 by introducing either Nb_2O_5 or Li_2CO_3 to stoichiometric $LiNbO_3$. Dielectric constants were measured as a function of temperature using a capacitance bridge at a frequency of 1 kHz. Crystal birefringence was determined with prisms and a quartz compensator at a wavelength $\lambda = 6328$ Å.

It has been established that the Curie temperature varies from 1190 °C for maxium lithium content to 1170 °C for samples pulled from a niobium-rich melt composition. Figure 7.6 shows the Curie temperature versus melt composition. These data are consistent with the earlier results of Smolensky *et al* (1966) on the Curie temperature as well as dielectric anomaly along the *c* axis. The same figure also presents the

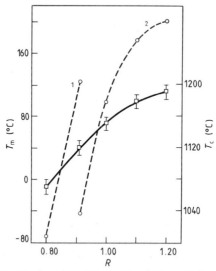

Figure 7.6. The Curie temperature T_C (full curve) and the phase-matching temperature T_m (broken curves) of lithium niobate versus melt composition from which the crystals have been grown: $\lambda = 1.15 \, \mu m$ (curve 1) and $\lambda = 1.08 \, \mu m$ (curve 2) (according to Bergmann *et al* 1968).

phase-matching temperature versus melt compositions for wavelengths 1.08 and 1.15 μm. It is evident from the plots that the higher the content of lithium ions in the melt, the higher the phase-matching temperature.

It is remarkable that the composition of a pulled crystal differs from that of the melt. For instance, as the R value in the melt varies from 0.8 to 1.2, the value of this ratio in the crystal changes from 0.96 to 1.04. This fact is consistent with the variations in the lithium niobate lattice parameter.

The change in the phase-matching temperature T_m may be related to the melt composition $R = \text{Li}/\text{Nb}$ as follows. As already mentioned, for the primary wave incident normal to the crystal optic axis to be phase-matched with the excited second harmonic wave, the crystal birefringence $B = n_2^o - n_2^e$ should be equal to the dispersion $D = n_2^o - n_1^o$, i.e. we should have $B - D = 0$. The variations in the value $(B - D)$ depending on the composition (R) should be compensated by the variations in the phase-matching temperature T_m, so we can write

$$\frac{\partial(B - D)}{\partial T}\Delta T = -\frac{\partial(B - D)}{\partial R}\Delta R. \qquad (7.34)$$

Under the assumption that the birefringence B is determined solely be the melt composition (i.e. $\partial D/\partial R = 0$), the plots of figures 7.6 and 6.3 enable us to determine the slope in the vicinity of the composition $R = 1$ (for $\Delta R = 0.1$), $\Delta B = 0.0072$, and $T_m = 122\,°C$:

$$\frac{\partial(B - D)}{\partial T} = \frac{0.0072}{122} = -5.9 \times 10^{-5}\ \text{K}^{-1}.$$

This value compares fairly well with the above value $\partial(B - D)/\partial T = -5.7 \times 10^{-5}\,°C^{-1}$ obtained for other crystals using another method. The linear relation $T_m = f(R)$ is violated at $R > 1.1$, maybe because of an increase $\partial D/\partial R$. Besides, the ratio R may have become saturated in a crystal with an increasing Li content in the melt. These findings are significant from another standpoint as well. It is common knowledge that lithium niobate crystals exposed to laser radiation may suffer from an inhomogeneous refractive index, a photo-refraction phenomenon to be discussed in chapter 8. It is also known that induced Δn is effectively annealed at 170 °C. Thus, by varying the melt composition we can specify a temperature $T_m > 170\,°C$ at which laser-induced inhomogeneities of the refractive index will disappear.

Fay *et al* (1968) examined a series of melts with the following equimolar ratios of Li_2O to Nb_2O_5: $R = 0.95, 1.00, 1.05, 1.10, 1.15$ and 1.20, in terms of the Li_2O molar fraction in the melt 0.487, 0.500, 0.512, 0.535 and 0.546 respectively.

Rectangular parallelepipeds were cut from crystals with their faces normal to the x, y and z axes. The radiation source was provided by a pulsed YAG:Nd laser with a pulse repetition frequency of 1200 Hz, with a mechanical Q-switch for enhancing the peak power at $\lambda = 1.06\ \mu m$ and producing a phase-matching signal. The output laser beam was focused onto a lithium niobate crystal which was oriented in a manner that the primary radiation propagated as an ordinary ray. The phase-matching temperatures versus the molar fraction of $Li_2O(x)$ in the liquid phase are plotted in figure 7.7. The dependence is roughly linear, increasing with the content of Li_2O in the melt. The straight line was obtained by the least-squares method and it can be approximated by the equation

$$T_m = 50.5 + 32 \times 10^2(x - 0.50). \tag{7.35}$$

The standard deviation from the straight line is $7\,^\circ C$.

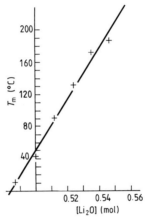

Figure 7.7. Phase-matching temperature versus melt composition for lithium niobate crystals grown under controlled conditions ($\lambda = 1.06\ \mu m$) (according to Fay *et al* 1968).

The solid phase, being at equilibrium with the melt, is not stoichiometric but instead has a variable composition that cannot be specified in the phase diagram. Negligible as they are, the departures in the solid phase cannot be identified in the course of chemical analysis, although the change in the phase-matching temperature may be significant. The latter circumstance is attributed to crystal defects (see chapter 3).

Thus the exact mechanism underlying phase-matching temperature variability with melt composition is not yet understood entirely. It is worthwhile noting that the extraordinary refractive index is more liable to variations with any departure from lithium niobate stoichiometry than the ordinary index. This may be due to the relationship of the extraordinary refractive index to the lithium ion vacancy concentration. The lower the vacancy concentration upon introduction of excess Li_2O into the melt, the higher the value of n_e and hence the lower the value of birefringence ($n_o - n_e$).

Midwinter and Warner (1967) studied the effect of the melt chemical composition on refractive indices and SHG of lithium niobate single crystals grown from these melts. Measurements were taken on crystals grown from four melt compositions—stoichiometric lithium niobate and those containing 0.5 wt.% Li_2CO_2, Nb_2O_5 and MgO respectively.

The crystal plates which were refractive-index measured were used subsequently for second harmonic generation. The light source was a He–Ne laser tuned to a wavelength of 1.08 μm. The SHG power was plotted as a function of temperature with an x–y recorder, so the phase-matching temperature T_m was easy to register.

Midwinter came to the conclusion that crystals containing minimum impurities and exhibiting exact stoichiometry have a minimum phase-matching temperature.

Table 7.4 cites calculated values of T_m for various wavelengths, drawing on the refractive indices values presented in table 6.1. The data of table 7.1 and the values of dn/dT, dB/dT, $dn/d\lambda$ calculated therefrom suggest that the variations due to variable melt compositions are complex and cannot be described solely in terms of crystal birefringence variability.

Table 7.4. Calculated phase-matching temperatures for second harmonic generation in lithium niobate (according to Midwinter and Warner 1967).

λ_1 (μm)	λ_2 (μm)	λ_3 (μm)	T_m (°C)
1.0840	1.0840	0.5420	90.8
1.0580	1.0580	0.5290	26.8
1.0590	1.0590	0.5295	29.6
1.0600	1.0600	0.5300	32.3
1.1520	1.1520	0.5760	209.3
1.1620	1.1620	0.5810	223.2

Byer *et al* (1970) announced that in their lithium niobate crystals the temperature of annealing-induced variations in the refractive index is reduced from 180 to 110 °C. Lower annealing temperatures permit growth of warm phase-matching and good optical quality crystals of lithium niobate suitable for efficient second harmonic generation at

$\lambda = 1.064\ \mu$m of a YAG:Nd laser. The mean SHG efficiency of such crystals reaches 40%.

The temperature of annealing optical inhomogeneities can be lowered by varying the crystal composition. Figure 7.8 is a plot of the phase-matching temperature T_m versus contents of the Li_2O and MgO oxides in the melt, where the iron impurity was carefully reduced prior to crystal growth.

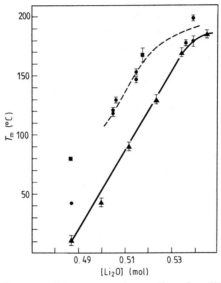

Figure 7.8. Phase-matching temperature T_m of a lithium niobate crystal for $\lambda = 1.06\ \mu$m versus molar fraction of Li_2O and MgO in the melt. The crystals have been grown from: ■, 2% MgO; ●, 1% MgO; ▲, pure lithium niobate (according to Byer *et al* 1981).

7.4 Criteria for nonlinear optical quality of crystals

Lithium niobate crystals exhibit, along with random variations in the refractive indices due to temperature and thermal stress fluctuations, regular birefringence gradients along the crystal length. This circumstance is responsible for a 1–3 °C change in the phase-matching temperature per cm of crystal length. This effect is presumed to result from the separation of the melt components at the melt–solid interface, which contributes to crystal inhomogeneity.

As already mentioned in the foregoing discussion, variations in the refractive index with crystal composition provide a very sensitive method for checking slight variations in the composition of lithium niobate crystals. For instance, a change of birefringence by 2×10^{-5} is easy to define by viewing a 1 cm long sample between crossed polarizers. This corresponds to a 10^{-3} wt.% change in one component of the melt. The sensitivity of chemical analysis is usually two orders of magnitude lower. However, the optical method yields no information directly on which component is varying, i.e. it allows measurement of effects only as a whole.

In chapter 2 it was already mentioned that congruently grown crystals of lithium niobate have bright prospects. Under the congruent growth conditions, there is no constant gradient of excess lithium ions along the crystal length and temperature fluctuations at the interface have a smaller bearing on the occurrence of local inhomogeneities in a growing crystal.

Birefringence induced by the defects occurring in the course of crystal growth is typically studied with an arrangement similar to that of figure 7.9. A 6328 Å laser beam polarized at an angle of 45° to the crystal optic axis excites both an ordinary and an extraordinary wave which undergo a relative phase shift in proportion to the path and birefringence at each point of the crystal in the plane normal to the beam. The resultant elliptically polarized light is passed through an analyser in order to reproduce the field with intensity variations corresponding to the birefringence changes in the crystal. To carry out a quantitative analysis, the surface of a studied sample is polished to have a gentle slope. In this case, the light traversing the crystal provides an interference pattern in the form of dark fringes corresponding to the contours of the surfaces of equal birefringence spreading normal to the wedge plane. Slight variations in the position and direction of displacement of

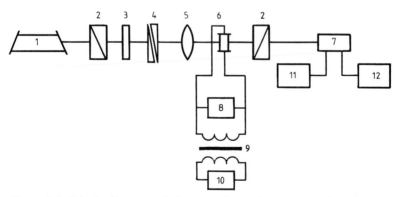

Figure 7.9. Block diagram of the experimental set-up used to examine crystal quality: 1, 0.63 μm He–Ne laser; 2, polarizers; 3, $\frac{1}{4}\lambda$ wave plate, 4, compensator; 5, lens; 6, crystal studied, 7, radiation detector; 8, kilovolt-meter; 9, high-voltage auto-transformer; 10, voltmeter; 11, recorder; 12, detector supply unit.

the interference fringes are indicative of birefringence inhomogeneities in the crystal. A 5×5 mm crystal with a slope of $10'$ displays two interference fringes. The lithium niobate crystals examined by Byer *et al* (1970) had been grown along the *a* axis and the crystal orientation permitted estimation of birefringence variations along the crystal growth.

The obtained data are very useful for establishing variations in the growth conditions that may occur during crystal pulling.

Figure 7.10 gives an interferogram of variations in the birefringence of crystals grown from stoichiometric and congruent melts. The stoichiometric crystal has a highly chaotically variable birefringence along the growth direction. As discussed earlier, these inhomogeneities are the result of slight temperature fluctuations at the solid–melt interface during crystal growth. The resultant birefringence in-homogeneities render this crystal unsuitable for SHG phase-matching. The same figure also shows an interference pattern for a congruently grown crystal. Although the birefringence testing is helpful to qualify the extent of optical inhomogeneity and its distribution in the bulk of a crystal, it cannot really determine a definitive answer about crystal applicability for SHG purposes.

Twyman–Green interferometric investigation of a crystal in monochromatic light yields information about its total optical length (Zernike and Midwinter 1973) defined by

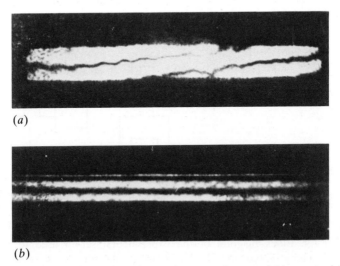

(*a*)

(*b*)

Figure 7.10. Birefringence of crystals grown from stoichiometric (*a*) and congruent (*b*) melts; at a length of 40 mm the birefringence remains at a constant level (according to Byer *et al* 1970).

$$l_{opt} = \int_0^L n(z)\,dz \tag{7.36}$$

where $n(z)$ is the refractive index as a function of the z coordinate in the observation direction and L is the crystal length.

If the crystal is uniaxial and observation is made in the plane normal to its optic axis, the optical length is determined by the polarization of light used for measurement. For ordinary–extraordinary interaction to occur, the phase-matching conditions of SHG require that the refractive indices of the fundamental and second harmonic waves of mutually perpendicular polarizations be equal. It is anticipated therefore that the relations

$$l_{opt}^o(\omega) = \int_0^L n^o(z,\,\omega)\,dz \tag{7.37}$$

$$l_{opt}^e(2\omega) = \int_0^L n^e(z,\,2\omega)\,dz \tag{7.38}$$

$$l_{opt}^o(\omega) = l_{opt}^e(2\omega) \tag{7.39}$$

are satisfied.

Consequently, a given crystal does not meet the requirement of phase-matching if the interferograms produced for ordinary and extraordinary rays for this crystal are different.

It is noteworthy, however, that if a crystal displays similar interferograms for ordinary and extraordinary rays with the two light polarizations, the crystal is not sure to be of a good optical quality even if it fits the linear optical requirements. The specific requirements imposed on crystals applied to nonlinear optics follow from the nature of the phase-matching process. Nonlinear interaction (second harmonic generation, parametric amplification or generation) is a process of travelling wave interaction where the energy is transferred in the direction determined by the interaction wave phase relationship at each point of the crystal. Phase-matching anticipates fine tuning of refractive indices for various frequencies and polarizations. In other words, if a desired phase relation is found at the input to a crystal, this relation should not be violated in subsequent wave propagation through the crystal. Once a refractive index for an ordinary or extraordinary wave at some point of a crystal is somewhat different from an exact phase-matching value, the phase relation of interacting waves will be violated and, for example, in SHG there will be generated a second harmonic wave with a phase differing from that of the primarily generated wave. As a result, the total second harmonic wave will prove attenuated. If the refractive index at a point is changed drastically and the phase shift of interacting waves runs to π, the direction of energy transfer will be changed too.

Hence, crystals employed in nonlinear optics should meet another, more stringent requirement—constancy of birefringence at each point of a crystal, i.e. $\partial(n^\circ - n^e)/\partial z = 0$. Along with condition (7.39) calling for a certain specific relationship of phases at the input and exit of a crystal, we have once more

$$\int_0^{L'} n^\circ(z, \omega)\mathrm{d}z = \int_0^{L'} n^e(z, 2\omega)\mathrm{d}z \qquad (0 \leqslant L' \leqslant L) \qquad (7.40)$$

requiring that the interacting wave phases be constant at each intermediate point.

This latter condition is difficult to comply with in practice since, as shown in chapter 3, the composition of lithium metaniobate crystals differs as a rule, from that of the melt they are grown from. That is why, to determine optical quality of lithium niobate crystals employed in nonlinear optics, the first step is to produce Twyman–Green interferograms for ordinary and extraordinary rays to check whether they are similar and, if they are, to undertake further interferometric studies to see if condition (7.40) is satisfied.

A more reliable way to check the crystal quality is to measure the second harmonic generation efficiency. This method has been most advantageously employed by several researchers (Midwinter 1968, Kovrighin *et al* 1972) and it relies on the shape of the curve plotting the second harmonic power versus temperature, the latter varying within the limits including the exact phase-matching temperature. For a perfect crystal, this curve virtually coincides with that of the function $(\sin^2 x)/x^2$.

If some regions of a crystal differ in the phase-matching temperature, this indicates that they also differ in composition and hence in birefringence. In that case, the central maximum of this curve will broaden and become lower, while the nearest secondary maxima will prove distorted. In this way, one can gain information about the mass value of birefringence inhomogeneity along the beam.

The technical aspects of SHG measurement aimed at determining crystal quality, as well as the experimental apparatus, are discussed in chapter 9.

To achieve a 90° phase matching in lithium niobate, the SHG power is defined (Byer *et al* 1970) by

$$P_{2\omega} = \frac{2\omega^2}{\pi} \frac{l^2}{r} \left(\frac{\mu_0}{\varepsilon\varepsilon_0}\right)^{3/2} d_{31}^2 P_\omega^2 \left(\frac{2n-1}{n}\right) \sin^2(\tfrac{1}{2}\Delta kl) \qquad (7.41)$$

where P_ω is the fundamental wave power, ω is the fundamental angular frequency, r is the beam radius, n is the number of arbitrary modes in laser pumping, Δk is the instantaneous mismatch factor and l is the crystal length or the coherence length. Equation (7.41) applies to wide focusing, $l/b < 0.4$, where b is the confocal parameter.

The width of a maximum of $[\sin^2(\Delta kl/2)]/(\Delta kl/2)^2$ enables the phase-matching length (coherence length) to be determined without measuring an absolute value of primary radiation to second harmonic conversion. It is problematic to measure the absolute conversion ratio, since one has to know an absolute energy of incident radiation, laser modal structure and radiation source stability, while the beam parameters should be measured.

To estimate an optical inhomogeneity of crystals from the temperature dependence of SHG, we will present some useful relations. Let us expand $D = (n_2^e - n_1^o)$ in the vicinity of the phase-matching temperature T_m into the Taylor series and confine ourselves to two terms

$$(n_2^e - n_1^o)_T = (n_2^e - n_1^o)_{T_m} + (T_m - T)d(n_2^e - n_1^o)/dT \quad (7.42)$$

where n_1^o is the ordinary refractive index for the fundamental radiation and n_2^e is the extraordinary refractive index at a double frequency. At the phase-matching temperature, we have $(n_2^e - n_1^o)_{T_m} = 0$. Hence, it follows that

$$T_m - T = -\frac{(n_2^e - n_1^o)_T}{d(n_2^e - n_1^o)/dT}. \quad (7.43)$$

The mismatch vector

$$2k_1 - k_2 = \Delta k$$

is related to

$$D = (n_2^e - n_1^o) = -\Delta k\lambda_1/2\pi \quad (7.44)$$

where λ_1 is the fundamental wavelength.

The previous expressions yield

$$T_m - T = \frac{\lambda_1 \Delta k}{2\pi d(n_2^e - n_1^o)/dT}. \quad (7.45)$$

For a perfect crystal, the second harmonic power is fitted by the function

$$[\sin^2(\Delta k L/2)]/(\Delta k L/2)^2 \quad (7.46)$$

which becomes twice as small at $\Delta k L/2 = \pi/2.25$. Thus the total width at a half maximum is given by

$$\Delta T = \lambda_1/[2.25 L d(n_2^e - n_1^o)/dT]. \quad (7.47)$$

For lithium niobate, $d(n_2^e - n_1^o)/dT = 8.06 \times 10^{-5} \ K^{-1}$, $\lambda_1 = 1.15 \ \mu m$; L is the crystal length along the beam propagation.

Thus, for a crystal with no induced variations in the refractive index, we have the following relations for $T \ (K \, cm^{-1})$:

$$\Delta T = 0.63/L. \quad (7.48)$$

This suggests that the value $(L\Delta T)$ remains at a constant level (ΔT

being the temperature difference at a half SHG power per unit crystal length). Knowing the value of ΔT for a perfect crystal, we can determine the effective length L_{ef} of a studied crystal, which determines the SHG power

$$L_{ef} = (0.63/\Delta T_{ex})L_{ex}. \qquad (7.49)$$

Theoretical values of ΔT have been calculated from the lithium niobate refractive indices found by Hobden and Warner (1966) and are presented in table 7.5. We should like to note here that Hobden and Warner studied lithium niobate crystals grown from stoichiometric melts. Table 7.5 also contains phase-matching temperatures for congruently grown lithium niobate crystals for which calculated values of ΔT coincided with experimental ones.

Table 7.5. SHG parameters for a 90° phase matching in LiNbO$_3$ (according to Hobden and Warner 1966).

Pumping wavelength (μm)	$(\partial n_1^o/\partial T)_{T_m}$ $\times 10^{-5}$	$(\partial n_2^e/\partial T)_{T_m}$ $\times 10^{-5}$	T_m (°C) Stoichiom.	T_m (°C) Congruent.	ΔT (°C)
1.064, YAG:Nd laser	0.24	6.08	47	−8	0.81
1.084, He–Ne laser	0.27	6.88	92	42	0.73
1.152, He–Ne laser	0.29	8.35	210	172	0.64

Crystal quality can be also characterized by the ratio of the first secondary maximum of a phase-matching curve to the central maximum, and by the effective phase-matching width governed by the ratio of the area under a phase-matching curve to the central maximum amplitude (Voronov *et al* 1974).

The criterion based on the ratio of the first secondary maximum to the central one loses sense where the phase-matching curve is asymmetric. Occurrence of additional substantial maxima and asymmetry of the phase-matching curve are accounted for by the presence of a longitudinal optical inhomogeneity. Various regions of a crystal inhomogeneity have different phase-matching temperatures.

Within an approximation of a specified pumping field, Ivanova (1979) has succeeded in relating L_{ef} (L, ΔB) to an optimized crystal length L_{opt} (ΔB) for second harmonic generation, ΔB being the variation in birefringence along the entire crystal. The specified-field approximation is valid at low parametric conversion coefficients. Since the parametric gain is roughly equal, in a small-gain approximation, to the efficiency of coversion to a second harmonic at the frequency of a parametrically amplified signal (Harris 1969), this calculation holds good also for a parametric amplifier.

If the amplitude of the second harmonic wave is represented in the form (Nash *et al* 1970)

$$E^{2\omega} \sim \int_0^L \exp\left(\int_z^L - i\Delta k \, dz'\right) dz \qquad (7.50)$$

then the intensity of the second harmonic leaving the crystal, in the presence of a wave mismatch Δk, is written as follows:

$$S^{2\omega}(L) \sim \left[\int_0^L \exp\left(\int_z^L - i\Delta k \, dz'\right) dz\right]\left[\int_0^L \exp\left(\int_z^L i\Delta k \, dz'\right) dz\right]. \quad (7.51)$$

Assuming that B is a linear function of the crystal length, the phase mismatch Δk is

$$\Delta k = \Delta k' + (\delta k/L)z' \qquad (7.52)$$

where

$$\Delta k = 4\pi B/\lambda \qquad (7.53)$$

λ being the pump wavelength.

Integration in expression (7.51) yields

$$S^{2\omega}(L) \sim \frac{\pi L^2}{4\varphi(L)}\left\{ \mathrm{erf}\left[(1 - i)\left(\sqrt{\frac{\phi(L)}{2}} + \frac{\varphi(0)}{\sqrt{2\varphi(L)}}\right)\right]\right.$$

$$\left. - \mathrm{erf}\left((1 - i)\frac{\varphi(0)}{\sqrt{2\varphi(L)}}\right)\right\} \qquad (7.54)$$

where

$$\mathrm{erf}\, x = \frac{2}{\pi}\int_0^x e^{-t^2} dt$$

and

$$\varphi(0) = \Delta k' L/2$$

$$\varphi(L) = \delta k L/2 = (2\pi L/\lambda)\Delta B.$$

$\varphi(0)$ and $\varphi(L)$ stand for the initial and the phase mismatch due to the nonlinear crystal inhomogeneity.

It is evident from equation (7.54) that for $\varphi(L) = 0$

$$S^{2\omega} \sim L^2(\sin^2\varphi(0))/\varphi^2(0)$$

i.e. with no birefringence variations along the crystal ($\Delta B = 0$), the second harmonic intensity curve will have an idealized form. In the presence of inhomogeneities in a crystal, resulting from birefringence variations along its length, the shape of the curve is radically changed. Figure 7.11 gives the plots of normalized intensity $S^{2\omega}_{\max}[L, \varphi(L)]/S^{2\omega}_{\max}[L, 0]$ versus initial phase mismatch $\varphi(0)$ for various extents of the phase mismatch $\varphi(L)$ caused by the crystal birefringence change

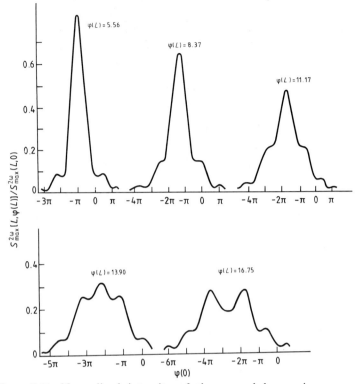

Figure 7.11. Normalized intensity of the second harmonic versus $\varphi(0)$ for various values of $\varphi(L)$ (according to Ivanova 1979).

ΔB. One can see that the larger the crystal inhomogeneity, the higher the distortions of the curve $S^{2\omega}$.

The maxima of the curves plotted in figure 7.11 were used to relate the normalized intensity $S^{2\omega}_{max}[\varphi(L)]/S^{2\omega}_{max}(0)$ to the value $\varphi(L)$.

Since $S^{2\omega} \sim L^2$, the following relation

$$S^{2\omega}(L_{ef})/S^{2\omega}(L) = L^2_{ef}/L^2 \tag{7.55}$$

is the case.

Figure 7.12 shows the L dependence of L_{ef} obtained from relation (7.61) with the use of $\varphi(L)$:

$$\varphi(L) = \frac{2\pi}{\lambda} \frac{\partial B}{\partial L} L^2 \tag{7.56}$$

where $\varphi = 1.06 \ \mu$m and $\partial B/\partial L$ is the time rate of birefringence change along the crystal, which serves as a parameter of the curves and which is commonly termed a parameter of inhomogeneity.

It is evident from figure 7.12 that at a certain specific crystal length the effective length attains its maximum value L^{max}_{ef} and then begins to

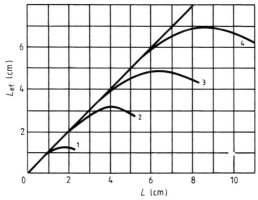

Figure 7.12. Effective length L_{ef} of a crystal versus its geometric length L for $\lambda = 1.06\ \mu m$ and various values of the inhomogeneity parameter $\partial B/\partial L$: 6×10^{-5}; 1×10^{-5}; 4×10^{-6}; $2 \times 10^{-6}\ cm^{-1}$ (curves 1–4 respectively) (according to Ivanova 1979).

reduce. So, to achieve a specified effective length, the inhomogeneity parameter and the crystal length should exceed certain values. The curves of figure 7.13 show that the ratio L_{ef}^{max}/L is virtually the same and equal to 0.82 for various inhomogeneity parameter values. Hence, we can optimize the length L_{opt} of a crystal possessing some inhomogeneity $\partial B/\partial L$:

$$L_{opt} = 1.22\ L_{ef}^{max}. \tag{7.57}$$

The relationships of L_{ef}^{max} and L_{opt} to the inhomogeneity parameter are depicted in figure 7.13. To make an efficient light parametric oscillator,

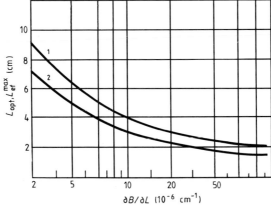

Figure 7.13. Optimal (curve 1) and maximal (curve 2) effective lengths of a crystal versus inhomogeneity parameter (according to Ivanova 1979).

one needs crystals of 4 to 6 cm in length. Proceeding from this length, one can estimate a maximum value of the inhomogeneity parameter. Figure 7.13 implies that for $L_{ef} \geqslant 4$ cm the inhomogeneity parameter should be $\partial B / \partial L \leqslant 6 \times 10^{-6}$ cm^{-1} and L_{opt} is 5 cm.

For relatively high optical quality single-domain crystals having no obvious defects, such as growth layers, fractures or striae, the inhomogeneity parameter may function as a convenient criterion for nonlinear crystal applicability to second harmonic generation and parametric generation.

It is remarkable that in lithium niobate type crystals, along with longitudinal inhomogeneity of birefringence, we may also deal with a transverse one characterized by an inhomogeneity length R_i (Midwinter 1967)

$$R_i = \frac{\lambda}{\pi[\partial(n_2^e - n_1^o)/\partial r]R_0}. \qquad (7.58)$$

The value $[\partial(n_2^e - n_1^o)/\partial r]R_0 = (\partial B/\partial r)R_0$ is roughly equal to the change of birefringence B at a distance equal to the radius R_0 of the pump beam cross section. In the case of a small transverse inhomogeneity, $R_i/R < 1$, where R is the crystal radius. It follows from the work by Butyaghin (1973) that

$$\frac{S^{2\omega}(R)}{S^{2\omega}(R_i)} \approx 1 - \frac{R^2}{6R_i^2}. \qquad (7.59)$$

For poor optical quality crystals of lithium niobate, the transverse inhomogeneity $\partial B/\partial r \sim 10^{-3}$–$10^{-4}$ cm^{-1} can be disregarded, if the pump beam diameter $R_0 \leqslant 200$ μm.

7.5 Enhancement of SHG in lithium niobate crystals with periodic laminar ferroelectric domains

In 1966 Bloembergen proposed a scheme for quasi-phase-matching by means of one-dimensional spatial periodic modulation of nonlinear susceptibilities, with a period just equal to $2l_c$, where l_c is the coherence length (Armstrong *et al* 1962, Bloembergen and Sievers 1972, Somekh and Yariv 1972). This scheme may be applied to non-birefringent crystals as well as to some birefringent crystals with nonlinear optical coefficients which cannot be phase-matched. Lithium niobate is a commonly used nonlinear optical crystal. However, its largest nonlinear coefficient d_{33} is not phase-matched, so it has not been utilized at all. Since d_{33} is about 7.5 times larger than d_{31}, which is the coefficient ordinarily used, if one fabricates a LiNbO$_3$ crystal with periodic laminar ferroelectric domains, alternate in polarization directions, and of a layer thickness equal to l_c, one may expect enhancement of SHG in a crystal quasi-phase-matched for d_{33}, compared to a single-domain crystal phase-matched for d_{31} of the same length. So there is great potential in exploiting quasi-phase-matching for d_{33} for applications in practical nonlinear optical devices.

In general the width of positive domains l_p and that of negative domains l_n are unequal; however, by adjusting the rotation and the pulling rates, the condition $l_p + l_n \simeq 2l_c$ may be achieved (Feng *et al* 1980).

To make d_{33} accessible, propagation and hence the stacking direction of the domains perpendicular to the c axis are required. The Bloembergen quasi-phase-matching (QPM) condition is met if the domain thickness d is an odd multiple of the coherence length l_c (Feisst and Kaide 1985):

$$d = (2m + 1)l_c \qquad (7.60)$$

giving the theoretical nonlinear optical efficiency, as compared with the 90° phase-matched (PM) process for the same interaction length, of

$$I_{QPM}/I_{PM} = [(2/\pi)(d_{33}/d_{31})]^2(l_c/d)^2. \qquad (7.61)$$

The nonlinear optical applicability of domain structures was tested by studying 1.06 μm pumped SHG. The crystal was irradiated by an acousto-optically Q-switched Nd:YAG laser beam under extraordinary polarization. The harmonic signal was recorded again under e polarization. The (e + e → e) process has a coherence length of 3.4 μm. It is driven by d_{33}. To generate a reference signal, part of the pump beam was split off and frequency-doubled in a temperature-controlled single-domain lithium niobate crystal under 90° type I phase matching. This reference signal was used to normalize the second harmonic intensity (SHI) of the single-domain crystal to correct for pump laser fluctuations.

Crystals with domain thicknesses slightly less than an odd multiple of the coherence length were prepared. Condition (7.60) was fulfilled by angle-tuning the crystal around the c axis. No temperature stabilization was required.

Figure 7.14 shows the SHI for two sets of crystals with $d/l_c = 5$ and 11 as a functional of the crystal length. It is seen that the SHI varies as the square of the domain number N, up to $N \approx 50$. This result indicates the high periodicity of the domain superstructures. The dependence of the nonlinear optical efficiency on domain thickness is shown in figure 7.15, where the SHI of a multidomain crystal, normalized to the 90° phase-matched SHI of a crystal of the same thickness, is plotted versus the domain thickness in units of l_c. A quadratic dependence on l_c/d is observed, as expected from (7.61). An extrapolation to an optimal case, $d = l_c$, yields a nonlinear optical efficiency 15 times larger than that of the 90° phase-matched process. This result is in agreement with equation

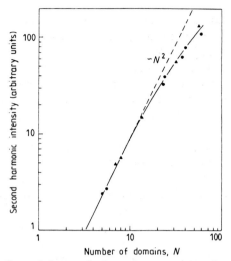

Figure 7.14. Second harmonic intensity versus number of domains of crystals with $d/l_c = 5$ (●) and 11 (▲) respectively (according to Feisst and Kaide 1985). $l_c = 3.4\,\mu$m.

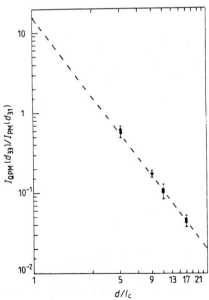

Figure 7.15. Second harmonic intensity of crystals with domain thickness d normalized to SHI of a 90° phase-matched single-domain crystal of equal length versus d/l_c (according to Feisst and Kaide 1985).

(7.61) which, based on the nonlinear optical coefficients of Miller *et al* (1971), predicts a theoretically possible efficiency enhancement by a factor of $[(2/\pi)d_{33}/d_{31}]^2 = 16 \pm 6$.

Lithium niobate is the dominating material for YAG:Nd laser parametric generation of near infrared radiation. The coherence lengths of these processes ($> 10\mu$m) are well within the region of possible domain thicknesses available with present-day techniques. For instance, parametric generation of a subharmonic at $2.12\,\mu$m (under $e \rightarrow e + e$ polarization) has a coherence length of $16\,\mu$m. Thus, a $1.06\,\mu$m pumped optical parametric oscillator would be well suited to use the possible gain enhancement offered by lithium niobate domain superstructures.

Blistanov *et al* (1986) studied light conversion into the second harmonic in lithium niobate crystals with a regular domain structure at a 90° phase-matching temperature for light propagating along the domain walls. Their experimental arrangement is similar to that used by Butyaghin (1973). A $1.06\,\mu$m 2 W single-mode CW YAG:Nd^{3+} laser was used. Measurements were taken of the phase-matching temperature T_m and of the temperature width of the phase-matching curve at a half-maximum level. For single-domain lithium niobate crystals, these values proved to be 45.4 and 0.9 °C respectively, while for regular domain structure crystals of the same dimensions and orientation they were 45.7 and 8.5 °C. The second harmonic power was observed to reduce by about 25%.

Temperature dependences of the SHG efficiency are plotted in figure 7.16. A comparison of the obtained results to the known relationship of the temperature width of the phase-matching curve to the optical quality of an inhomogeneous crystal permits a claim that the observed broadening of the temperature width of phase-matching is hardly related to the

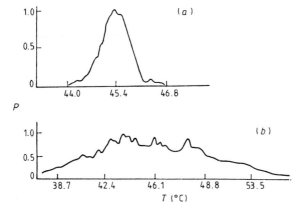

Figure 7.16. SHG efficiency versus temperature for a single-domain lithium niobate crystal (*a*) and for a crystal of a 40 μm regular domain structure (*b*) (according to Blistanov *et al* 1986).

transverse inhomogeneity of a regular domain structure lithium niobate crystal. The investigation of the peculiarities of light propagation in the vicinity of domain walls has revealed that transverse scanning of a regular domain structure lithium niobate crystal with a 0.63 μm helium–neodymium laser beam, focused onto the surface of the crystal by means of a micro-objective, made the light intensity near the domain interface at the exit face of the crystal jump suddenly by a factor of 7 to 10. The spatial period of such jumps corresponded to that of a regular domain structure. Thus it is presumed that the region adjacent to the domain walls possesses the properties of a planar waveguide. For unannealed lithium niobate crystals of a regular domain structure, the transverse dimension of the waveguide was measured indirectly and proved to be 8 μm at a refractive index jump $\Delta n \approx 10^{-3}$.

The marked enhancement of thermal stability of SHG observed in regular domain structure lithium niobate crystals is attributable to the joint effect of refractive index transverse inhomogeneity and waveguide properties of the domain walls. In the latter case, waves can be phase-matched thanks to the modal dispersion in a waveguide (Suematsu et al 1972). By confining the radiation near the domain walls, we can virtually retain the conversion efficiency.

8 Photoelectrical and Photo-refractive Properties

8.1 Model representations of the photo-refractive effect

It has already been mentioned that lithium niobate and lithium tantalate crystals exposed to laser radiation experience local variations in their refractive index Δn. After the irradiation has been discontinued, the crystal persists to have a region with an altered refractive index. This region is in the form of a 'track' and it may exist in the crystal for a long time. In the literature, this phenomenon goes under the names of induced optical inhomogeneity, optical distortion or photo-refractive effect (Ashkin *et al* 1966, Chen *et al* 1968).

Recently the phenomenon of optical distortion in crystals has received considerable attention of researchers in conjunction with its possible application to holographic recording of information.

In the suggested models, optical distortion is attributed to the migration of photo-excited electrons from an illuminated region and to the establishment of a space charge field which affects the refractive index due to the electro-optical effect. The electron migration from an illuminated region is related to the internal electric fields existing in ferroelectrics (Chen 1969), or to the field set up due to the variations of spontaneous polarization in an illuminated region of a crystal (Johnston 1970, Levanyuk and Osipov 1975), to the carrier diffusion (Amodei 1971a,b) to the photovoltaic effect (Glass *et al* 1974). Ferroelectric crystals fall under a class of materials registering the phase of a light wave. The virtues of holographic recording in ferroelectric crystals on the basis of the photo-refractive effect are the high diffraction efficiency and the possibility of re-recording (Amodei 1971a,b). As a rule, ordinary ferroelectric crystals have a low illumination responsivity. But their sensitivity can be enhanced markedly by doping, say, of Fe, Mn, Cu, Rh, Ce, V (Ishida *et al* 1972, Okamoto *et al* 1975). It is noteworthy that in most studies published thus far, optical distortion is investigated in Fe-doped crystals of lithium niobate with a view to their application in holography. Meanwhile the interaction of undoped lithium niobate crystals with large-diameter light beams extensively used in quantum electronics has remained largely unstudied. This is to be explained partially by the fact that, to clarify the mechanism of optical distortion, one has to carry out experiments on photoconductivity, photovoltaic current and other photoelectric phenomena. These tasks are rather problematic in undoped crystals exhibiting low dark conductivities (Bollman and Gernand 1972).

If lithium niobate crystals are used as nonlinear elements for second harmonic generation at $\lambda = 1.06\ \mu$m, SHG introduces an optical distortion in the element. The variations in the refractive index of an extraordinary ray will violate the phase-matching conditions and reduce abruptly the SHG efficiency of a laser (which was discussed in chapter 7). Optical distortions also give rise to distorted distributions of the second harmonic radiation both in time and space, thereby impairing the parameters of the converted laser radiation.

When lithium niobate crystals are employed as electro-optical elements of modulators and light deflectors, optical distortions enhance the residual luminous flux in an uncontrollable manner and reduce the contrast ratio. The effect of optical distortion upon the characteristics of nonlinear optical elements is made less pronounced by decreasing the laser radiation intensity and raising the operating temperature of an element.

Thus, optical distortions in lithium niobate crystals impose limitations on the application of this material as radiation frequency converters and electro-optical elements in modulators and light deflectors: they reduce the conversion efficiency in harmonic generators and contract the range of operating laser radiation intensities in light modulators and deflectors. By making ferroelectric crystals less sensitive to optical distortions one will be able to extend the limits of applicability of this material to quantum electronic devices and improve their efficiency.

Prior to taking up various models of the optical damage effect, we think it is good to consider electric fields that can exist in an idealized ferroelectric (Ohmori *et al* 1975). As already mentioned in the introduction, the internal field in a crystal is in the opposite direction to the spontaneous polarization vector. In an idealized ferroelectric crystal, the depolarizing field is $P_s/\varepsilon\varepsilon_0$, where P_s and ε, ε_0 are the spontaneous polarization and the dielectric constants of a crystal and vacuum respectively. Taking $P_s = 70\ \mu$C and $\varepsilon = 80$ for lithium niobate, we expect the depolarizing field to be of the order of 10^6 V cm^{-1}. In actual ferroelectric crystals, however, dielectric polarization at the surface is smaller due to the surface conductivity during high-temperature polarization, finite bulk conductivity and damage of surface domains and crystal surface during machining.

Figure 8.1(*a*) shows the configuration of a crystal and electrodes. Assume that in the surface layer $d/2$ spontaneous polarization behaves linearly (see figure 8.1(*b*)). We will introduce the following designations: $d/2$ is the surface layer depth, where spontaneous polarization undergoes variations, l is the sample thickness, $P_s^o(x)$ and P_s^s stand for spontaneous polarization in the bulk and at the surface of a crystal respectively, E is the electric field, $\rho_{c\,ch}$ is the compensating charge at the electrode surface, $\rho_{p\,ch}(x)$ is the polarization charge in the bulk of a crystal, and $\rho_{s\,p\,ch}$ is the surface polarization charge.

It follows from the Poisson equations that

$$\varepsilon_0(\operatorname{div} E) = \rho_{c\,ch} - \operatorname{div} P_s(x) = \rho_{c\,ch} + \rho_{p\,ch}(x)$$

$$\operatorname{div} P_s(x) = -\rho_{p\,ch}(x). \tag{8.1}$$

Equation (8.1) yields expressions for $\rho_{p\,ch}$, $\rho_{s\,p\,ch}$ and $\rho_{c\,ch}$:

$$\rho_{c\,ch}(x) = -\operatorname{div} P_s(x)$$

$$= \begin{cases} (P_s^o - P_s^s)/(d/2) & (l - d)/2 \leqslant x \leqslant l/2 \\ -(P_s^o - P_s^s)(d/2) & -l/2 \leqslant x \leqslant (l - d)/2 \end{cases} \tag{8.2}$$

$$\rho_{s\,p\,ch} = \mp \frac{P_s^s}{\varepsilon_0} \delta(x \pm l/2) \tag{8.3}$$

$$\rho_{c\,ch} \approx \rho_{c\,ch}\delta(x \pm l/2) \tag{8.4}$$

where $\delta(x \pm l/2)$ is the δ function with centres at $+l/2$ and $-l/2$. Let $P_s(x)$ change as shown in figure 8.1(*b*); the induced charge of a sample can be obtained from equations (8.3) and (8.4) (see figure 8.1(*c*)). For closed electrodes the potential difference between two electrodes is zero

$$\int_{-l/2}^{+l/2} E(x)\,\mathrm{d}x = 0. \tag{8.5}$$

The internal field in the bulk of a ferroelectric crystal E_i and the one in the surface layer E_b for shorted electrodes are obtainable from equations (8.2) and (8.5):

$$E_i = \frac{d}{\varepsilon_0(l - 2d)} (P_s^s - P_s^o) \tag{8.6}$$

$$E_b = -\frac{l - d}{\varepsilon_0(l - 2d)} (P_s^s - P_s^o) \tag{8.7}$$

where $P_s^o > P_s^s$, $l \gg d$. The fields E_i and E_b are schematically shown in figure 8.1(*d*), while the potential distribution in a sample is given in figure 8.1(*e*). The internal field E_i in equation (8.6) is directed opposite to the polarization P_s^o, since P_s^s is negligible compared to P_s^o. For a

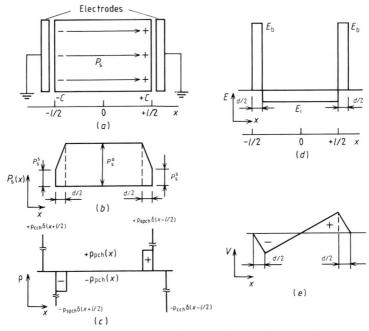

Figure 8.1. (*a*) Configuration of a crystal and electrodes. Distributions of spontaneous polarization (*b*), charge (*c*), electric field strength (*d*) and potential (*e*) in a ferroelectric crystal (according to Ohmori *et al* 1975).

lithium niobate crystal having thickness $l = 1$ mm, $d \approx 1\ \mu$m and $P_s^o \approx 70\ \mu$C cm^{-2}, we have $E_i \approx 10^5$ B cm^{-1}. This value is in reasonable agreement with experiment (~ 180 kV cm^{-1}) for an internal field of an undoped lithium niobate crystal. Thus, internal fields can exist in the lithium niobate crystal prior to its exposure to light.

The first model of the optical distortion phenomenon has been suggested by Chen (1969). He observed a steady-state current in a shorted crystal on its exposure to light with no external electric field applied, which is indicative of the presence of a constant internal electric field along the optic axis. Thus, the electron migration within Chen's model is determined by the drift of electrons in a constant internal electric field. Chen has established that for the refractive index to change by $\Delta n \sim 10^{-3}$ the space charge field should be around 6.7×10^4 V cm^{-1}. For a 2×10^{-2} cm diameter laser beam this electric field is provided by an electron density of 4×10^{14} cm^{-3}. Photo-excited electrons from impurity levels or from the valence band drift by the action of the internal electric field (migration process). The space charge field set up by the drift and subsequent electron retrapping introduces changes in the refractive index due to the electro-optical effect (see chapter 6). Chen's model is represented schematically in figure 8.2.

Johnston (1970) proposed a mechanism where variations in the refractive index are related to those of spontaneous polarization in an illuminated crystal due to a changed density of free carriers. A similar conclusion follows from the far earlier study by Fridkin (1966).

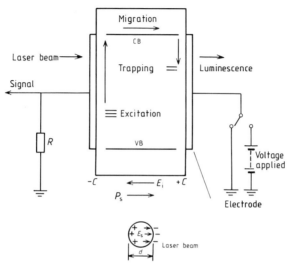

Figure 8.2. Schematic diagram of a change occurring in birefringence Δn in a ferroelectric crystal exposed to laser radiation: CB, conduction band, VB, valence band (according to Ohmori *et al* 1975).

The main claim of Johnston's model is that in a pyroelectric crystal the macroscopic polarization ΔP_s changes to the ionization of some traps only along the direction P_s.

Hence, photo-excitation of defects is responsible for spatial variations in macroscopic polarization in an illuminated region. The resultant polarization charge acts as a source of electric field forcing photo-excited electrons to drift from an illuminated region to an unilluminated one. In the steady state, an equilibrium establishes itself between the field related to the space charge and the field of the polarization charge. The refractive index is made to vary by the resultant field due to the electro-optical effect.

Later on Levanyuk and Osipov (1975, 1977) examined Johnston's model critically. Johnston relates ΔP_s to the variations in the impurity charge state, while it is common knowledge (Burfoot 1967) that the dipole moment of a system of impurities having the same sign charge is governed by the choice of the coordinate origin and hence is determined ambiguously.

Levanyuk and Osipov consider, in a sequential manner, the mechanisms basic to refractive index variations due to the changes in spontaneous polarization. These mechanisms include a charge exchange of impurity centres as well as a photo-induced transition between an equilibrium and a metastable ion configuration of the impurity centre.

The mechanism related to the charge exchange of impurity centres comes about as follows. Wide-band materials including lithium niobate are known to contain both donor- and acceptor-type defects (chapter 3). When subject to light, donor–acceptor pairs experience charge exchange: an electron from a deep negatively charged acceptor goes over to a positively charged donor (Yunovich 1972). The net charge of the pair is preserved but the charge exchange will affect the optical polarizability of a system and hence the refractive index of a medium. Since in a ferroelectric each defect or impurity centre is polarized by its environment and lies in the macroscopic field of spontaneous polarization $E = fP_s$ (f being the Lorentz factor), the defect dipole moment is also changed by $\Delta P_s = \chi f P_s$ (χ being the polarizability). Thus in an illuminated region the polarizability ΔP_s is changed and the electro-optical effect causes variations in the refractive index. Once a crystal is illuminated along the x or y axis, in which case the light spot is in the form of a strip perpendicular to P_s, a macroscopic depolarizing field arises, making photo-excited carriers drift into an unilluminated region of the crystal. In the steady state, the space charge field and the depolarizing field are in equilibrium. The refractive index variations are proportional to the density of charge exchange centres N_g^o, and the relationship of Δn to the radiation intensity and temperature is determined by the dependence of N_g^o upon these parameters.

In their theory, Levanyuk and Osipov have established relationships of optical damage to the laser radiation intensity and crystal temperature observed in experiment.

Amodei and Staebler (1971) applied Chen's model directly to explain the process of hologram recording. They took account not only of the carrier drift in the internal field but also of the diffusion of free electrons formed in photo-ionization. Calculations were performed for a simple interference pattern—sinusoidal distribution of light intensity produced by the superposition of two plane waves. The authors supposed that the carrier path was short and constant compared to the holographic pattern, and that the number of photo-excited electrons in the conduction band is far less than the total number of donor centres.

A comparison of the drift and carrier diffusion effects on the spatial distribution of the space charge resulting from the sinusoidal light intensity distribution in a crystal has revealed that for more than 100 cycles per mm of a diffraction grating it is the diffusion that is decisive in the formation of a holographic image.

Deighen *et al* (1975) and Vinetsky *et al* (1977) consider holographic recording on the basis of a steady-state solution of the system of equations describing the process. The solution is accomplished in a diffusion approximation and the spontaneous polarization variations are discarded.

The diffusion model approach to the solution has proved very fruitful and has enabled its authors to explain a number of interesting phenomena arising in hologram recording (Markov *et al* 1977). Here the authors would only like to stress that the diffusion model approach seems to be justified only in hologram recording for an illuminated region of several micrometers in size. If the illuminated region is around $100\,\mu$m or more in diameter the diffusion mechanism does not play a decisive role. A similar situation is encountered in quantum electronic devices where ferroelectric crystals are used as electro-optical elements and radiation frequency converters.

We will estimate the internal field induced by the diffusion mechanism of optical distortion (Ohmori *et al* 1975). For the sake of simplicity, assume that the light intensity distribution in a one-dimensional case along the axis is gaussian. It is well known that the density of the charge carriers formed is proportional to the light intensity and is defined by

$$\rho = \rho_0 \exp\left(-2x^2/r_0^2\right) \tag{8.8}$$

where ρ_0 is the maximum carrier density at the centre of a laser beam while r_0 is the laser beam radius.

The density of the diffusion current J_{dif} (and of the drift current J_{dr} equivalent thereto) can be expressed in the form

$$J_{\mathrm{dif}} = eD(\mathrm{d}\rho/\mathrm{d}x) = e\mu\rho E_{\mathrm{i}} = J_{\mathrm{dr}} \tag{8.9}$$

where D is the diffusion coefficient. Employing the Einstein relation

$$D = kT\mu/e \tag{8.10}$$

we will write an expression for a maximum field E_{i} for $x = \pm r_0/2$

$$E_{\mathrm{i}} = 2D/(\mu r_0) = 2kT/(er_0). \tag{8.11}$$

Taking $r_0 = 0.1$ and $T = 300$ K we find a maximum electric field intensity to be $E_{\mathrm{i}} \approx 0.3$ V cm^{-1}.

This value is negligible compared to that of an observed internal crystal field. Thus the diffusion current is not significant in the formation of an optical inhomogeneity in lithium niobate crystals, except for the case of very small r_0. This fact supports the claim of Amodei that the holographic grating is formed by the diffusion mechanism.

While studying the photovoltaic effect in iron- and copper-doped lithium niobate, Glass *et al* (1974) have established that the density of a steady-state photocurrent J is proportional to the laser radiation intensity I:

$$J = K_G \alpha I \qquad (8.12)$$

where K_G is the Glass constant and is dependent upon the nature of absorbing centres and wavelength and independent of the crystal geometry, electrode configuration or dopant content; α is the absorption coefficient.

The existence of a steady-state photocurrent with no external electric field applied is attributed by the authors to the different probabilities for charge carrier migration in the positive and negative directions of spontaneous polarization, which is due to the local asymmetry of the ferroelectric structure. The predominant migration of charge carriers in a certain direction results in charge separation and establishment of a space charge field which changes, through the electro-optical effect, the refractive index.

In an open steady-state crystal, the resultant current in an illuminated region is zero, i.e.

$$J = K_G \alpha I + \sigma E_i = 0 \qquad (\sigma = \sigma_{ph} + \sigma_d) \qquad (8.13)$$

where σ_d and σ_{ph} are the dark conductivity and the photoconductivity of a crystal respectively and E_i is the electric field strength in an illuminated region. Hence the steady-state value of an electric field in an illuminated region is defined by

$$E_i = K_G \alpha I / \sigma. \qquad (8.14)$$

For iron ion-doped lithium niobate crystals, the field strength is $E_i \approx 0.9 \times 10^5 \, \text{V cm}^{-1}$. Proof of the fact that optical distortion occurs due to the photovoltaic effect is provided by the relationship of experimentally measured factors of proportionality between the optical damage Δn and density of absorbed radiation energy, on the one hand, and the photovoltaic effect and absorbed energy density on the other.

Fridkin *et al* (1977) suggest two possible mechanisms of the photovoltaic effect. The first one is that the spontaneous polarization screening length may be comparable to the crystal length along its optic axis due to the low density of equilibrium carriers in an unilluminated region of a crystal. In this case the internal electric field in the bulk of a crystal is nonzero and it may cause, once the crystal is illuminated, a photovoltaic effect and optical damage. This mechanism of optical damage seems to make sense with thin plates of lithium niobate cut normal to the crystal polar axis.

The second mechanism relates the photovoltaic effect in a crystal to the influence of non-equilibrium carriers on spontaneous polarization due to the Jahn–Teller effect (Fridkin 1976).

Fridkin has also elaborated a fluctuation model of the photovoltaic effect (Fridkin and Popov 1978).

By summing up, all the models assume that the effect of laser radiation upon a crystal lies in exciting electrons into the conduction band. Since optical damage is observed to take place on laser irradiation at wavelengths corresponding to the transmittance range of lithium niobate crystals, this suggests the appearance of discrete energy levels in the gap.

8.2 Occurrence of optical distortion in lithium niobate crystals exposed to cw laser radiation

The phenomenon of optical distortion in lithium niobate and lithium tantalate crystals was first observed by Ashkin *et al* (1966).

As long as lithium niobate crystals were used as harmonic generators, the second harmonic beam spread out further and the conversion efficiency decreased. A crystal subject to laser radiation experiences local variations in the refractive index in an illuminated region, the variations persisting after the illumination has been removed. The region of a changed refractive index is perceivable by eye and is in the form of a track. If the irradiated lithium niobate crystal is raised to a temperature around 200 °C or if it is illuminated uniformly by a mercury lamp, the change in the refractive index disappears. But its subsequent irradiation by laser light will cause an optical distortion again.

Thus the optical distortion phenomenon was first observed as the one impairing the operating characteristics of electro-optical elements and imposing limitations on ferroelectrics applications in quantum electronic devices.

Optical distortions are also found to take place in other ferroelectric crystals, e.g. in $KNbO_3$ (Gunter and Micheron 1977), $KTaO_3$ (King *et al* 1972), $BaTiO_3$ (Townsend and La Macchia 1970), $Ba_xSr_{1-x}Nb_2O_6$ (Volk *et al* 1978), KH_2PO_4 (Wax *et al* 1970), $KTa_xNb_{1-x}O_3$ (Chen 1967), $NaBa_2Nb_5O_{15}$ (Voronov *et al* 1976a,b). It is noteworthy that in ferroelectrics at temperatures above the Curie point optical distortion occurs only with an external electric field applied.

Chen (1969) was the first to study the optical distortion effect under a 0.48 μm Ar laser radiation. He measured a change in birefringence $\Delta(n_e - n_o)$ by the polarization optical technique. These experiments have established that the extraordinary refractive index changes much more than the ordinary one does. So it can be claimed that $\Delta(n_e - n_o) \approx \Delta n_e$. The author also measured the distribution of Δn_e across the laser beam along the polar axis of a crystal and in the plane normal to the radiation propagation. The distribution of Δn_e along the polar axis virtually follows the radiation intensity distribution, while this value varies more smoothly along the x axis. The experimentally determined dependence of Δn_e of the irradiation time increases linearly with time and emerges onto a steady-state level Δn_{st}.

When the laser radiation intensity increases, the value of Δn_{st} is proportional to the square root of this intensity. At high intensities, however, the value of Δn_{st} is intensity-independent. For lithium niobate at room temperature a limit of Δn_{st} is 10^{-3}.

Chen was also the first to observe a steady-state current in a shorted crystal exposed to the argon laser radiation with no external electric field applied. But the external electric field does affect optical distortion. The latter may either increase or decrease, depending on the external field direction relative to spontaneous polarization in a crystal. However, the effect cannot be compensated completely by the external electric field.

The external electric field compensating the optical damage in a crystal has been evaluated by extrapolating the dependence of Δn upon the external electric field to zero. In undoped lithium niobate crystals the compensating field E_{com} is about $180 \, \text{kV cm}^{-1}$ and is independent of the radiation intensity.

However, under roughly the same conditions induced birefringence can be compensated by an external electric field (Augustov and Gotlib 1976, Augustov *et al* 1977). In the works of these authors the compensating field E_{com} was dependent upon the laser radiation intensity and for $I = 25 \, \text{W cm}^{-2}$ it was $11 \, \text{kB cm}^{-1}$.

To clarify the mechanism of optical distortion it is important to know whether or not the strength of a compensating electric field is related to the radiation intensity. That is why the inconsistent experimental findings require additional verification.

Serreze and Goldner (1975) studied spectral dependence of Δn_{st} at wavelengths ranging from 300 to 800 nm. As they extended to the short-wavelength region, the sensitivity to optical damage increased. Near $\lambda \sim 415 \, \text{nm}$, the sensitivity was at its maximum.

The phenomenon of optical distortion can be used for holographic recording of information, which is particularly attractive for researchers. Townsend and La Macchia (1970) have put forward a holographic method for measuring an optical distortion Δn. A diffraction grating resulting from the superposition of two plane waves is recorded in a crystal. When the hologram is read out by an extraordinary ray at a wavelength λ, the diffraction efficiency of a sinusoidal grating is determined according to Kogelnik's (1969) formula

$$\eta = \sin^2 \left[\pi l \Delta n_e / \lambda \cos (\theta/2) \right] \qquad (8.15)$$

where Δn_e is the variation in amplitude of the extraordinary refractive index and l is the optical path in a crystal.

Thus by measuring the diffraction efficiency of a grating we can find the value of Δn_e. Holographic investigations of the photo-refractive effect provided additional experimental data indispensable for obtaining insight into the physics of the phenomenon. For instance, the holographic method enables the changes of an extraordinary and an ordinary refractive index to be measured separately. For $BaTiO_3$ it has been established that

$$\Delta n_e / \Delta n_o \sim r_{33} n_e^3 / (r_{13} n_o^3) \tag{8.16}$$

where r_{33} and r_{13} are the electro-optic coefficients. This expression attests to the electro-optical nature of the photo-refractive effect.

The holographic method has also been employed to measure temperature dependence of the time constant of hologram disappearance in the darkness. The τ value was shown to decrease exponentially with increasing temperature at an activation energy of about 1.1 eV (Amodei and Staebler 1972a,b). However, Belobaev *et al* (1976) have obtained data different from the previous findings: the optical recording time constant proved to decrease exponentially with increasing temperature at an activation energy of about 0.25 eV, while the time constant of hologram thermal erasure decreases with an activation energy around 0.66 eV.

The spectral dependence of lithium niobate photoconductivity coincides with spectral dependence of optical distortion. After illumination has been discontinued, an optical distortion relaxes due to the Maxwellian relaxation of a space charge (Pestryakov and Entin 1976, Barkan *et al* 1977a). This is confirmed by the fact that the activation energy calculated from the temperature dependence of optical distortion relaxation time coincides with the activation energy of thermal conductivity.

Working with congruent nominally pure crystals of lithium niobate, Pashkov *et al* (1979a) related refractive index variations to the radiation time and power as well as to the crystal temperature and chemical composition. They used a 0.63 μm He–Ne laser.

Figure 8.3 plots a change in the refractive index Δn versus time of exposure of crystals for various component ratios $R = Li/Nb$ (Anghert *et al* 1972). As a glance at the figure shows, Δn first increases with time and then emerges onto a steady-state level. The experimental dependence can be expressed as

$$\Delta n = \Delta n_{st}[1 - \exp(-t/\tau)]. \tag{8.17}$$

The process studied is characterized by two independent parameters: a steady-state value Δn_{st} and a growth time constant τ. It follows from figure 8.3 that the higher the ratio R in a crystal, the larger the value Δn_{st} and the steeper the slope of the curve $\Delta n(t)$, which is initially determined by the ratio $\Delta n/\tau$.

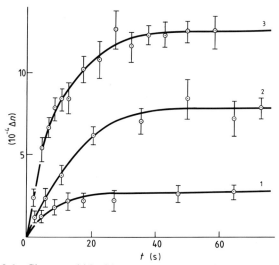

Figure 8.3. Change of birefringence Δn versus time of 0.63 μm laser irradiation of lithium niobate crystals of various chemical compositions $R = \text{Li}/\text{Nb}$: 0.925, 0.957, 0.984 (curves 1–3) (according to Anghert *et al* 1972).

Figure 8.4 plots Δn_{st} versus radiation power for a constant beam diameter in the focus of a lens with a spot diameter $d = 4.5 \times 10^{-3}$ cm (the focal length of a lens being $f = 4.5$ cm). Here one can see that the value Δn_{st} increases linearly with radiation power. The dependence of $1/\tau$ upon laser radiation intensity is depicted in figure 8.5. It follows from experimental data that the time constant τ is inversely proportional to the laser radiation intensity.

Figure 8.6 is a plot of Δn_{st} versus light beam diameter for a constant radiation power. The light beam diameter was made to vary from 4.5×10^{-3} to 17×10^{-3} cm by changing the lens focal length from 4.5 to 17 cm. Experiments have revealed that Δn_{st} is proportional to $1/d^2$. Thus the above results on optical distortions versus radiation power and laser beam diameter yield the conclusion that Δn_{st} depends linearly on the laser radiation power.

The temperature dependence of Δn was studied over a temperature range 300 to 430 K. A crystal was thermostatically controlled and the temperature was taken by a copper–constantan thermocouple. Over this temperature range the variations in Δn_{st} are fitted by the exponential

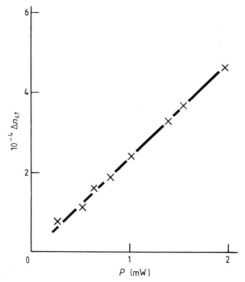

Figure 8.4. Steady-state value Δn_{st} in a congruently grown lithium niobate crystal versus laser output power at $\lambda = 0.63$ μm (according to Anghert *et al* 1972).

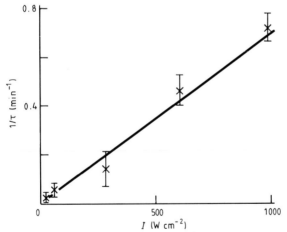

Figure 8.5. Inverse time of relaxation $1/\tau$ of an induced change in birefringence in a lithium niobate crystal versus intensity of laser radiation at $\lambda = 0.63$ μm (according to Anghert *et al* 1972).

dependence $\Delta n_{st} \sim \exp \Delta \varepsilon_{ef}/kT$), where $\Delta \varepsilon_{ef}$ is the effective activation energy, k is the Boltzmann constant and T is the thermodynamic temperature. The slope of the curve $\ln \Delta n = f(1/T)$ (figure 8.7) was used to calculate the value $\Delta \varepsilon_{ef}$. The latter proved to be 0.3 ± 0.06 eV

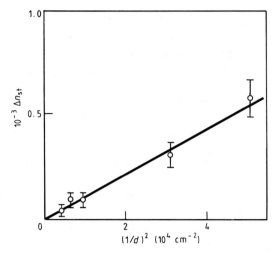

Figure 8.6. Steady-state value of induced birefringence Δn_{st} versus diameter of a laser spot acting on a lithium niobate crystal (according to Anghert *et al* 1972).

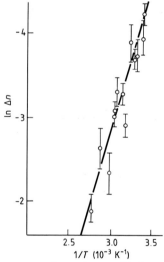

Figure 8.7. Temperature dependence of induced birefringence Δn in a lithium niobate crystal (according to Anghert *et al* 1972).

and remained at a constant level for various ratios $R = \mathrm{Li/Nb}$ ($R = 0.92$–0.98).

The role of an external electric field applied to a crystal in the behaviour of the photo-refractive effect was studied by Pashkov *et al* (1979a). The laser beam was made to propagate along the crystal optic axis and the field was applied along the x axis. Variations in the refractive index of a crystal exposed to an external electric field under laser radiation were measured by the polarization optical technique.

Samples were polished single-domain slugs having dimensions 4, 6 and 16 mm along the x, y and z axes respectively. The faces perpendicular to the crystal axis carried glued electrodes made of silver-plated brass. The source was a He–Ne laser with an output power up to 40 mW in a single mode. The laser beam intensity was lowered by calibrated neutral filters and focused into a sample by a lens whose focal length exceeded the length of a crystal. A sample was placed between two cross polaroids; the incident beam was polarized along the x axis of a crystal. The birefringence induced by an applied external electric field was compensated by a quartz wedge. When the applied voltage was a multiple of the half-wave voltage, compensation was accomplished by turning the analyzer through an angle of 90°. Occurrence of an optical inhomogeneity versus laser radiation intensity was examined in the interval from 40 to 1000 W cm^{-2} for a constant lens focal length and external electric field strength.

Figure 8.8 plots Δn_{st} versus irradiation time t for various laser radiation intensities and a constant field strength corresponding to a half-wave voltage (2.5 kB cm^{-1}). It is evident from the figure that $\Delta n(t)$ increases until it reaches a steady-state value Δn_{st}. Occurrence of inhomogeneities is also characterized by two independent parameters: a

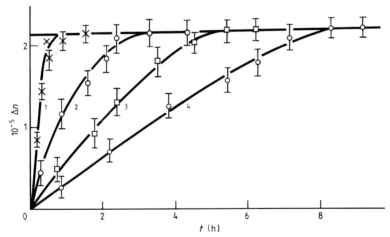

Figure 8.8. Induced birefringence versus time of irradiation of a lithium niobate crystal for a constant strength of external electric field $E_0 = 2.5$ kV cm^{-1} and various intensities of laser radiation at $\lambda = 0.63$ μm: $I_0 = 1000$ W cm^{-2}, $0.3I_0$, $0.075I_0$ and $0.04I_0$ (curves 1–4 respectively) (according to Pashkov *et al* 1979a).

steady-state value Δn_{st} and a growth time constant τ. This relationship is fitted by an expression of the type (8.17).

In the above range of laser power densities a steady-state change of birefringence Δn_{st} is independent of radiation intensity and corresponds to a value induced by an external field with a strength equal to a half-wave voltage $V_{\lambda/2}$. Thus, induced birefringence can be compensated by an external electric field.

Once the electric field is off, the induced birefringence will relax in the centre of the beam at a time constant equal to that of growth Δn. The crystal regions lying near the beam boundaries or off the beam persist in having a changed refractive index. After the electric current has been switched off and the laser radiation removed, induced bi-refringence exists for many days at room temperature.

8.2.1 Dependence of Δn upon external electric field strength

The value Δn was measured for various strengths of the external electric field ranging from 1.2 to $30\,\mathrm{kV\,cm^{-1}}$ at a constant laser radiation intensity $I \sim 10^3\,\mathrm{W\,cm^{-2}}$. To prevent an electric breakdown at the surface under a strong field, a crystal was placed into a glass cell filled with an organosilicon liquid. Experimental data indicate that Δn_{st} increases linearly with external electric field strength (figure 8.9), while the time constant is independent of the latter.

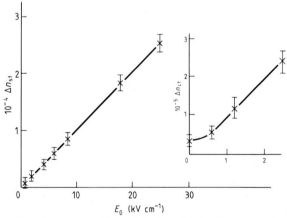

Figure 8.9. Steady-state value of induced birefringence Δn_{st} in lithium niobate versus strength of an external electric field E_0 (the light propagates along the z axis, the field is applied along the x axis) (according to Pashkov *et al* 1979a).

The temperature dependence of Δn_{st} was studied for crystal temperatures ranging from room temperature to $300\,°\mathrm{C}$ at a constant laser radiation intensity $I \sim 10^3\,\mathrm{W\,cm^{-2}}$ and external electric field strength $E_0 \sim 2.5\,\mathrm{kV\,cm^{-1}}$.

Over a certain specific temperature range the value Δn_{st} was found to remain constant and equal to the birefringence induced by an external electric field; subsequently Δn_{st} decreases with increasing temperature. The temperature range over which the value Δn_{st} is maintained at a constant level extends as the laser radiation intensity is increased (figure 8.10).

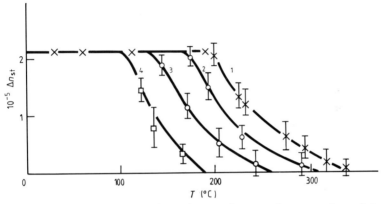

Figure 8.10. Temperature dependence of a steady-state value of induced birefringence Δn_{st} in congruently grown lithium niobate crystals for various laser radiation intensities: $I_0 = 1000\,\mathrm{W\,cm^{-2}}$, $0.5I_0$, $0.1I_0$ and $0.01I_0$ (curves 1–4 respectively) (according to Pashkov *et al* 1979a).

Over a range where Δn_{st} does depend on temperature, the former is defined by an expression $\Delta n_{st} \sim \exp(\Delta \varepsilon_{ef}/kT)$, where $\Delta \varepsilon_{ef}$ is the effective activation energy determined experimentally to be $0.5 \pm 0.1\,\mathrm{eV}$.

For lithium niobate crystals containing 5×10^{-3} wt.% Fe the segment of the curve where Δn_{st} is temperature independent is longer than that for undoped crystals, the light intensities being the same.

Occurrence of optical distortion for light propagation normal to the crystal polar axis has been investigated by Pashkov *et al* (1979a).

An external electric field was applied along the z axis of a sample. Figure 8.11 plots a steady-state value Δn_{st} versus external electric field strength for two laser radiation intensities (100 and 200 W cm^{-2}). It is an experimentally established fact that Δn_{st} is a linear function of the external electric field strength. If the direction of an electric field coincides with that of spontaneous polarization in a crystal, as long as the strength E_0 increases, the value Δn_{st} goes down to zero in some field $E_0 = E_c$ and then increases in magnitude again. The obtained relationship of Δn_{st} to the external electric field strength can be written as follows:

$$\Delta n_{st} = aE_0 + \Delta n_{st}^o \qquad (8.18)$$

where Δn_{st}^o is a steady-state induced birefringence for a given laser radiation intensity at $E_0 = 0$ and a is a factor of proportionality equal to $(1.57 \pm 0.1) \times 10^{-8}$ cm V^{-1}. The experimentally determined value of the coefficient a coincides with that of the effective electro-optic coefficient defining the electro-optical effect in a crystal for light propagating normal to the z axis of a crystal under a field applied along the x axis. The compensating field E_c is defined by

$$E_c = \Delta n_{st}/a = 2\Delta n_{st}^o/(r_{13}n_o^3 - r_{33}n_e^3). \qquad (8.19)$$

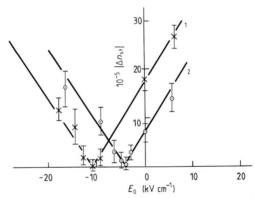

Figure 8.11. Steady-state value of induced birefringence Δn_{st} versus external electric field strength (the light propagates along the y axis, while the field is applied along the z axis): 1, 200 W cm^{-2}; 2, 100 W cm^{-1} (according to Pashkov *et al* 1979a).

Since, with light propagating normal to the polar axis and with no electric field applied, the value Δn_{st}^o increases linearly with laser radiation intensity and decreases exponentially with a rising temperature of a crystal, this suggests that the compensating field E_c is also dependent upon the laser radiation intensity and crystal temperature.

Figure 8.12 plots the strength of a compensating field E_c versus laser radiation intensity. It is evident from the figure that E_c increases linearly with the laser radiation intensity. The linear dependence of E_c was observed by Augustov *et al* (1977). It has been experimentally established that E_c decreases exponentially with increasing temperature of a crystal for an activation energy of 0.3 eV. Thus both Δn_{st}^o and E_c

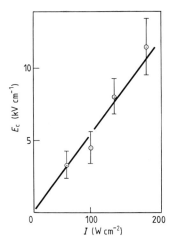

Figure 8.12. Strength of an external electric field E_c compensating induced birefringence versus laser radiation intensity I ($\lambda = 0.63 \ \mu m$) (according to Pashkov *et al* 1979a).

depend in a similar way upon laser radiation intensity and crystal temperature.

It is remarkable that the dependence of Δn_{st} upon laser radiation intensity and crystal temperature differs markedly for various values of an external electric field. If the latter is insufficient for compensating induced birefringence, Δn_{st} depends on laser radiation intensity and crystal temperature just as it does in the absence of an external electric field. But if the electric field intensity exceeds a value required for compensation, Δn_{st} is independent of laser radiation intensity or crystal temperature, as is the case with light propagation along the polar axis of a crystal.

It follows from the above results that: (i) at the same intensity an optical distortion in an external electric field is observed at a higher temperature than it is in the absence of an electric field; (ii) once a crystal is raised to a temperature at which an optical distortion is vanishingly small and then an external electric field is applied along the polar axis, the illuminated region experiences a change in the refractive index. In doing so, the value Δn_{st} is determined by the strength of an external electric field, while the sign is dictated by the electric field direction relative to spontaneous polarization. These conclusions are supported by experiment. When a crystal was heated to 60 °C and the radiation intensity was 200 W cm^{-2}, no optical distortion was observed. It occurred in an illuminated region of a crystal when an electric field $E_0 = 6$ kV cm^{-1} was applied. If the direction of the electric field coincided with that of spontaneous polarization the laser beam traversing the optical distortion region became narrower, while in the opposite direction it spread out along the optic axis. As the temperature of the crystal was raised further, the value Δn_{st} remained constant up to 160 °C. At lower radiation intensities, Δn_{st} was dependent upon the radiation intensity, just as is the case of radiation propagation along the polar axis in an external electric field.

Pashkov *et al* (1979a) interpret the obtained results as follows. One deals with an optical distortion occurring in lithium niobate for light propagating along the crystal polar axis in an electric field applied normal to this axis. It is assumed that a crystal contains impurity centres with a density M. Some of them (M_t) are occupied and can serve as donors, while the others ($M - M_t$) are free and act as traps, i.e. carrier trapping centres. If a crystal is exposed to a laser beam of a small diameter, compared to the crystal transverse dimensions, photo-excited carriers drift in the electric field to the light beam periphery and are trapped there. This sets up a space charge field E_i which makes the refractive index of an electro-optical crystal change:

$$\Delta n = 2n_o^3 r_{22} E_i \qquad (8.20)$$

where n_o is the ordinary refractive index and r_{22} is the electro-optic coefficient.

A steady-state value Δn_{st} can be evaluated by proceeding from the following considerations. In a steady state the density of a current J flowing through a crystal should be the same in any cross section perpendicular to the direction of an external electric field. Then the dark current in the external electric field E_0 in an unilluminated region will be equal to that in an illuminated region in the resultant field

$$e\mu N_{st}(E_0 - E_i) = e\mu N_d E_0 \qquad (8.21)$$

where e and μ are the electron charge and mobility respectively, N_{st} is the steady-state electron density in the conduction band in an illuminated region and N_d is the dark electron density in the conduction band in an unilluminated region.

The density of free electrons in an illuminated region can be determined from the continuity equation (Ryvkin 1963)

$$\frac{dN}{dt} = (\alpha I + \beta)M_t - BN_{st}(M - M_t) - \frac{\text{div } J}{e} \qquad (8.22)$$

where α is the photo-ionization cross section, β is the probability for thermal ionization and B is the recombination cross section.

The experimentally determined linear dependence of Δn_{st} on the external field strength E_0 (see figure 8.11) shows that no appreciable detrapping occurs during optical distortion. Hence the density of filled centres in an illuminated region is supposed to be constant throughout the process of irradiation (M_t = const.) Taking account of this circumstance as well as the stationarity conditions $dN/dt = 0$ and $\text{div}\, J = 0$ and using equation (8.22), we derive an expression for N_{st} in the form

$$N_{st} = (\alpha I + \beta)M_t/[B(M - M_t)]. \tag{8.23}$$

The probability for thermal excitation β is found from the principle of detailed balance

$$\beta = BN_d(M - M_t)/M_t. \tag{8.24}$$

Substituting (8.24) and (8.23) into (8.21) and employing E_i^{st} and hence (8.20), we obtain Δn_{st}:

$$\Delta n_{st} = 2r_{22}n_o^3E_0\{\alpha IM_t/[\alpha IM_t + N_dB(M - M_t)]\}. \tag{8.25}$$

For high radiation intensities or rather low temperatures, where the condition $\alpha IM_t \gg N_dB(M - M_t)$ is fulfilled, using also expression (8.25) we derive

$$\Delta n_{st} = 2r_{22}n_o^3E_0. \tag{8.26}$$

Indeed, at room temperature and radiation intensities used in the experiment the value Δn_{st} is independent of the intensity and is governed by the external field E_0 (see figure 8.9).

For low intensities (or high temperatures) the condition $\alpha IM_t \ll N_dB(M - M_t)$ is fulfilled; then (8.25) yields

$$\Delta n_{st} = 2r_{22}n_o^3E_0\{\alpha IM_t/[N_dB(M - M_t)]\}. \tag{8.27}$$

Allowing for the fact that dark equilibrium density of free carriers in the conduction band in the case of donor impurity is defined by (Kireev 1969)

$$N_d = M_t^{1/2}(2\pi m^*kT/h^2)^{3/4}\exp(-\Delta/2kT) \tag{8.28}$$

where Δ is the trap thermal activation energy, m^* is the effective electron mass and h is Planck's constant, we obtain a final expression for Δn_{st}:

$$\Delta n_{st} = 2r_{22}n_o^3E_0 \frac{\alpha IM_t^{1/2}}{B(M - M_t)}\left(\frac{2\pi m^*kT}{h^2}\right)^{-3/4}\exp\left(\frac{\Delta}{2kT}\right). \tag{8.29}$$

Since $\Delta \gg kT$, the temperature dependence of Δn_{st} is fitted by an exponential term. Thus at rather high temperatures the value of Δn_{st} decreases exponentially with increasing temperature. A comparison of expression (8.29) and experimentally determined temperature dependence of Δn_{st} gives a trap thermal activation energy in lithium niobate of $\Delta = 2\Delta\varepsilon_{ef} \approx 1$ eV. Additional experiments have revealed that over a rather high temperature range (above 200 °C), where expression (8.29) holds good, the value of Δn_{st} is proportional to the light intensity and the external electric field strength.

The kinetics of the refractive index behaviour at rather low temperatures, where $\alpha I M_t \gg N_d B(M - M_t)$, can be determined by proceeding from the following considerations. Since $\Delta n = r_{ef} E_0$, the kinetics of the refractive index are governed by the kinetics of the behaviour of E_i. A space charge field establishes itself in an illuminated region thanks to the drift of photo-excited carriers in an external electric field and the formation of space charge on the boundaries of the illuminated region. Variations in the density of space charge dQ in time dt are determined by the density of current flowing per unit area in an illuminated region in the time dt. Thus

$$dE_i = dQ/(\varepsilon_0 \varepsilon) = (\varepsilon_0 \varepsilon)^{-1} J(t) \, dt. \tag{8.30}$$

The current density $J(t)$ in an illuminated region is determined by the conduction current in the resultant field $E_0 - E_i$ and, on condition that $\sigma_{ph} \gg \sigma_d$, it is

$$J(t) = \sigma_{ph}[E_0 - E_i(t)]. \tag{8.31}$$

Substitution of (8.31) into (8.30) yields

$$\frac{dE_i}{E_0 - E_i(t)} = \frac{\sigma_{ph}}{\varepsilon_0 \varepsilon} dt. \tag{8.32}$$

Integrating this equation we obtain an expression

$$E_i = E_0\{1 - \exp[-\sigma_{ph}t/(\varepsilon_0 \varepsilon)]\} \tag{8.33}$$

or

$$\Delta n = r_{ef} E_0\{1 - \exp[-\sigma_{ph}t/(\varepsilon_0 \varepsilon)]\}. \tag{8.34}$$

Expression (8.34) coincides with (8.17) derived experimentally for τ equal to the Maxwellian relaxation time constant $\tau = \varepsilon_0 \varepsilon/\sigma_{ph}$.

The above considerations imply that the basic features of optical distortion observed experimentally in lithium niobate with light propagating along the polar axis under an external electric field directed normal to this axis are described by a simple model with a single-type impurity centre.

Thus, a conclusion can be drawn that crystal photoconductivity exceeds its dark conductivity in undoped lithium niobate under typical experimental conditions (at a radiation wavelength $\lambda = 0.63 \, \mu m$) at intensities $I \geqslant 1 \, W \, cm^{-2}$ and temperatures not higher than $100 \, °C$. Besides, here photoconductivity σ_{ph} increases linearly with radiation intensity and grows exponentially with temperature (figures 8.13 and 8.14). This behaviour is particularly pronounced in crystals exposed to argon laser radiation.

The experimental results obtained are explained within the polariza-

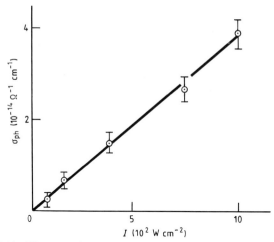

Figure 8.13. Photoconductivity σ_{ph} of a lithium niobate crystal versus laser radiation intensity ($\lambda = 0.63 \, \mu m$) (according to Pashkov *et al* 1979b).

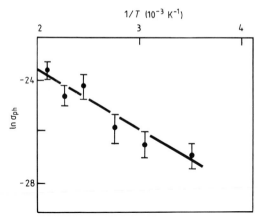

Figure 8.14. Temperature dependence of photoconductivity σ_{ph} of a lithium niobate crystal (according to Pashkov *et al* 1979b).

tion model developed by Levanyuk and Osipov (1975). It follows from their works that the charge exchange of donor–acceptor pairs causes a change in the polarization ΔP_s in an illuminated region and a macroscopic depolarizing field $E = -4\pi\Delta P$. Considering that $\Delta P = \Delta P_s + \chi E$ we derive $E = -4\pi\Delta P_s/\varepsilon$ (ε being the dielectric constant). This field introduces an additional change in the lattice polarization $\Delta P_1 = \chi E = -(\varepsilon - 1)/(\varepsilon\Delta P_s) \approx -\Delta P_s$. Consequently, the variations in the polarization induced by a depolarizing field are virtually compensated by the polarization change ΔP_s due to the charge exchange of donor–acceptor pairs. Free carriers formed in an illuminated region under laser radiation tend to drift in the depolarizing field E_{dep} to the laser beam periphery and a space charge field is set up on the boundaries of the illuminated region. In a steady state the space charge field compensates the depolarizing field, the change of the refractive index is determined by the value ΔP_s and hence it is proportional to a steady-state density of donor–acceptor pairs that have exchanged their charges. Since the steady-state density of excited centres is determined by the radiation intensity I and crystal temperature T, the steady-state variations in the refractive index will be governed in a similar way by these parameters. The polarization model suggested by Levanyuk and Osipov implies that at low laser radiation intensities Δn_{st} increases linearly with laser radiation and decreases exponentially with increasing crystal temperature, which is observed experimentally (see figures 8.4 and 8.7). At high radiation intensities Δn_{st} is proportional to the square root of the radiation intensity ($\Delta n_{st} \propto \sqrt{I}$), which is also confirmed by experiment (Chen 1969).

Since charge exchange of donor–acceptor pairs seems to take less time than the carrier drift does, a steady-state depolarizing field will establish itself before a stationary space charge does. Thus the kinetics of induced inhomogeneity occurrence will be largely determined by the process of free carrier migration from an illuminated region.

The compensation of an optical distortion by an external electric field is also explainable within the polarization model where Δn_{st} is determined by a change in spontaneous polarization ΔP_s in an illuminated region of a crystal due to the photo-induced charge exchange of donor–acceptor pairs. The condition of optical distortion compensation by an external field can be written as follows:

$$(\sigma_{ph} + \sigma_d)E_1 = \sigma_d E_2 \qquad \Delta P_s + \chi E_1 = \chi E_2. \qquad (8.35)$$

Solution of system (8.35) yields expressions for E_1 and E_2:

$$E_1 = \frac{\sigma_d}{\sigma_{ph}} \frac{\Delta P_s}{\chi} \qquad E_2 = \left(1 + \frac{\sigma_d}{\sigma_{ph}}\right) \frac{\Delta P_s}{\chi}. \qquad (8.36)$$

The compensating voltage V_c is defined by

$$V_c = E_1 l + E_2(L - l) = \frac{\Delta P_s}{\chi} \left(\frac{\sigma_{ph} + \sigma_d}{\sigma_{ph}} L - l\right). \qquad (8.37)$$

Under practicable conditions $L \gg 1$ and $\sigma_{ph} \gg \sigma_d$ we have

$$V_c = \Delta P_s L/\chi \qquad E_c = \Delta P_s/\chi. \qquad (8.38)$$

Within the model in question $\Delta n_{st} = r_{ef}\Delta P_s/\chi$. Therefore the relation between E_c and Δn_{st}, with no external field applied, coincides completely with expression (8.19). The compensating field E_c has a straightforward physical meaning: this is a field in an illuminated region of a crystal which is close in magnitude but opposite in direction to a depolarizing field, forcing the carriers to drift towards the boundary of the illuminated region.

It follows from (8.38) that the I and T dependences of E_c (as well as the dependences of Δn_{st} upon these parameters) are determined by the I and T dependences of ΔP_s. As the temperature rises, ΔP_s decreases exponentially. At low radiation intensities the I dependence of ΔP_s is linear, while at fairly high intensities it is square root, which is observed in experiment.

The analysis of mechanisms of optical distortion for light propagating along the polar axis of a crystal under an external electric field and normal to this axis with no electric field applied provides an explanation of the different temperature dependences and activation energies in the two cases. In the absence of an electric field an induced inhomogeneity forms due to the change in spontaneous polarization in an illuminated region as a result of electron excitation from deep to shallow levels. The value of Δn_{st} is proportional to the density of excited centres. The temperature dependence of Δn_{st} is determined by the probability of thermal ionization of shallow levels that have trapped electrons whose activation energy is 0.3 eV (figure 8.7).

In an external electric field an optical distortion occurs due to the drift of free electrons which go from a deep level in the gap over to the conduction band under laser radiation. When photoconductivity is comparable to dark conductivity Δn_{st} is temperature-dependent. Since dark conductivity is determined by the probability of thermal excitation of electrons from the same deep levels ($\Delta \sim 1$ eV), it becomes comparable to photoconductivity only at sufficiently high temperatures. This explains in particular why, at the same laser radiation intensity in an external electric field, induced inhomogeneities are observed at higher temperatures than those in the absence of an electric field, with light propagating normal to the optic axis.

Belincher *et al* (1977) have measured internal crystal fields, temperature, refractive indices and diffraction efficiencies of undoped lithium niobate crystals. In the opinion of the authors, their experimental data (partly depicted in figure 8.15) fit the hypothesis about the pyroelectric nature of an internal crystal field which is set up by laser heating of a crystal.

8.2.2 Lithium niobate crystals doped with iron ions and partially reduced
In congruently grown crystals of lithium niobate reduced partially in the hydrogen atmosphere as well as doped with iron ions, Belobaev *et al* (1978) have observed a linear dependence of photovoltaic current on the light intensity up to a power density around 1 W cm^{-2}. Current–voltage

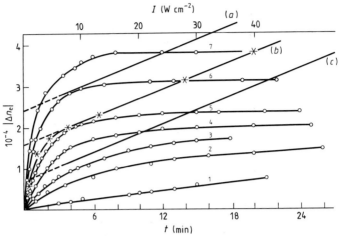

Figure 8.15. Time dependence of Δn_e for various constant intensities of light illumination I: 0.5, 1.5, 2, 7.5, 13, 28 and 40 W cm^{-2} (curves 1–7 respectively) as well as the relationship of Δn_e to the radiation intensity I for various applied fields: (*b*) limiting Δn_e with no external field applied; (*a*) and (*c*) limiting Δn_e with external fields of 6 and -6 kV cm^{-1} respectively. The exposure time in these experiments was 18 min (according to Belincher *et al* 1975).

characteristics of crystals exposed to uniform illumination were also linear, which enabled the crystal photoconductivity σ_{ph} and the compensating field E_c to be calculated. These data are presented in table 8.1 for various iron contents and degrees of reduction.

As the iron content in crystals increases the photovoltaic current increases linearly with absorption, which is in agreement with the data obtained by other authors (Glass *et al* 1975, Von der Linde *et al* 1974). The compensating field also increases linearly over the above iron content range (figure 8.16), whereas crystal conductivity is hardly variable. The Glass coefficient K_G (see equation (8.12)) remains constant for any reduced crystal and is $K_G = (2.5–2.7) \times 10^{-9}$ A cm W^{-1}. This value coincides with the data obtained by other authors (Glass *et al* 1975, Volk *et al* 1977) within an uncertainty up to 30%. Deviations from this value are observed only in highly absorbing samples prepared by long reducing annealing in hydrogen. This permits a conclusion that reduction of crystals does not affect, to a certain specific limit, the threshold for occurrence of a photovoltaic current but instead merely changes the density of active centres involved in the process.

8.2.3 Dependence of the photo-refractive effect on the shape of a light spot

As shown earlier, the photo-refractive effect is determined by depolarizing fields set up in non-uniformly illuminated ferroelectrics. That is why

Table 8.1. Parameters of lithium niobate crystals and characteristics of the photovoltaic effect $\lambda = 0.44\ \mu m$, $T = 20\ °C$ (according to Belobaev *et al* 1978).

Annealing time in H_2(min)	Fe content (wt.%)	I (W cm^{-2})	α (cm^{-1})	σ_d ($10^{-15}\ \Omega^{-1}$ cm^{-1})	$\sigma_{ph'}$ ($10^{-14}\ \Omega^{-1}$ cm^{-1})	I (10^{-10} A cm^{-2})	E_c (10^3 V cm^{-1})	K_G (10^9 A cm W^{-1})
—	—	1	0.12	1.5	1	0.4	4.5	2.7
5	—	1	0.34	20.0	16	1.2	0.66	2.7
20	—	1	0.95	220	180	3.2	0.12	2.6
120	—	1	1.6	1000	600	5.4	0.08	2.6
—	0.005	1	0.6	1.5	2.8	1.8	8.6	2.6
120	0.005	1	10.9	8	29	12	4.7	0.9
—	0.02	1	1.2	1.3	3.6	3.4	9.0	2.6
—	0.03	1	2.0	2.0	4.5	5.0	11.0	2.5
—	0.05	1	4.4	3.0	5.5	11.0	20.0	2.5

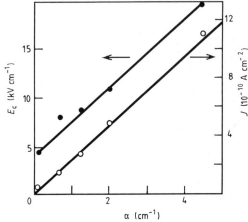

Figure 8.16. Photovoltaic current J and compensating field E_c versus coefficient of absorption α at a wavelength of $0.44\,\mu m$ in a series of unreduced crystals with an increasing content of iron (according to Belobaev *et al* 1978).

photo-refraction is expected to depend on the shape and orientation of the region at the crystal surface illuminated by a 'damaging' light (Volk *et al* 1977).

If the light spot is in the form of a narrow strip parallel to the axis of spontaneous polarization P_s, the macroscopic field E_i will be near zero and variations in birefringence $\delta\Delta n$ should be determined solely by those in optical polarizability of impurity centres. If the light spot is a narrow strip normal to the axis of P_s, a crystal should develop a macroscopic field $4\pi\Delta P$ inducing further variations in the lattice polarization. The corresponding change in birefringence will be given by a complex expression, with the total value $\delta\Delta n$ much higher than in the former case due to the appreciable macroscopic field formed.

In the work by Volk *et al* (1977) these conclusions were verified experimentally. The authors examined photo-refraction versus orientation of a crystal-damaging light strip relative to the C polar axis (axis of P_s) in lithium niobate crystals containing $0.03\,\text{mol.\%}$ Fe_2O_3 (figure 8.17(*a*)). They took three cases: (i) damaging light strip perpendicular to the C axis; (ii) former parallel to the latter; (iii) an intermediate case of $45°$ orientation of the two. A source of damaging light was provided by a He–Cd laser (440 nm, 15 mW). The probing focused beam of a He–Ne laser (633 nm, 1 mW) was oriented normal to the crystal plane. Induced photo-refraction was measured by the compensation technique.

It has been established that the observed change in birefringence is at its maximum where the light strip is perpendicular to the C axis. It is far

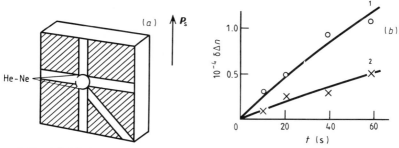

Figure 8.17. (*a*) Various conditions of photo-refraction excitation and (*b*) the photo-refraction effect in a LiNbO$_3$: Fe crystal versus illumination conditions: 1, the damaging light strip is perpendicular to the *C* axis; 2, it is at an angle 45° to the *C* axis (according to Volk *et al* 1977).

less in the intermediate case and at least two orders of magnitude smaller with a strip parallel to the *C* axis (the sensitivity of the method is not worse than 5×10^{-6}) (figure 8.17(*b*)). The relationship obtained of photo-refraction to the orientation of a damaging light strip is evidence of a certain contribution of a macroscopic depolarizing field to the photo-refraction effect in a non-uniformly illuminated ferroelectric. The photovoltaic model put forward by Glass and Von der Linde (1976) gives no explanation to the observed relationship, since it does not assume formation of macroscopic fields dependable upon the light spot orientation and shape.

Kanaev and Malinovsky (1982) interpret the independence of induced Δn under illumination of a light strip parallel to the crystal optic axis as follows. The value of light-induced Δn can be represented as the sum

$$\Delta n = \Delta n_1 + \Delta n_2 = \Delta n(E) + \Delta n(A). \qquad (8.39)$$

Here $\Delta n_1 = \Delta n(E)$ is the change introduced by a photo-induced electric field E; $\Delta n_2 = \Delta n(A)$ stands for a non-field change in the refractive index. To separate the terms experimentally, we can draw on the boundary-condition dependence of the field. Near the boundary on a conducting medium the tangential component is known to be zero. Obviously, if the first term is determined only by one component of a photo-induced field, near the boundaries for which this component is tangential the value Δn is defined by non-field mechanisms. In lithium niobate the extraordinary refractive index n_e depends only on the field component E_z. Hence near the boundaries parallel to the *C* optic axis the total change Δn will be $\Delta n_e(A)$, while at a distance from them it is a sum $\Delta n_e(E) + \Delta n_e(A)$.

An examination of the above experimental facts argues, first of all, in favour of the predominant role of the field photo-refraction mechanisms including the photovoltaic one. Second, using a strip-like light spot we can arrange it relative to the electrodes in such a manner that the photo-refraction effect might be determined solely by the component $\Delta n(A)$, which is small in undoped lithium niobate crystals.

8.3 Occurrence of optical distortion in lithium niobate exposed to pulsed laser radiation

Studies of optical distortion arising under pulsed laser radiation is a matter of interest from the standpoint of its effect upon second harmonic generation in a lithium niobate crystal.

This section presents the findings on the dynamics of occurrence of residual optical distortion Δn_{res} in undoped lithium niobate crystals under pulsed laser radiation normal to the crystal polar axis, depending on the energy emitted in a pulse, the pulse width and the number of pulses (Pashkov and Solovyeva 1978). The radiation source was provided by a neodymium garnet laser ($\lambda_1 = 1.06\ \mu m$) operated in a pulsed mode.

The laser radiation was converted into the second harmonic ($\lambda_2 = 0.52\ \mu m$) in a 10 mm long lithium niobate crystal at a conversion efficiency of about 10%. The laser parameters were as follows: pulse width ~10 ns; divergence $\alpha' = 0.6 \times 10^{-3}$ minutes of arc; the energy emitted in a pulse ~6×10^{-4} J. The radiation at $\lambda_2 = 0.53\ \mu m$ was focused into a studied lithium niobate sample by a lens with $f = 10$ cm. The optical distortion in a sample that has transmitted a laser pulse was measured by the polarization optical technique using a probe beam from a He–Ne laser with an output power around 1 mW. The probe beam coincided with the pulsed laser beam. The compensating device was a quartz wedge. A signal from a photoreceiver was recorded by an oscillograph. The time constant of the recording system was 10^{-6} s. Oscillograms of the changing power of the probe beam passing through the aperture were photographed. A change in the refractive index was determined according to the expression

$$\Delta n = \frac{D^2}{16f^2 n_{o,e}} \left(\frac{P_o}{P} - 1 \right) \qquad (8.40)$$

where D is the laser beam diameter, $n_{o,e}$ stands for the ordinary and extraordinary refractive indices of a crystal, f is the lens focal length, P_o is the output power with no variations in the crystal refractive index and P is the output power with a changed refractive index of a crystal.

After the laser pulse has traversed a crystal, the latter displays a region of a changed refractive index Δn_{res}. Experiments have revealed that the value Δn_{res} is determined by the energy of a pulse and the number of pulses. Figure 8.18 plots Δn_{res} versus the number of pulses N for various values of the energy emitted in a pulse. It is evident from the figure that Δn_{res} grows rather rapidly and attains saturation n_{res}^{st}; all the subsequent laser pulses will not change the value Δn_{res}^{st}. The value of Δn_{res} was also measured in the frequency mode at a constant energy emitted in a pulse. The pulse repetition frequency was made to vary from 0 to 10 Hz. Over this frequency range n_{res} did not prove to depend upon the pulse repetition frequency.

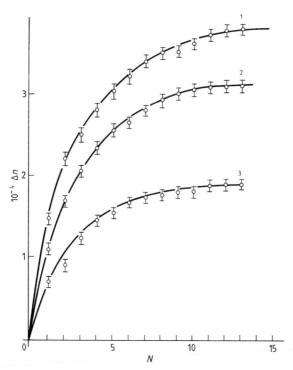

Figure 8.18. Induced birefringence Δn in lithium niobate versus the number of laser pulses N emitting various energies: 1.8, 1.3 and 0.85 J cm^{-2} (curves 1–3 respectively) (according to Nechaev *et al* 1972).

Changes in birefringence after the first pulse traversal Δn_1 and Δn_{res} versus pulse energy are plotted in figure 8.19. The values Δn_1 and Δn_{res} are a linear function of the pulse energy density within the interval 0.3 to 1.6 J cm^{-2}.

The dynamics of optical damage are represented in figures 8.20(a) and (b), which show the time dependences of Δn. As a glance at figure 8.20(a) shows, the time behaviour of the refractive index $\Delta n(t)$ under

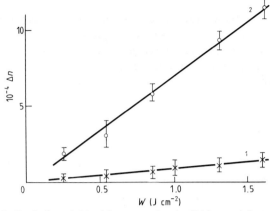

Figure 8.19. Induced birefringence Δn in lithium niobate after the first pulse traversal N_1 (curve 1) and Δn_{res} (curve 2) versus energy density in a pulse (according to Nechaev *et al* 1972).

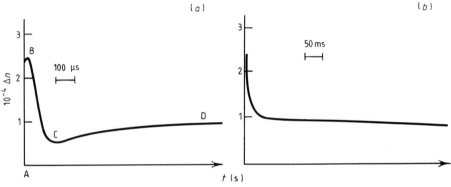

Figure 8.20. Change of induced birefringence Δn in a lithium niobate crystal at the centre of a probe beam: (*a*) on a short time base; (*b*) on a long time base (according to Nechaev *et al* 1972).

the pulsed lasing conditions is as follows: during transmission of a light pulse Δn increases (segment AB); after its transmission it decreases down to a certain specific value at a time constant $\tau_1 \approx 10^{-4} – 10^{-5}$ s (segment BC); then it rises slightly (segment CD) and then goes down again to a residual value of Δn_{res} (figure 8.20(*b*)) at a time constant $\tau_2 \approx 10^{-2}$ s.

The change of the refractive index Δn during laser pulse transmission makes the transmitted laser beam spread along the z axis.

Figure 8.21 shows the dynamics of the refractive index behaviour when the same crystal region is exposed to a sequence of laser pulses following in 1 min intervals. During each successive laser pulse, the value of Δn changes starting from the Δn_{res} that has remained in the crystal after the previous transmitted pulse. After a steady-state value of Δn_{st} has been achieved, all the subsequent laser pulses make Δn increase further during pulse passage. But after the pulse has been transmitted, this increased value relaxes again to the Δn_{st} level at a time constant 10^{-2} s.

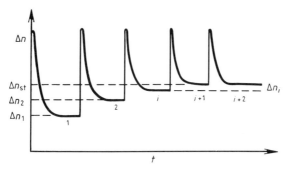

Figure 8.21. Dynamics of induced change of birefringence Δn in a lithium niobate crystal with an increasing number of light pulses: $\Delta n_1 - \Delta n$ after the first pulse; $\Delta n_2 - \Delta n$ after the second pulse; $\ldots \Delta n_i - \Delta n$ after the ith pulse (according to Nechaev *et al* 1972).

Pashkov and Solovyeva (1978) have also endeavoured to get insight into the dynamics of optical distortion versus crystal temperature ranging from room to 500 °C. It is evident from the oscillograms that a change in the refractive index is observed to increase during a laser pulse at temperatures up to 500 °C, its maximum value remaining virtually at the same level. The residual value of Δn_{res} after the laser pulse decreases with increasing temperature and even at ~ 100 °C Δn relaxes to zero (figure 8.22). The relaxation time constant τ_1 is independent of temperature, while τ_2 decreases from 10^{-2} to 10^{-3} s.

If the pulsed radiation propagates along the polar axis of a crystal, the refractive index is also found to change during the lasing pulse. However, after the pulse has been transmitted n relaxes at time constants $\tau_1 \sim 10^{-4}$ s and $\tau_2 \sim 5 \times 10^{-3}$ s down to a value roughly two orders of magnitude smaller than Δn_{res} observed under the same conditions but with the radiation propagating normal to the polar axis. The experimental results obtained are attributable to changed spontaneous polarization in an illuminated region of a crystal. Within this model the dynamics of Δn behaviour can be interpreted as follows. During transmission of a laser pulse the radiation will excite electrons to the conduction band and cause a charge exchange of donor–acceptor pairs. As a result, spontaneous polarization is changed by ΔP_s and a

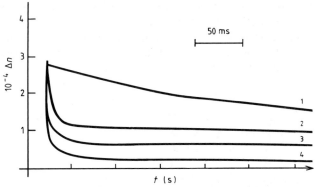

Figure 8.22. Dynamics of induced change of birefringence Δn in lithium niobate under pulsed radiation for various crystal temperatures T: 23, 35, 65 and 157 °C (curves 1–4 respectively) (according to Nechaev *et al* 1972).

depolarizing field E_{dep} establishes itself. The depolarizing field is responsible for medium depolarization, thereby reducing ΔP_s. These processes seem to take place rather fast. The resultant change of polarization in an illuminated region gives rise to Δn, which is observable in the segment AB (figure 8.20(a)). By the action of the depolarizing field electrons drift towards the boundaries of an illuminated region, thereby forming a space charge and increasing Δn further. On the other hand, recombination of photo-excited electrons to acceptor levels is likely to result in relaxation of ΔP_s in an illuminated region and, accordingly, in reduction of Δn. It is also noteworthy that relaxation of ΔP_s reduces the depolarizing field E_{dep} and hence slows down the process of space charge formation. Reduction of Δn due to recombination is faster than increase of Δn due to enhancement of the space charge field. This fact accounts for the decrease of Δn in the segment BC at a time constant $\tau \sim 10^{-4}$ s (figure 8.20(a)). A slower decrease of n at a time constant $\tau \sim 10^{-2}$ s (figure 8.20(b)) can be attributed to the relaxation of ΔP_s due to the reverse charge exchange of donor–acceptor pairs.

When a laser pulse propagates along the C polar axis of a crystal, the value of E_i is small. The space charge on the boundaries of an illuminated region is therefore also small. Hence the value of Δn_{res} in a crystal after laser pulse transmission will also be small. The dynamics of Δn behaviour during a laser pulse will be determined by an increase of the value ΔP_s and subsequently by its relaxation. Presumably, relaxation of ΔP_s will first proceed through electron recombination and later on as a result of reverse charge exchange of donor–acceptor pairs. Reduction of the time τ_2 may be explained by the absence of the competing process of space charge formation in the case of radiation propagation along the polar axis of a crystal.

Each successive laser pulse makes the process of variation in Δn repeat itself (see figure 8.21). When the radiation is normal to the crystal polar axis, the specificity lies in the fact that by the time of onset of any following mth pulse the crystal already exhibits a change Δn_{m-1} produced by the previous $(m - 1)$ laser pulses. Hence it follows firstly that the change Δn_m starts from the level Δn_{m-1}, and secondly that electron drift is already observed in the field $E_{dep} - E_{i(m-1)}$, where $E_{i(m-1)}$ is the space charge field set up in a crystal by the action of the $(m - 1)$th pulse. If E_i is smaller than E_{dep} and Δn_{res} in a crystal is determined only by E_{dep}, the value of Δn_{res} depends linearly on the number of laser pulses (see figure 8.18). As E_i is enhanced, the value of E_{dep} becomes compensated.

It is also remarkable that the resultant field $E_{dep} - E$ in which electrons drift will reduce in time as a result of relaxation of ΔP_s after light pulse transmission. The above circumstances are responsible for the fact that Δn_{res} no longer increases in a crystal. During a light pulse the value of Δn increases and then relaxes to the initial value of Δn_{res}.

With increasing temperature, the probability for thermal excitation of an electron from the donor level is greater. This may be the cause of the reduced time of ΔP_s relaxation. This in turn leads to a shorter time of action of the depolarizing field E_{dep} and smaller value Δn_{res}. Besides, the higher the crystal temperature the higher its conductivity, and accordingly the shorter the time of space charge screening. Evidently, these factors may explain the experimental fact that at about 180 °C no Δn_{res} is observed any longer in a crystal after laser pulse transmission. Meanwhile the change Δn during the laser pulse is observed up to 500 °C. The change Δn during the radiation pulse at high temperatures has been pointed out also in the work by Barkan *et al* (1978).

Variations in the pulse width can be explained as follows. The residual value Δn_{res} is proportional to the number of photo-excited electrons. Using the equation

$$dN/dt = \alpha I M_t \tag{8.41}$$

where N is the density of photo-excited electrons, M_t is the density of filled traps and I is the radiation intensity, we derive

$$N = \alpha I M_t \tau \tag{8.42}$$

where τ is the laser pulse duration. Thus the value of n_{res} is proportional to the product $I\tau$, i.e. it is determined by the laser pulse energy and is independent of its width. It is worthwhile noting that the above consideration took account of only one-photon processes of electron photo-excitation. Allowance for multiphoton processes in the photo-excitation of electrons for high radiation intensities yields a nonlinear dependence of Δn_{res} upon the radiation intensity (Von der Linde and Glass 1975, Dmitriev *et al* 1979). Thus the experimental results obtained are qualitatively interpreted within the polarization model. This permits some conclusions regarding permissible operating conditions of undoped lithium niobate crystals as nonlinear optical elements.

The dynamics of variations in the refractive index of a lithium niobate crystal exposed to pulsed irradiation ($\tau \simeq 3 \times 10^{-8}$ s) has been also studied using the holographic technique (Kanaev and Malinovsky 1974). It is established experimentally that a change in the diffraction efficiency (see formula (8.15)) reproduces within a fairly good accuracy the profile of the function

$$F(t) = \int_0^t I(t)\,\mathrm{d}t \qquad (8.43)$$

where $I(t)$ is the intensity of light incident onto a crystal. According to Kanaev and Malinovsky, this is possible if the dominant role in the mechanism of optical damage belongs to the effect of a changed spontaneous polarization, whereas the features of electron relaxation in the conduction band show up only in insignificant details of the phenomenon. The characteristic lifetime of electrons in the conduction band at a temperature of 20 °C is 10^{-7} s; at -50 °C it is of the order of 10^{-8} s (Hordvik and Schlossberg 1972). The features of optical damage of iron-doped lithium niobate induced by pulsed laser radiation are as follows (Schwartz 1977).

(i)   Optical distortion occurs at an early point of the light pulse action (the measurement error is under 5×10^{-9} s).

(ii)   No optical distortion threshold has been discovered.

(iii)   Optical distortion relaxes in a time characteristic of the Maxwellian relaxation $\varepsilon\rho$ (ε being the dielectric constant, ρ being the crystal resistivity).

(iv)   The dependence of diffraction efficiency upon the total energy of a light pulse is of the second-power nature.

(v)   Diffraction efficiency by the latest point of the laser pulse is independent of the crystal temperature (over the temperature range 20–250 °C).

(vi)   Optical distortion in a lithium niobate crystal is due to the photovoltaic effect.

The above research work provides several practical recommendations. The characteristic time of amplitude modulation of light should exceed markedly the relaxation time of optical distortion in a heated lithium niobate crystal. On the other hand, we can employ lithium niobate crystals in dynamic holography in a wide range of durations (10^4–10^2 s) without impairing the crystal sensitivity to holographic recording.

In iron ion doped crystals of lithium niobate the form of the functional dependence $\Delta n_{st}(I)$ is dictated to a great extent by the degree of doping (Kanaev and Malinovsky 1982). Note that the characteristics of crystals containing under 0.01 wt.% Fe (pure) and over 0.01 wt.% Fe (doped) differ very much. The typical form of $\Delta n_{st}(I)$ for doped samples is given by curve (*a*), while that for pure samples is represented by curve (*b*) in figure 8.23. The dependence $\Delta n(I)$ does not tend to saturation, as follows from all the models, but instead increases with light intensity. For $I \geqslant 10^7$ W cm^{-2} the strength of photo-induced fields runs to 160–200 kV cm^{-1}. These are the values at which a breakdown occurs in the bulk of a material. Breakdown takes place in chaotically distributed micro-regions of about 2 μm in size, rather than in the whole of the illuminated region. For instance, at a pulse width of 5×10^{-8} s and beam intensity of 10^8 W cm^{-2} irreversible damage occurs in the irradiated volume at a density of 10^4 cm^{-3}.

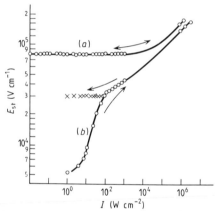

Figure 8.23. Relationship of $\Delta n_{st} \propto E_{st} = K_G/\beta$ to light intensity I for two lithium niobate crystals of various absorption coefficients: (*a*) 0.5; (*b*) 0.04 cm^{-1} ($\lambda_0 = 0.69$ μm) (according to Kanaev and Malinovsky 1982).

We should like to point out that the value of $\Delta n_{st}(I)$ is multiply reversible, i.e. at an increasing or decreasing light intensity I the value of $\Delta n_{st}(I)$ repeats itself. The reversible portion of Δn_{st} runs to 50% in doped samples and to 80% in pure samples.

It is evident from figure 8.23 that the dependences $\Delta n_{st}(I)$ for pure and doped crystals only differ considerably in the intensity range $I \leqslant 10^3$ W cm^{-2}. At $I \geqslant 10^6$ W cm^{-2} the photo-refractive characteristics of pure and doped samples are virtually the same.

In the assumption that behind the photo-refraction mechanism is the photovoltaic effect, Kanaev and Malinovsky were led to conclude that the behaviour of $\Delta n_{st}(I)$ stems from the relationship of the Glass coefficient K_G and quantum yield β to light intensity, and that the absorption coefficient α is independent of light intensity. We should like to stress here that the conclusions of Kanaev and Malinovsky refer to crystal exposure to radiation with a high power density reaching 10^6 W cm^{-2}.

8.4 Laser-induced physical effects in lithium niobate

8.4.1 Short-circuit currents

When a shorted ferroelectric is illuminated by photo-active light, the crystal passes a short-circuit current I_{sc} consisting of a transient and a stationary component (Volk *et al* 1975, Voronov *et al* 1976a,b). Occurrence of photo-refraction in a crystal is always accompanied by variations in the local transient current I_{sc} (Ginzberg *et al* 1975) induced by the pyro-charge formed in an illuminated region of a crystal through its heating by absorbed radiation as well as screening space charge. On subsequent illuminations the transient portion of I_{sc} differs from an initial one by an amount determined by the space charge field E_i.

Figure 8.24 (Volk *et al* 1977) plots the photocurrent I_{sc} in iron ion doped lithium niobate crystals under various illumination conditions. A change in the transient current ΔI_{sc}, and hence the space charge field E_i, are at their maximum when illumination is carried out by a light strip normal to the C axis of a crystal. In this case the space charge field strength amounts to $E_i \approx 1800$ V cm^{-1}.

Occurrence of a substantial internal field in the bulk of a ferroelectric crystal gives rise to some characteristic ferroelectric effects.

When Fe-doped lithium niobate single-domain crystals are exposed to a rather high-intensity (~ 10 W cm^{-2}) laser beam incident normal to the C axis, the curve of decay of I_{sc} displays anomalies accompanied by virtually synchronous discontinuities of birefringence (figure 8.25). The discontinuity Δn ('quasi-breakdown' according to Augustov *et al* 1977) takes place when a certain critical value Δn_{cr} has been attained. Under continuous illumination the discontinuities n are periodically repeated but the mean level Δn_{cr} is preserved. The space charge field E_i set up by the instant of breakdown initiation ($\Delta n_{cr} = 1.7 \times 10^{-3}$) runs to a value of 200 kV cm^{-1}.

Volk *et al* (1977) have shown experimentally that the observed effect is related to local repolarization of a crystal due to an increasing internal field, while current anomalies may be Barkhausen discontinuities. The

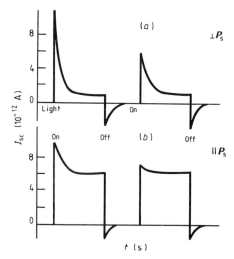

Figure 8.24. Time behaviour of the short-circuit photocurrent I_{sc} in a LiNbO$_3$: Fe crystal accompanying the photo-refraction effect: (a) the damaging light strip is normal to the C axis; (b) it is parallel to the C axis (according to Volk *et al* 1977).

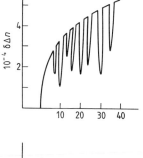

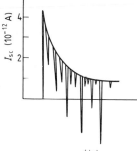

Figure 8.25. 'Discontinuities' in the curve of birefringence $\delta\Delta n$ and anomalies of the short-circuit photocurrent I_{sc} for a LiNbO$_3$: Fe crystal exposed to a He–Cd laser (12 W cm^{-2}) (according to Volk *et al* 1977).

density of discontinuities is determined only by the power density rather than by the damaging light energy. The short-circuit current observed for the first time on lithium niobate crystals by Chen was interpreted by Glass as a photovoltaic current. Initially the photovoltaic current was described by scalar quantities, although the effect was observed in certain crystallographic directions. Belincher *et al* (1977, and Belincher and Sturman 1980) suggested that the photovoltaic current in non-centrosymmetric crystals should be given by a third-rank tensor α_{ijk}. The tensor relationship between the short-circuit current density and the illumination can be written as

$$J_i = \sum \alpha_{ijk} E_j E_k^* \qquad (8.44)$$

where J_i is the current density vector, E_j and E_k^* are components of the light complex electric field vector, α_{ijk} is the third-rank photovoltaic tensor (27 elements) and the asterisk represents complex conjugation. Since $E_j E_k^* = E_j^* E_k$, the α_{ijk} tensor can be described by only 18 independent elements and can be written as a 3×6 matrix using the reduction of jk as given by Nye (1964). The application of the Neumann principle to the α_{ijk} tensor followed by the use of the reduced-subscript notation gives

$$\alpha_{ijk} = \begin{bmatrix} 0 & 0 & 0 & 0 & \alpha_{15} & -2\alpha_{22} \\ -\alpha_{22} & \alpha_{22} & 0 & \alpha_{15} & 0 & 0 \\ \alpha_{31} & \alpha_{31} & \alpha_{33} & 0 & 0 & 0 \end{bmatrix}. \qquad (8.45)$$

Note that $\alpha_{15} = \alpha_{24}$, $\alpha_{22} = -\alpha_{21} = -\alpha_{16}/2$ and $\alpha_{31} = \alpha_{32}$. Thus the bulk photovoltaic effect in lithium niobate can be described by four independent coefficients α_{15}, α_{22}, α_{31}, and α_{33}.

Fridkin and Magomadov (1979) were the first to relate the photovoltaic current in $LiNbO_3$: Fe to the light polarization direction and to establish components of the photovoltaic tensor α_{ijk}. The crystal was illuminated by linearly polarized light at a wavelength $\lambda = 500 \ \mu m$ corresponding to the edge of the absorption band of Fe^{2+} in $LiNbO_3$: Fe. All measurements were carried out at a constant light intensity $I = 2.3 \times 10^{-3} \ W \, cm^{-2}$ and room temperature. The photovoltaic current was measured both in the direction of spontaneous polarization (C axis) and along the x and y axes for various orientations of the light polarization planes. Figures 8.26(*a*), (*b*) present experimental curves for the photovoltaic current J_z, J_y, J_x versus the angle β made by the light polarization plane with the corresponding crystal axis. Expressions for the photovoltaic current in crystals with point group symmetry 3 m are cast in the form

$$J_z = \alpha_{31} I + (\alpha_{33} - \alpha_{31}) I \cos^2 \beta \qquad (8.46)$$

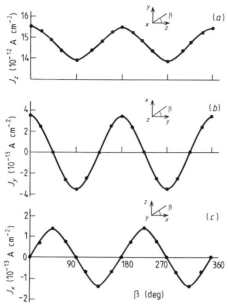

Figure 8.26. Photovoltaic current components J_x, J_y and J_z versus orientation of the light polarization plane in a LiNbO$_3$: Fe crystal. The direction of light propagation is shown in the upper inserts (according to Fridkin and Magomadov 1979).

$$J_y = \alpha_{22}I(1- \sin^2 \beta) \qquad (8.47)$$

$$J_x = \alpha_{15}I \sin 2\beta \qquad (8.48)$$

where I is the light intensity. A comparison of the experimental orientation dependences (figure 8.26) with expressions (8.46)–(8.48) reveals their fairly good agreement. According to theory, the currents J_x and J_y are non-zero only for polarized light and they reverse sign twice as the polarization plane is turned through 360°. The current flowing in the direction of spontaneous polarization has a component independent of the light polarization direction. Perhaps this may explain why the effect of light polarization on the photovoltaic current J_z has not been observed earlier. The amplitudes of the currents J_y and J_x are more than an order of magnitude lower than that of the current J_z. Accordingly, the generated fields E_y and E_x proved also to be an order of magnitude lower than E_z and they did not exceed 200 B cm^{-1}. These fields produced a photo-refractive effect along the z axis: $\delta\Delta n = 10^{-6}$ is more than an order of magnitude smaller than in the case of the transverse photo-refractive effect. The values of the photovoltaic coefficients (Glass coefficients) $K_{ijk} = 1/(\gamma\alpha_{ijk})$, where γ is the absorption coefficient (in lithium niobate $\gamma \approx 4.5$ cm^{-1} at $\lambda = 500$ nm), were found to be as follows: $K_{31} \approx 1.4 \times 10^{-9}$; $K_{33} \approx 1.5 \times 10^{-9}$; $K_{22} \approx 0.5 \times 10^{-10}$; $K_{15} \approx 1.0 \times 10^{-11}$ A cm W^{-1}. The values of K_{31} and K_{33} are close to that of the photovoltaic coefficient determined earlier for LiNbO$_3$: Fe in unpolarized light (Glass and Von der Linde 1976).

8.4.2 Space-oscillating currents

Carrying out experiments on recording dynamic phase gratings by cross-polarized beams in iron-doped crystals of lithium niobate, Odulov (1982) discovered a new mechanism of space charge formation in a crystal, namely space-oscillating currents described by non-diagonal components of the photovoltaic effect tensor. When a lithium niobate crystal is subject to the light wave field

$$E = e_o E_o \exp(ik_o \cdot r) + e_e E_e \exp(ik_e \cdot r) \qquad (8.49)$$

(e_o and e_e being the polarization unit vectors, and E_o and E_e the complex amplitudes of ordinary and extraordinary light waves), an oscillating current arises in the crystal and flows normal to the C axis (Belincher and Sturman 1980):

$$J_{phg} = \alpha^s_{15} E_o E_e \cos(k \cdot r) + \alpha^a_{15} E_o E_e \sin(k \cdot r) \qquad (8.50)$$

$$= [(\alpha^s_{15})^2 + (\alpha^a_{15})^2]^{1/2} E_o E_e \cos[k \cdot r - \tan^{-1}(\alpha^s_{15}/\alpha^a_{15})]$$

and consists of two parts related to the symmetric α^s and antisymmetric α^a components of the photogalvanic tensor (figure 8.27). The oscillating current results in the appearance of a stationary grating of space charge with a wavevector k. The space charge field E_i determines the depth of refractive index modulation

$$E_i = J_{phg}/\sigma_{ph} \qquad (8.51)$$

where σ_{ph} is the crystal photoconductivity and the condition $\sigma_{ph} \gg \sigma_d$ is fulfilled. The resultant grating is detectable by the diffraction of one of the recording beams whose polarization is orthogonal relative to the read-out beam. No diffraction is observed if the polarization plane of a read-out beam is turned through 90°. All this argues in favour of the fact that the observed effect is actually related to space charge redistribution due to the space-oscillating currents.

The efficiency of recorded gratings was low, as compared to that with similarly polarized beams. Thus, in iron-doped lithium niobate crystals exposed to cross-polarized light beams, an oscillating photovoltaic current is excited whose amplitude is of the order of the scalar photovoltaic current.

8.4.3 Photo-Hall effect

Figure 8.28 shows Hall mobility μ_H and dark conductivity σ_d in the

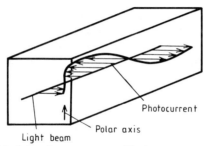

Figure 8.27. Model of a space-oscillating current arising in a $LiNbO_3$: Fe crystal exposed to cross-polarized light beams (according to Belincher *et al* 1977).

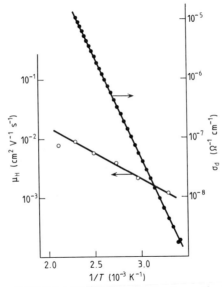

Figure 8.28. Hall mobility μ_H of electrons (the scale on the left) and dark conductivity σ_d (the scale on the right) in a highly reduced crystal of lithium niobate versus inverse temperature (according to Jösch *et al* 1978).

temperature interval from room temperature to $200\,°C$ in reduced lithium niobate crystals (Jösch *et al* 1978). At room temperature the electron mobility is $1.1 \times 10^{-3}\,cm^2\,(V\,s)^{-1}$. It increases with temperature for an activation energy of $0.18\,eV$. These data imply that the charge transfer in lithium niobate is accomplished through the flip-flop mechanism, presumably by small-radius polaritons, as was supposed earlier (Callaerts *et al* 1972). We can expect the same mobility of charge carriers in unreduced crystals of lithium niobate. The basis for this claim is provided by the close value of the photoconductivity activation energy which is around 0.16 eV. The photo-Hall effect in reduced lithium niobate crystals was measured earlier (Ohmori *et al* 1976), but their results sharply differ from those presented above: at room temperature $\mu = 0.8$ cm^2 V^{-1}s^{-1}, i.e. 10^3 times larger. Because of this circumstance it is worthwhile dwelling upon the photo-Hall mobility measurement technique. The lithium niobate crystal was reduced by annealing in a vacuum of ~10^{-5} Torr at 750 °C for 10 h. In the course of reduction there appeared oxygen vacancies in the crystal with an absorption band near 450 nm. Photoelectrons were excited from these levels by light pulses from a xenon lamp at a repetition frequency of 30 Hz. A delayed pulse of voltage V synchronized with the exciting light was applied to a sample along the x axis, while the magnetic field B was applied along the y axis. The accompanying pyroelectric or photovoltaic signal, independent of the magnetic field, was completely compensated by means of a potentiometer. The Hall mobility μ_H (in cm^2 V^{-1} s^{-1}) was calculated according to the formula

$$\mu_H = \frac{c}{H} \frac{\Delta V}{V} \frac{l}{d} \qquad (8.52)$$

where c is a constant (10^8), H is the magnetic field strength (in G), ΔV is the Hall voltage, V is the electrical potential, l is the crystal length along the electric field and d is the crystal thickness along the z axis of the Hall field. The experimental results of the Hall mobility as a function of inverse temperature are depicted in figure 8.29.

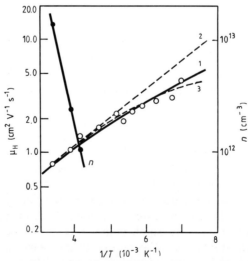

Figure 8.29. Temperature dependence of Hall mobility μ_H of electrons in reduced lithium niobate (according to Ohmori *et al* 1976).

The sign of the Hall voltage indicates that the charge carriers are electrons. The Hall mobility is independent of the magnetic field within 5–15 G and at room temperature it is $\mu_H = 0.8$ cm^2 V^{-1}s^{-1}.

The temperature dependence of the Hall mobility μ_H can be represented by an empirical formula

$$\mu_H = 0.21[\exp(450/T) - 1] \qquad (8.53)$$

(curve 1 in figure 8.29).

Comparing equation (8.53) and the theoretical formula

$$\mu_H = \mu_o[\exp(\theta/T) - 1], \qquad (8.54)$$

one can establish the Debye temperature $\theta_D = 450$ K, which differs from that obtained earlier ($\theta = 560$ K: curve 2 of figure 8.29).

The difference between the Debye temperatures may be attributed to two causes. (i) The phonons by which conduction electrons are scattered differ from those arising upon crystal heating. (ii) At low temperatures conduction electrons may be scattered by impurities. Over the temperature range where optical scattering and scattering by impurities are important, the total mobility μ can be represented in the form

$$1/\mu = 1/\mu_{oph} + 1/\mu_{im} \qquad (8.55)$$

(curve 3 of figure 8.29). Here μ_{oph} is the mobility of electrons scattered by optical phonons, while μ_{im} designates the mobility of electrons scattered on impurities. Substitution of experimental data into formula (8.54) yields the expressions for these mobilities (in cm^2 V^{-1}s^{-1})

$$\mu_{oph} = 0.14[\exp(560/T) - 1] \qquad (8.56)$$

$$\mu_{im} = 9 \qquad (8.57)$$

(curves 2 and 3 in figure 8.29). The same figure also shows the density of charge carriers (curve n). At the present stage of experimental technique the Hall mobility can be measured only in reduced lithium niobate crystals because in unreduced crystals the density of donor impurities is low.

Summing up the results on the Hall effect in lithium niobate crystals, Ohmori *et al* (1976) conclude that: (i) the carriers of conductivity in lithum niobate are electrons; (ii) electrons are scattered by phonons in the optical mode over a temperture range from 143 to 300 K. The difference between the transport mechanisms in the cases of short-circuit photocurrent and photoconductivity leading to a striking difference in mobilities is confirmed by three additional observations. First, the short-circuit photocurrent I_{sc} increases with temperature far more slowly than photoconductivity does. Second, the short-circuit photocurrent is proportional to the light intensity for any photon energy, while photoconductivity varies according to the law $\sigma_{ph} \propto I^{0.6}$ at photon energies within 3.5 to 4.7 eV. Third, the short-circuit photocurrent responds promptly to light illumination, whereas photoconductivity increases for several hours, which is due to the slow growth of carrier density.

The high mobility determined from measured short-circuit photocurrents yields a conclusion about a short path s of an excited electron in the direction of spontaneous polarization commensurable with the lattice parameter

$$I_{sc} = e(\alpha I/h\nu)\beta s. \tag{8.58}$$

This circumstance renders unacceptable the theory of the photovoltaic effect in lithium niobate crystals, developed by Glass and Von der Linde (1976).

The model describing the generation of macroscopic electric fields and explaining the high mobility of photo-excited electrons relies on fluctuations of photo-induced polarization. This model was suggested earlier by Koch *et al* (1976) and developed further by Fridkin (1977). Fluctuations of photo-induced polarization have been demonstrated experimentally by Chanussot *et al* (1977). The fluctuating field affects photo-generated carriers only for a short time comparable to their lifetime prior to recombination, unlike carriers responsible for photoconductivity. The fluctuation model provides a better explanation for the photovoltaic effect in lithium niobate crystals than the asymmetric emission of electrons suggested by Glass.

An alternative explanation of the abnormally high mobility of electrons causing a bulk photovoltaic effect was given by Belincher and Sturman (1980), who relate it to the excess momentum acquired by a non-equilibrium electron in the conduction band. The modulus and direction of the momentum are determined by the asymmetry of excitation, recombination and scattering of a carrier in a non-centrosymmetric crystal. The photo-excited electron contributes to the photovoltaic current in a time shorter than the relaxation time before pulse averaging.

Thus the observed abnormally high effect of the magnetic field upon the photovoltaic current in a lithium niobate crystal is related to the mechanism of the photovoltaic effect, in which case the expression for a transverse Hall current (8.52) includes high mobility of a non-equilibrium electron before it becomes isotropic.

8.5 Photo-induced distortion of the crystal structure in lithium niobate

Lithium niobate is one of the best studied ferroelectrics exhibiting a photo-refraction effect. However, light illumination of this crystal is not only responsible for variations in the refractive index or photocurrent. For instance, Phillips *et al* (1972) observed light scattering in a

$LiNbO_3$:Fe crystal exposed to a single or to two interacting beams (in hologram recording). Shwartz (1977) reports discontinuities in birefringence induced by a focused laser beam. Electric noise arising in illuminated crystals of $LiNbO_3$:Fe was detected by Volk *et al* (1977) and Kovalevich *et al* (1978). The authors reach the conclusion that the noise is related to the partial inversion of spontaneous polarization within the illuminated region. These data suggest that the light affects not only the spatial distribution of electric charge but also the crystal structure of lithium niobate. To answer this question, Abramov *et al* (1978) and Ohnishi (1977, 1978) endeavoured to study the light effect on the $LiNbO_3$:Fe crystal using X-ray diffraction.

Variations in the crystal structure were examined in experiments of two types: (i) half-width measurement of the diffraction maximum deviation curve; (ii) production of X-ray topograms. In the two cases, a two-crystal spectrometer and Cu K_α radiation were used. Samples were 120 and 190 μm thick plates of pure lithium niobate or ones doped with Fe (containing 0.05 and 0.15 wt.% Fe) ($\mu t = 5.1$ and 8.1 respectively). The plates were cut normal to the direction [110], which enabled high-intensity topograms of crystals for interferences (006) and (110) to be produced. The plates were ground, polished and etched in a mixture of HF and HNO_3 acids.

The light effect on crystals was studied as follows. Prior to illumination, deviation curves were plotted and X-ray topograms were produced. The crystals were then exposed to light. This was achieved using a 0.63 μm He–Ne laser, a 0.44 μm He–Cd laser and a mercury lamp. During light illumination the intensity of transmitted X-rays was recorded and the deviation curves were plotted. After the illumination had been discontinued, X-ray topograms for interferences (006) and (110) were taken.

In the two types of experiment, the local effect of light on a crystal reduced the anomalous transmission of X-rays in an illuminated region. This indicated a distortion of the crystal structure. The distortions observed in topograms were the result of deformed planes $(00h)$ perpendicular to the ferroelectric axis. The planes $(hh0)$ parallel to the ferroelectric axis were deformed to a far lesser degree.

Figure 8.30(*a*) is a direct-light topogram of $LiNbO_3$:Fe crystals prior to illumination. The light lines in the topogram are due to the distortions of the crystal lattice on dislocations whose density is around 10^3 cm^{-2}, which is evidence of high-quality samples.

The topogram of figure 8.30(*b*) was produced after the crystal had been irradiated by a focused beam from a He–Ne laser. Here the region of distortions consists of two semicircles where there is no effect of anomalous transmission of X-rays. A narrow strip is seen between the semicircles, where the Borrmann effect persists. In the topogram taken

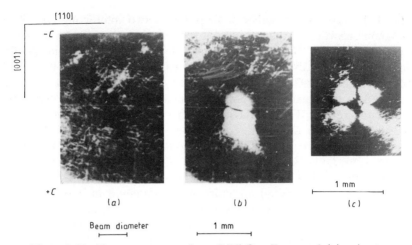

Figure 8.30. X-ray topograms for a LiNbO$_3$: Fe crystal (*a*) prior to and (*b, c*) after the irradiation. The X-ray interferences are (006) (*a, b*) and (110) (*c*). Magnification ×20 (*a, c*) (according to Abramov *et al* 1978).

for interference (110), the Borrman effect disappears from four sectors separated by strips which are parallel and perpendicular to the ferroelectric axis of a crystal and which cross at the centre of an illuminated region (see figure 8.30(*c*)). In the topograms optical damage is largely located in the region of a light beam. In the case of interference (006) the distortions have a 'tail' in the +*C* direction from the light spot.

The distortion patterns observed in the case of a He–Ne laser, a He–Cd laser or a mercury lamp with a filter displayed no qualitative difference. In undoped lithium niobate crystals the effect is less pronounced.

The traces of the light effect on the crystal structure relax very slowly in time. They were still perceptible in topograms for samples stored in the darkness for a month. The distortions were completely removed by illuminating the whole of the sample with a mercury lamp for several minutes or by heating the crystal to 170 °C. In doing so, the initial topographic picture was restored completely.

Upon illumination ferroelectric macro-domains will appear near the +*C* face of a crystal, while near the −*C* face they do not arise. These macro-domains are not stable and disappear rather rapidly. For instance, no domains were seen in a topogram taken seven days after the illumination. The conclusion that the diffraction contrasts are ferroelectric domains has been drawn on the basis of the findings by Sujii *et al* (1973) and Krausslich and Mohring (1974), who also investigated the defect structure of lithium niobate by X-ray topography.

To get insight into the kinetics of formation of laser-induced defects in the lithium niobate crystal, a series of experiments have been performed in which, along with the behaviour of the Laue reflection maximum intensity I, the half-width of the deviation curve β and the diffraction angle 2θ have been measured (figure 8.31). At an early stage of crystal illumination the angle 2θ is observed to decrease slowly, which corresponds to an increasing unit cell dimension. It is remarkable that at this stage the width of the deviation curve hardly changes. After the angle 2θ has reached its minimum value corresponding to the change $\Delta c = 4 \times 10^{-3}$ Å and hence to a relative lattice deformation of 0.03%, the value of 2θ rapidly jumps nearly to its initial level. The point at which the angle 2θ starts to increase coincides, within experimental

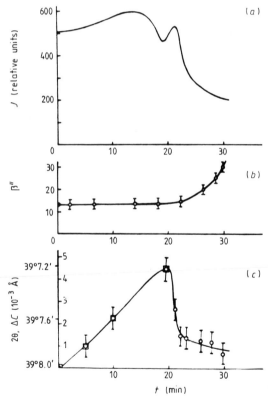

Figure 8.31. (*a*) Diffraction maximum intensity I; (*b*) width of the deviation curve β of this maximum; and (*c*) change of the unit cell dimension Δc (squares) and 2θ (circles) versus time of exposure to laser radiation (laser output power 0.2 W cm^{-2}, $\lambda = 0.44\ \mu$m) (according to Abramov *et al* 1978).

accuracy, with the starts of reflection intensity depletion and deviation curve broadening. The latter is indicative of a marked distortion of the crystal structure. As the laser output power runs to $16\,W\,cm^{-2}$, the deviation curve broadens from $14''$ to $140''$, which corresponds to a defective crystal. After the light is switched off, the deviation curve width relaxes in the first 5 h more than it does in the subsequent 150 h.

To ascertain the nature of processes responsible for diffraction maxima broadening, half-widths of X-ray interferences (006) and (00.12) were measured with a diffractometer for an unilluminated and illuminated samples. Data processing has revealed that the true broadening of X-ray lines spans, within an accuracy of 25%, over the curve $\beta \sim \tan\theta$. This is evidence that the broadening is mainly contributed to by the lattice micro-deformations. The reorientation and micro-deformation effects have been separated so that reorientation is $\Delta\theta = 30''$, while the relative micro-deformation is

$$(\Delta d)/d = 6 \times 10^{-4}. \tag{8.59}$$

To clarify the mechanism of photo-induced distortion of the crystal structure, parallel investigations were made of the polarization optical image of the light-affected region (figure 8.32). The optic axis of a crystal was at an angle of $45°$ to the polarizer and analyser axes. Photo-refraction was induced by a He–Cd laser with a radiation intensity of $100\,mW\,cm^{-2}$. The radiation was incident onto a crystal through a 0.3 mm aperture located at the crystal. The boundary of the light beam coincides with the central circle but the region exhibiting variability of birefringence exceeds considerably the size of the light spot, particularly

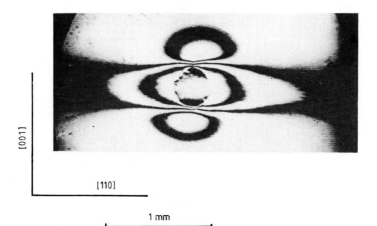

[001]

[110]

1 mm

Figure 8.32. Photomicrograph of a $LiNbO_3$: Fe (0.15 wt.%) crystal between cross polarizers after the action of laser radiation. Magnification $\times$ 52 (according to Voronov 1980).

in the direction normal to the C axis. Figure 8.33 shows the behaviour of birefringence $\Delta(n_e - n_o)$ along the direction [001] calculated from the data of figure 8.32 according to the formula (Mustel and Paryghin 1970)

$$I = I_0 \sin^2 [2\pi\Delta(n_e - n_o)l/\lambda] \qquad (8.60)$$

where l is the crystal thickness along the beam and I and I_0 are the intensities of the transmitted and incident light. The rest of the designations are the same as before.

A change of birefringence in a crystal can be expressed either in terms of variations of the internal electric field or through variations of spontaneous polarization P_s. However, light-induced changes in P_s can take place only in an illuminated region, whereas the space charge

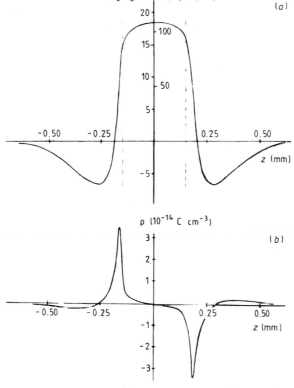

Figure 8.33. Variations of (a) birefringence $\Delta(n_e - n_o)$ and (b) space charge density ρ along the ferroelectric axis [001]. The broken lines show the boundaries of a light spot on a crystal (according to Abramov and Voronov 1979).

electric field extends over a far larger area of a crystal. The experimentally observed change in birefringence spans far beyond an illuminated region, particularly in the direction normal to the ferroelectric C axis. Hence we can assume that birefringence variability is related to the space charge electric field. The distribution of the electric field E_z along the C axis in a crystal coincides with the distribution of Δn presented in figure 8.33. The values of electro-optic coefficients and refractive indices needed for distribution calculations have been borrowed from the book by Fridkin (1979). Allowing for the symmetry of a crystal and action we find the distribution of an electric charge density in a plate

$$\varepsilon_{33} \mathrm{d}E_z/\mathrm{d}z = \rho(z) \tag{8.61}$$

given in figure 8.33. We see a maximum value $\rho = 3.5 \times 10^{14}\,\mathrm{cm}^{-1}$, which is close to the data of Chen (1969).

This distribution is distinguished by a symmetry of the density of positive and negative charges about the centre of an illuminated region, in contrast to Chen's scheme constructed in the assumption that photo-excited electrons leave the illuminated region, thereby rendering it positively charged. It is evident from figure 8.33 that the charge density at the centre of the light spot is zero.

Photo-induced distortion of the crystal structure can be explained as follows. An increase of the unit cell dimension c observed at the initial stage of illumination is related to the macro-deformation of the central part of an illuminated region. One of the causes for the observed macro-deformation may be the inverse piezoelectric effect brought about by the electric field which is set up as a result of the photovoltaic effect (Glass *et al* 1974). A maximum field due to the photovoltaic effect in LiNbO$_3$:Fe is around $10^5\,\mathrm{V\,cm}^{-1}$. Owing to the piezo-effect it produces a deformation $\Delta c/c = 1 \times 10^{-4}$, which is close to the experimentally observed value.

A comparison of figures 8.30(b) and 8.33(b) leads to the conclusion that the disappearance of the Borrmann effect is related to the crystal lattice deformation, which results from micro-deformations around charge-exchanged centres in an illuminated region of a crystal.

As shown by Amodei and Staebler (1972a,b), when LiNbO$_3$:Fe crystals are illuminated by light, Fe ions exchange their charges. Electrons are photo-excited from the Fe^{2+} ions and transferred, through the photovoltaic effect, in the direction of P_s, and are trapped by the Fe^{3+} ions. In a piezoelectric medium charged point defects produce deformations diminishing according to the law $1/r^2$ (Krivoglaz 1967). This in turn causes experimentally observed broadening of diffraction maxima. According to the same author, neutral defects do not invite diffraction maxima broadening.

In the electric field of charge-exchanged centres, atoms are mainly displaced along the C crystal axis. This is related to the ratio of the tensor components of the piezoelectric coefficients $d_{33} \gg d_{13}$ (Fridkin 1979).

In figure 8.30(b) in the direction [001] away from the light spot, one can see a marked distortion of the structure that may be attributed to the onset of micro-domains of opposite signs in a crystal. The existence of the latter was mentioned already by Ohnishi (1977, 1978). However, the author of these works claimed incorrectly that inside an illuminated region the area breaks down into domains. In fact, according to Fridkin *et al* (1974), inside the illuminated region the resultant space charge field E_i is directed along P_s, while beyond this region E_i and P_s are in opposite directions. That is why in these parts acicular domains may arise, whose formation proceeds in fields below threshold ones (Evlanova 1978). They are $\leqslant 1 \ \mu$m in transverse dimension and are not, therefore, resolvable in a topogram. The latter displays only a region of mechanical stress caused by the existence of 180° domain walls. The model of photo-induced distortion of the crystal lattice is shown schematically in figure 8.34.

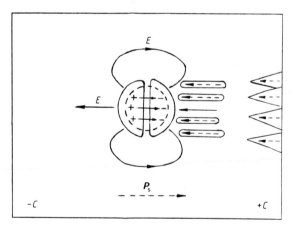

Figure 8.34. Schematic model of photo-induced distortion of the crystal lattice in iron-doped lithium niobate (according to Voronov 1980).

9 Optical Inhomogeneity of Crystals and Methods of its Investigation

9.1 Nature of optical inhomogeneity

In practical applications of lithium niobate crystals to laser radiation control systems, of eminent concern and importance are not only certain specific standard values of optical constants but also the stability of the latter in the bulk of a crystal, i.e. its optical homogeneity.

An optical inhomogeneity may be either relatively stable or variable according to the external effect on a crystal. A stable inhomogeneity of the lithium niobate crystal arises from structural defects and internal stresses that develop in the course of crystal growth and its subsequent technological processing. It is related primarily to the degree of purity and homogeneity of raw materials and to the instability of growth parameters or departure from optimized conditions of crystal thermal treatment. This nonlinearity has its origin in a variable chemical composition (non-uniform impurity distribution, foreign matter) or in various non-uniformities of the crystal lattice (non-uniformly distributed dislocations, growth striations, cellular structure, mosaic structure, twins, schlieren (see chapter 3)). Unlike a congruently grown crystal, a stoichiometric crystal exhibits significant chaotic variability of birefringence in the growth direction (Byer *et al* 1970). Inhomogeneities in the refractive index n may be the result of a changed crystal composition during growth or of internal thermal stresses. Limited variations in this inhomogeneity may arise in the course of long-term high-temperature annealing, providing mass transfer in the anion and cation sublattices. Methods used to remove stable optical inhomogeneities boil down to optimizing the technological conditions of crystal growth and processing (see chapter 2).

328

The other type of optical inhomogeneities in lithium niobate depends strongly on external (light, thermal, electrical, radiation) effects. It shows up particularly in the effect of local reversible change of birefringence in a laser-irradiated lithium niobate crystal, the so-called optical distortion treated in chapter 8.

The effects of temperature and external electric field upon optical inhomogeneities in lithium niobate have been studied less thoroughly. Thermally induced optical inhomogeneity in lithium niobate has been examined by Ivleva and Kuz'minov (1971a, b). When a multidomain crystal is heated up to $T = 120\,°C$, its anomalous biaxiality decreases and approaches the values obtained for single-domain crystals (figure 9.1) (Solov'yeva 1969). Upon cooling the anomalous biaxiality is restored. The author of the above mentioned work believes that the thermally induced optical inhomogeneity is due to the electric field applied perpendicular to the optic axis of a crystal. The existence of electric fields normal to the z axis is confirmed by experiments on annealing the y-cut of a crystal, indicating domain walls parallel to the C axis.

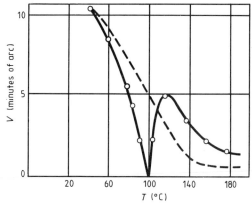

Figure 9.1. Temperature dependence of anomalous biaxiality V of a multidomain lithium niobate crystal on heating (full curve) and cooling (broken curve) (according to Solov'yeva 1969).

Temperature dependence of anomalous birefringence in the lithium niobate crystal was studied using the polarization optical technique (Belobaev *et al* 1973). Figure 9.2 plots typical temperature dependences of anomalous birefringence in single- and multidomain crystals for the same time rates of temperature change. The temperature dependence of lithium niobate birefringence is distinguished by the following features:

(i) the maximum of the $\Delta n(T)$ curve for a multidomain sample is much higher than that for single-domain one;

(ii) the maxima of the $\Delta n(T)$ curves for single-domain samples lie within 60–80 °C for any time rate of temperature decrease;

(iii) the temperature at which Δn disappears is around 150 °C and is independent of the degree of sample poling.

The above features have been verified by optical observations of the behaviour of growth layers having domains which are fixed with changing temperature. Striations present in a crystal vanish at about 120 °C and appear again as the temperature is lowered.

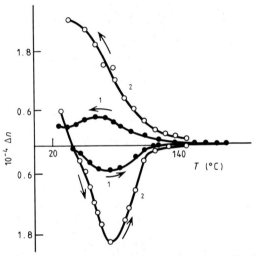

Figure 9.2. Temperature dependence of anomalous birefringence n: (1) for a single-domain; (2) for a multidomain lithium niobate crystal (according to Belobaev *et al* 1973).

Belobaev *et al* (1973) relate thermally induced inhomogeneity of the refractive index to stresses arising in the course of crystal growth and annealing.

This claim is, however, in conflict with the strong non-monotonic temperature dependence at $T < 150$ °C and the practically complete disappearance of n at temperatures lower than the lithium niobate melting point ($T \geq 150$ °C), since these temperatures are too low for stresses to be annealed. A maximum value of the temperature dependence of residual luminous flux (RLF) is also dependent on impurities in a crystal (figure 9.3). The residual luminous flux is determined not only by surface pyro-charges, as was supposed by Belobaev *et al* (1973), but also by pyro-charges arising in bulk inhomogeneities. The effects of surface

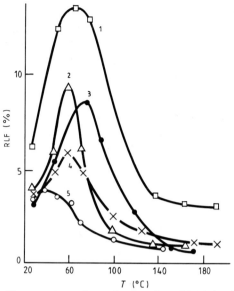

Figure 9.3. Temperature dependence of residual luminous flux in lithium niobate crystals: 1, undoped crystal; 2, 0.5 wt.% WO_3; 3, 0.5 wt.% MgO; 4, 0.1 wt.% CuO; 5, 0.1 wt.% MoO_3 (according to Blistanov *et al* 1978).

and space charges are comparable, while surface conductivity of lithium niobate crystals is insufficient to ensure surface charge compensation during observation. The residual luminous flux is attributed to the electro-optical effect produced by the internal electric fields present in a crystal and normal to its optic axis. These fields are found in some regions of a crystal suffering from structural inhomogeneities that make the pyro-field non-uniform. The temperature dependence of a residual luminous flux over a temperature range 20–150 °C is governed by the competition between the pyro-effect and bulk conductivity compensating the pyro-charge field. Electric fields normal to the z axis were observed in a lithium niobate crystal heated or cooled within 120–430 K with a view to studying spontaneous electric breakdown. At the instant of breakdown the strength of the electric field and optical inhomogeneity induced by this field are changed discontinuously (Avakyan *et al* 1976a, b).

Occurrence of optical inhomogeneities in lithium niobate exposed to the electric field was also observed by Peterson *et al* (1964), who studied its electro-optical properties. They used samples having isolated partially transparent 'blocks' in parallel light between crossed polaroids. A constant electric field was applied along the y axis and the light was made to propagate along the z axis. As the external field is enhanced, birefringence of isolated blocks changes differently. After the field has been removed the induced birefringence is restored to its intial level in about 30 s.

An examination of the polariscopic pictures of relaxation of anomalous birefringence in electric field switching, and an analysis of the temperature dependence of the relaxation time constant of residual luminous flux, have revealed that the nature of electrically induced optical inhomogeneity in lithium niobate is related to the non-uniform extrinsic conduction in a crystal, which causes a space charge in a dielectric. The relaxation of optical inhomogeneities induced by the electric field is due to electron conductivity, while the time constant corresponds to the Maxwellian relaxation time $\tau = 8.8 \times 10^{-12}/\sigma$, where σ is the crystal conductivity.

Induced optical inhomogeneities of lithium niobate affect adversely the operation of electro-optical elements. So it is essential to seek ways of reducing them. By annealing lithium niobate crystals in hydrogen ($T = 500$–$700\,°C$) for times $t = 5$–7 min, we increase the conductivity of their skin, thereby reducing the adverse effect of the pyro-field. In this way the electrically induced residual luminous flux relaxes faster. Reducing annealing impairs sample transparency and the onset of colour centres in the visible spectral region increases photosensitivity and impairs laser damage stability of crystals. Moreover, it decreases the phase-matching temperature of second harmonic generation because of the refractive index inhomogeneity. Optical inhomogeneity of lithium niobate can be temperature-stabilized by doping the crystal with molybdenum, which does not affect sample transmission in the visible (up to 0.3 wt.% MoO_3 in the mixture) (Blistanov *et al* 1978).

9.2 Electrically induced optical inhomogeneity of crystals

Application of an external field to a crystal redistributes its internal electric field and changes the relevant optical inhomogeneity, i.e. it is responsible for optical instability of lithium niobate.

The effect of an external electrostatic field upon crystal optical inhomogeneity was studied by Blistanov *et al* (1981). On raising a sample to a specified temperature, they observed a typical temperature dependence of residual luminous flux: the ratio of the transmitted beam intensity in the case of crossed polarizers to that for parallel polarizers ($I_\perp/I_\parallel$) increases and at $T \approx 320$–340 K reaches its maximum value determined by the competition between the pyro-effect and bulk conduction compensating the pyro-charge field. The height of the maximum for a constant heating rate depends on the type of impurities contained in a crystal. By leaving a crystal standing for 2.5 to 6 h at a specified temperature, we establish a steady-state value of residual luminous flux $I_\perp/I_\parallel$ characteristic of this temperature (figure 9.4). In a steady state, an external electric field E_0 with a strength ranging from 100 to 500 kV m^{-1} was applied to the faces of a sample normal to the x axis. The residual luminous flux arising after the field has been switched on was calculated according to formula (9.1) for E_0 equal either to E_λ or to $E_{\lambda/2}$ and producing for an ordinary and an extraordinary ray in a crystal a phase difference $\Gamma = 2\pi$ or π due to the electro-optical effect:

$$I_\perp/I_\parallel = \frac{\sin^2 2\psi \sin^2(\Gamma/2)}{1 - \sin^2 2\psi \sin^2(\Gamma/2)} \tag{9.1}$$

where ψ is the angle made by the polarizer axis and the semi-axis of the cross-section ellipse of the crystal optical indicatrix normal to the system optic axis.

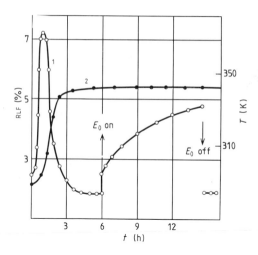

Figure 9.4. Time behaviour of residual luminous flux (1) and temperature (2) of a crystal. The arrows show the instants at which an external electric field E_0 is switched on and off (according to Kudasova 1980).

This method of measuring residual luminous flux excludes the effect of system transparency due to the electro-optical effect.

When the electric field is switched on, residual luminous flux jumps instantaneously. This jump is maximal at $E_0 = E_{\lambda/2}$. Further enchancement of the external electric field will reduce the jump value, its minimum being at $E_0 = E_\lambda$ (figure 9.5). The jump can be compensated partially by turning the crystal about its optic axis and minimized to a value $\text{RLF}_{\min}$, still exceeding the RLF value observed in lithium niobate ahead of field switching.

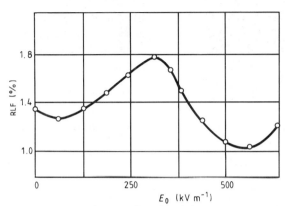

Figure 9.5. Residual luminous flux (RLF) jump versus external electric field E_0 applied to a crystal (according to Kudasova 1980).

Lithium niobate standing in an electric field usually increases the RLF value for any crystal. Figure 9.4 plots residual luminous flux versus the lithium niobate standing time in a field $E_0 = E_{\lambda/2}$ after a jump and further on to a value $\text{RLF}_{\min}$. The RLF value does not emerge onto a steady-state level even after crystals have been left standing in an electric field for long times (up to 50 h). After the field is switched off, the RLF value goes instantaneously to its initial level.

The temperature of a crystal and the field applied have a substantial bearing on optical destabilization of lithium niobate. The transparency distribution at the surface of a lithium niobate sample in parallel light between crossed polarizers changes sharply at the instant of field switching on at any temperature (figure 9.6). At $T < 350$ K, the nature of optical inhomogeneity is virtually the same during the action of an external electric field; the RLF value increases by about 3 to 6%. For $T > 350$ K the nature of optical inhomogeneity changes in time, with a general tendency to transparency of the picture as a whole. At stronger electric fields this process is accelerated. For instance, for $E_0 = 250 \, \text{kV m}^{-1}$, $T = 350$ K and $E_0 = 500 \, \text{kV m}^{-1}$, $T = 345$ K, the RLF value changes during experiment by 1.2–4.75% and 3–6.2% respectively. At high temperatures ($T > 470$ K) and in strong electric fields ($E_0 = 500 \, \text{kV m}^{-1}$) at the surface of the cathode clearly defined dark strips are formed, which gradually spread out and shift towards the

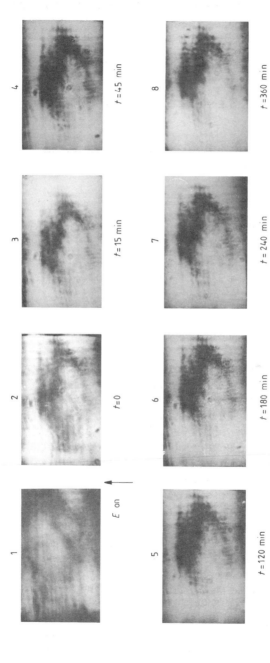

Figure 9.6. Polarization optical patterns characterizing the effect of an external electric field ($E_0 = 250 \text{ kV m}^{-1}$) on the nature of optical inhomogeneity in lithium niobate ($T = 310 \text{ K}$, t is the standing time of a crystal in an electric field) (according to Kudasova 1980).

E on

1   2   3   4

$t = 120$ min $t = 0$ $t = 15$ min $t = 45$ min

5   6   7   8

$t = 180$ min $t = 240$ min $t = 360$ min

anode. In these conditions an optical inhomogeneity consists of alternating light and dark strips. This manifestation of an optical inhomogeneity at high temperatures and in strong electric fields as well as the specific behaviour of residual luminous flux which begins to increase 4–5 h after a jump indicate that the mechanisms through which optical instabilities occur in lithium niobate at low and high temperatures are qualitatively different (figure 9.7) (Blistanov *et al* 1981).

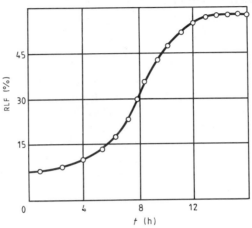

Figure 9.7. Time dependence of the residual luminous flux (RLF) in a lithium niobate crystal exposed to an external electric field $E_0 = 500 \, \text{kV m}^{-1}$ at $T = 480 \, \text{K}$ (according to Kudasova 1980).

The kinetics of occurrence of optical instability in lithium niobate subject to an electric field was studied at a low temperature, not involving the nature of optical inhomogeneity. Here the electric field facilitates manifestation of the optical inhomogeneity already present in a crystal, but does not cause any structural changes related to dielectric ageing or forming.

To eliminate optical instability of lithium niobate in an external electric field, imposing limitations on the application of this material as an electro-optical element in laser radiation control, we have to answer two main questions. (i) What is the cause of a RLF jump when an external electric field is switched on and why does the value of the jump decrease with increasing strength of the field (from $E_{\lambda/2}$ to E_{λ})? (ii) Why does the RLF value change in time when the crystal is exposed to an external electric field?

To explain variations in optical inhomogeneities by the action of an external electric field, Blistanov *et al* (1980) presumed the existence of an uncompensated electric field in a lithium niobate crystal, a field set up, say, by the pyro-effect due to a random change of the crystal temperature.

Structural inhomogeneities of a crystal may give rise to an electric field whose direction will not coincide with the third-order axis in some regions of a crystal. In a simple case, this kind of inhomogeneity may show up in the form of blocks, with individual crystal reorientation determined by the turning of the block crystallophysical axes through angles α_1, α_2 and α_3 (see figure 5.8) about the x, y and z axes of the matrix respectively. Spontaneous polarization in adjacent blocks produces an electric charge on their boundary and components of the internal electric field E_x^i, E_y^i normal to the mean direction of the crystal z axis.

Blistanov *et al* (1980) have examined expression (9.1) used to establish an RLF value with due account of the electro-optical effect brought about by both an external and an internal electric field. The contribution to crystal transparency owing to the deviation of light propagation from the system optic axis can be neglected.

Consider the case of a zero external electric field E_0. The crystal blocks are oriented relative to the x, y and z axes at angles α_1, α_2 and α_3 respectively. As the consequence of block orientation, internal electric field components

$$E_x^i = E^i \sin \alpha_2 \qquad (9.2)$$

and

$$E_y^i = E^i \sin \alpha_1 \qquad (9.3)$$

are produced.

The cross section of an optical indicatrix, perpendicular to the crystal optic axis, is an ellipse whose major semi-axis makes with the y axis an angle ψ defined by

$$\tan (2\psi) = E_x^i/E_y^i = \sin \alpha_2/\sin \alpha_1. \qquad (9.4)$$

The total electric field in a crystal is approximately equal to $(E_x^2 + E_y^2)^{1/2}$. Hence the residual luminous flux is

$$\text{RLF} = I_\perp/I_\|$$

$$= \frac{\sin^2 [\tan^{-1}(\sin \alpha_2/\sin \alpha_1)] \sin^2 (\pi l/\lambda) n_o^3 r_{22} E^i (\sin^2 \alpha_1 + \sin^2 \alpha_2)^{1/2}}{1 - \sin^2 [\tan^{-1}(\sin \alpha_2/\sin \alpha_1)] \sin^2 (\pi l/\lambda) n_o^3 r_{22} E^i (\sin^2 \alpha_1 + \sin^2 \alpha_2)^{1/2}}. \qquad (9.5)$$

Expression (9.5) yields a practically important conclusion: with no external electric field applied, the residual luminous flux in lithium niobate is determined by the block structure of a crystal, which turns the crystal lattice through angles α_1 and α_2 about the x and y axes and leads to non-zero field projections perpendicular to the z axis.

Consider the case where an external electric field $E_0 \parallel x$ is applied to a sample. Employing the same approach, we find electric field components normal to the optic axis

$$E_x = E_i \sin \alpha_2 + E_0 \cos \alpha_2 \cos \alpha_3 \qquad (9.6)$$

$$E_y = E_i \sin \alpha_1 + E_0 \cos \alpha_1 \cos \alpha_3 \qquad (9.7)$$

in lithium niobate. In the presence of the electric field components E_x and E_y the cross section of an optical indicatrix normal to the optic axis is an ellipse whose major semi-axis makes an angle

$$\psi = \tfrac{1}{2} \tan^{-1}(E_x/E_y) \qquad (9.8)$$

with the crystal x axis. The total electric field in a crystal is $(E_x^2 + E_y^2)^{1/2}$. In this case the residual luminous flux is (provided $\alpha_1 \ll \pi/2$)

$$\frac{I_\parallel}{I_\perp} = \frac{1 - \sin^2[\tan^{-1}(E_x/E_y)] \sin^2(\pi l/\lambda) n_o^3 r_{22}(E_x^2 + E_y^2)^{1/2}}{\sin^2[\tan^{-1}(E_x/E_y)] \sin^2(\pi l/\lambda) n_o^3 r_{22}(E_x^2 + E_y^2)^{1/2}} \qquad (9.9)$$

where E_x and E_y are fitted by equations (9.6) and (9.7) respectively.

A comparison of the expressions for the residual luminous flux with an external electric field applied (equation (9.9)) and with no field (equation (9.5)) shows that at $E_0 = 0$ the RLF value is determined by the crystal structural inhomogeneity produced by the rotation of the crystal blocks through angles α_1 and α_2 about the x and y axes. When the external electric field is on, the RLF value is also contributed to by the structural inhomogeneity making the crystal lattice turn through an angle α_3 about the z axis. This circumstance is evidently responsible for the instantaneous discontinuity of the RLF value observed experimentally at the point at which the external electric field is on and off.

For a wave voltage of an external electric field ($E_0 = E_\lambda$) the residual luminous flux is

$$\frac{I_\perp}{I_\parallel} = \frac{\sin^2[\tan^{-1}(E_x/E_y)] \sin^2(\pi l/\lambda) n_o^3 r_{22}(E_x^2 + E_y^2)^{1/2}}{1 - \sin^2[\tan^{-1}(E_x/E_y)] \sin^2(\pi l/\lambda) n_o^3 r_{22}(E_x^2 + E_y^2)^{1/2}} \qquad (9.10)$$

where E_x and E_y are again defined by equations (9.6) and (9.7). If RLF is calculated according to formulas (9.9) and (9.10) for $E_{\lambda/2}$ and E_λ respectively, for the same values of the internal electric field E_i and block reorientation angles α_1, α_2 and α_3, the RLF jump is much higher when the external electric field is equal to a half-wave voltage than when it is equal to the wave voltage. This has also been observed experimentally (figure 9.5).

Optical instability of lithium niobate, i.e. a change in RLF when the crystal is left standing in an external electric field, is due to the fact that the total electric field in a crystal may change as a result of the variability of the space charge density on crystal inhomogeneities owing to the current initiated by the external electric field applied.

Direction of the optical instability process (increasing or decreasing RLF) is determined by the sign and ratio of the external and internal electric fields. Once the components of the external electric field greatly exceed in magnitude those of the internal electric field $E_i \sin \alpha_1$ and $E_i \sin \alpha_2$, the RLF value is expected to go up when the lithium niobate crystal is exposed to the field. If the external electric field is equal to a half-wave one where

$$\frac{\Gamma}{2} = \frac{\pi l}{\lambda} n_o^3 r_{22} (E_x^2 + E_y^2)^{1/2} = \frac{\pi}{2} \qquad (9.11)$$

then any change of the total electric field in a sample, irrespective of the sign (decreasing or increasing), will reduce the function $\sin^2 (\Gamma/2)$ and enhance the residual luminous flux defined by expression (9.9).

If the external electric field is equal to a full-wave one,

$$\frac{\Gamma}{2} = \frac{\pi l}{\lambda} n_o^3 r_{22} (E_x^2 + E_y^2)^{1/2} = \pi \qquad (9.12)$$

then a decrease or increase of the total electric field in a sample will increase the function $\sin^2 (\Gamma/2)$ and the RLF value fitted by equation (9.10).

The residual luminous flux decreases in time only if the external electric field components $E_0 \cos \alpha_2 \cos \alpha_3$ and $E_0 \cos \alpha_1 \cos \alpha_3$ in expressions (9.6) and (9.7) are smaller in magnitude and of different sign, as compared with the internal electric field components. Then, for acute angles α, the value $\Gamma/2$ is near zero and an increase of the total electric field in a sample reduces the function $\sin^2 (\Gamma/2)$ and hence the residual luminous flux defined by formula (9.10). For equal magnitudes of the internal and external electric fields, the net field in a sample is zero and RLF is minimal.

Thus the electro-optical model of an optical inhomogeneity in lithium niobate crystals, suggested by Blistanov and Kudasova, provides an explanation for the instantaneous RLF jump on switching an external electric field on and off as well as for the change in the RLF value when a lithium niobate crystal is left standing in an external electric field.

The kinetics of an RLF change under an external electric field indicates the mechanism of electrical conduction that leads to electric charge redistribution at the interface of structural inhomogeneities, screening of an external electric field, and change of the total electric field in a lithium niobate crystal.

The experimental relationship of a residual luminous flux in undoped lithium niobate to the residence time of a crystal in an external electric field for various temperatures and field strengths is fitted fairly well by an exponential law

$$I_\parallel/I_\perp = (I_\parallel/I_\perp)_{t=0} \exp(-t/\tau). \tag{9.13}$$

The characteristic time τ of the optical instability process in lithium niobate exposed to an external electric field decreases with increasing crystal temperature and external electric field strength. Charge transfer resulting in redistributed electric fields in a crystal is presumably a thermally activated process. Conduction electrons stem from thermal activation rather than from light-induced photo-excitation. This is supported by the fact that the process of optical instability of lithium niobate in the electric field is hardly sensitive to intensity depletion by a factor of 40 of the light incident onto a sample, or to light turning on at an instant of RLF measurement. The dependence of the characteristic time τ upon the crystal temperature is also fitted by an exponential function

$$\tau = \tau_0 \exp(-H/kT). \tag{9.14}$$

The process activation energy H goes from 0.3 down to 0.1 eV as the strength of an external electric field increases from 125 to 500 kV m^{-1} (figure 9.8).

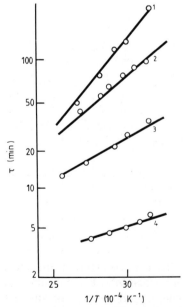

Figure 9.8. Temperature dependence of the characteristic time τ of the optical instability process in a lithium niobate crystal exposed to an external electric field E_0 equal to 125, 250, 375 and 500 kV m^{-1} (curves 1–4 respectively) (according to Blistanov *et al* 1980).

Thus the kinetics investigation of RLF changes due to the external electric field in undoped lithium niobate crystals has shown that electric charge redistribution affecting the total electric field in a crystal is the result of electrical conduction, which has a low activation energy, as compared to extrinsic conduction.

9.3 Doping and heat treatment effects on crystal optical inhomogeneity

It is common knowledge that cationic doping and annealing of lithium niobate in a reducing atmosphere affect profoundly the electro-physical properties and internal electric fields of crystals, and hence also their optical inhomogeneity and instability under an external electric field.

Figure 9.9 shows experimental time dependences of residual luminous flux in doped and vacuum-annealed lithium niobate for various temperatures and strengths of an applied electric field (Kudasova *et al* 1979). Neither introduction of the above dopants nor heat treatment of a crystal removes optical instability of lithium niobate in an external electric field. For instance, the RLF value increased from 1.3 to 2.7% for a hydrogen-annealed sample left in an external electric field for 6 h at $T = 330$ K.

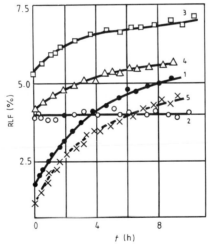

Figure 9.9. Residual luminous flux versus time of application of an electrostatic field $E = 2.5$ kV cm^{-1} in LiNbO$_3$ crystals: 1, pure; 2, molybdenum-doped; 3, iron-doped; 4, copper-doped; 5, magnesium-doped ($T = 60$ °C) (Kudasova *et al* 1979).

An increase with time of the residual luminous flux in an external electric field is described by the same relation (9.13) as for an undoped crystal. The shortest characteristic times have been observed with lithium niobate crystals annealed in vacuum.

The temperature dependence of the characteristic time is defined according to equation (9.14). The activation energy for the process of optical instability in lithium niobate under an external electric field depends on the dopant type and heat treatment of a given crystal. The highest activation energy is noted for vacuum-annealed crystals (table 9.1).

Table 9.1. Activation energy H for the optical instability process in lithium niobate crystals under an external electric field ($E_0 = 250 \, \text{kV m}^{-1}$)(according to Kudasova 1980).

Crystal	H (eV)
Undoped	0.17 ± 0.02
Fe-doped	$H_1 \; 0.22 \pm 0.02$
	$H_2 \; 0.70 \pm 0.05$
Cu-doped	0.20 ± 0.02
Mg-doped	0.25 ± 0.03
Annealed in vacuum	0.40 ± 0.04
Annealed in H_2	0.35 ± 0.03

The temperature curve $\ln \tau = f(T)$ for iron-doped lithium niobate crystals has a knee at about $T = 390 \, \text{K}$ (see chapter 5). Anomalies for these crystals have also been observed earlier; e.g. the temperature dependence of residual luminous flux has two peaks, one being near 380 K (Blistanov *et al* 1978). So the knee of the temperature curve is attributed to two processes of charge transfer differing in their activation energy over this temperature range. The charge transfer with an activation energy of 0.7 eV in lithium niobate is related to extrinsic conduction involving the Fe^{2+} levels.

Electrical conduction having a lower activation energy than extrinsic conduction was observed in lithium niobate by several workers. In particular, Hall mobility of electrons with a 0.18 eV activation energy was found to increase with temperature (Jösch *et al* 1978). The authors think that the low activation energy charge transfer in lithium niobate develops through the hopping mechanism of conduction involving small polarons that are formed, according to Zulberszteln (1976), under simultaneous ionization of atoms of the impurity (Fe^{3+}–Fe^{2+}) and of the host material (Nb^{5+}–Nb^{4+}) at a temperature higher than half the Debye temperature, i.e. at $T > 280 \, \text{K}$ (Lawless 1978). The electron conduction at an activation energy of 0.3–0.2 eV observed by Bollman and Gernand (1972) at room temperature is attributed to the F^1 centres. The screening of an external electric field in lithium niobate crystals at photoconductivity is the result of the space electric charge redistribution with an activation energy of 0.2 eV. This process is related to small trapping centres (see also chapter 8).

Thus, as is evident from the above considerations, there is no single model for low-activation electrical conduction observed in lithium niobate at 290–350 K. To obtain insight into the low activation mechanism of charge transfer and effects of various factors on this process in lithium niobate, indirect methods can be employed, including the crystal internal friction technique treated in chapter 5.

Optical instability of lithium niobate in an external electric field is due to electrical conduction characterized by a low activation energy, where the decisive role belongs to the energy levels related to oxygen vacancies. The levels corresponding to oxygen vacancies are deep localized levels lying 1.1–1.3 eV off the conduction band bottom. As a result, electric charge transfer involving these levels notable for low activation energy is similar to the hopping conduction process in deep localized levels in semiconductors.

Transition of an electron from the occupied level (F^1 centre) to a free one (F centre) occurring upon phonon absorption is made via tunnelling through the potential barrier separating the localized states inside the forbidden band. The electric charge moves along the crystal lattice, thereby redistributing the space charge and changing the electric fields in a crystal. The external electric field bends the energy bands, shifts the local levels and forces the Fermi level to incline. The spacing of energy levels in the forbidden band becomes shorter, while the activation energy for electrical conduction involving these levels decreases, which is observed experimentally.

A similar effect is observable where electrical conduction is due to small trapping centres. The claim that optical instability of lithium niobate at low temperatures is determined by hopping conductivity is supported by the results of the impurity effect upon the temporal behaviour of a residual luminous flux under an external electric field.

Hopping conductivity can be varied by doping a crystal with an electrically active substance. In particular, a donor impurity raises the Fermi level, thereby decreasing hopping conductivity at deep localized energy levels and hence stabilizing the optical properties of a crystal. For lithium niobate this impurity may be Mo^{6+}, whose donor properties were outlined earlier. Molybdenum doping of lithium niobate increases (by two orders of magnitude) the extrinsic conduction observed at $T > 350$–370 K and reduces its activation energy. This will raise the Fermi level in lithium niobate.

Incorporation of the Mo^{6+} dopant into lithium niobate crystals, replacing Nb^{5+} at lattice points, should give rise to an effective positive charge that would reduce the content of F centres and increase the content of F^1 centres occupied completely by electrons. Hence the probability for electron transition from the F^1 to the F level becomes smaller, i.e. hopping conductivity is blocked.

In fact, lithium niobate crystals grown from a Mo^{6+}-containing mixture exhibit no optical instability on exposure to an external electric field for 50 h (Kudasova 1980).

Doping of lithium niobate crystals with manganese should not increase optical instability, since in lithium tantalate and lithium niobate the present manganese is in the Mn^{2+} and Mn^{3+} states and does not increase conductivity markedly (Tsuya 1975).

The suggested mechanism provides one more way for improving optical stability of lithium niobate crystals. Hopping conductivity is blocked, not only by reducing the number of free energy states to which an electron may make a transition from occupied levels (Fermi level rise), but also by decreasing directly the density of the states corresponding to the oxygen vacancies in the forbidden band, i.e. by annealing lithium niobate in the oxygen atmosphere.

The contribution of hopping conductivity to charge transfer is smaller at higher temperatures and enhanced conductivity of crystals annealed in a reducing atmosphere. Here transitions of thermally generated electrons from impurity levels to the conduction band are prominent, as is the case with $LiNbO_3$:Fe at $T > 390$ K. For $T > 470$ K and strong control electric fields, electrical conduction is determined to a large extent by ageing and forming of a dielectric. The latter processes are obviously related to the sharp change of optical inhomogeneity under the electric field and to the change in the behaviour of the residual luminous flux.

Thus molybdenum-doped lithium niobate crystals are more stable optically in an external electric field.

9.4 Methods used to observe optical inhomogeneities in lithium niobate crystals

As mentioned above, determination of a change in the refractive index dependence on crystal composition is a sensitive method for checking slight deviations in the chemical composition of lithium niobate. For instance, a change of 2×10^{-5} in the refractive index is readily detectable in a 1 cm long sample placed between crossed polarizers. This corresponds to an equivalent change by 10^{-3} wt.% in a single component of the melt composition. Elementary chemical analysis is two orders of magnitude less sensitive. However, the optical method measures only integral effects and does not provide information about the nature of impurities.

The optical inhomogeneity B is characterized by the ratio of the difference between the ordinary and extraordinary refractive indices per 1 cm crystal length

$$B = d(n^\circ - n^e)/dl.$$

The ordinary refractive indxed n° is hardly variable with crystal composition, while the extraordinary one n^e changes markedly ($dn^e/dR = -1.63$, where R is the mole fraction of Li_2O: Lerner *et al* 1968, Bergmann *et al* 1968).

9.4.1 Polarization optical method

Optical inhomogeneity of lithium niobate crystals shows up in incomplete light extinction at the output of a system consisting of two crossed polaroids with a crystal in between them. Figure 9.10 is a block diagram of the experimental set-up used by Kudasova (1980) to study optical inhomogeneity of lithium niobate crystals over a temperature range 260–570 K at electrostatic field strengths up to 500 kV m^{-1}.

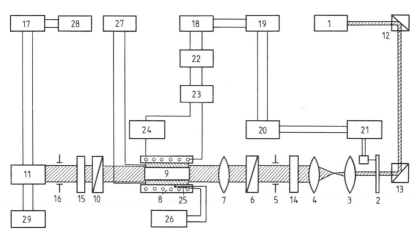

Figure 9.10. Block diagram of the experimental set-up used to measure the residual luminous flux in lithium niobate crystals (according to Kudasova 1980).

The beam from a He–Ne laser (1) is modulated by a mechanical Q-switch (2), which is a disk with openings mounted on the shaft of a micro-motor (21). The motor is supplied by a voltage stabilizer (18) via a step-down transformer (19) and rectifier (20). A telescopic system incorporating a 75 mm (3) and a 50 mm lens (4) spaced by 125 mm in conjunction with an aperture (5) and polaroid (6) creates a plane-polarized parallel light beam of 0.5 to 20 mm in diameter. The beam divergence does not exceed 30″. The rotation angle of a polarizer, which is provided by a polarizing prism, is measured using a graduated circle with a scale factor of 1°. A a rule, vertically polarized light is used. The beam shaping system includes a 160 mm focal length swivel lens (7) to study crystals in convergent light. The focal length of the lens is chosen so that the focus might be at the centre of a furnace (8) into which a crystal (9) is put by means of a sample holder. The design of the latter is such that the sample may be supplied with control voltage from a high-voltage stabilized rectifer (27). The furnace is energized by a voltage stabilizer (18) via an auto-transformer (22) and a rectifier (23). The current in the furnace heater is checked by an ammeter (24), while the crystal temperature is measured by a chromel–alumel thermocouple (25) connected to a potentiometer (26). The furnace is arranged on a special table movable in the directions normal to the light beam.

The light leaving the crystal gets to an analyser (10), which is a polarizing filter housed together with a graduated circle. The light intensity is registered by a photomultiplier (11) supplied by a high-voltage rectifier (29). An iris (16) is placed in front of the photocathode of the photomultiplier.

In the case of parallel polaroids, the light signal intensity was reduced by attenuating filters (14, 15) required for ensuring photomultiplier operation in the linear segment of the current–voltage characteristic.

For the convenience of system adjustment, visualization and photographing of optical inhomogeneities in crystals, a screen (16) may be incorporated in the system. All the above mentioned optical elements (1–16), together with image erecting prisms (12, 13), are mounted on a large optical bench.

A signal from the photomultiplier is fed to a selective amplifier (17) whose output is connected to an oscillograph (28) to display the shape of the signal studied. The selective amplifier is tuned to the light beam modulation frequency. This excludes the noise background in the measurement of the transmitted light intensity.

The intensity of light transmitted through this system is defined by (Sonin and Vasilevskaya 1971)

$$I = I_0 \cos^2(\beta - \psi) - I_0 \sin 2\psi \sin 2\beta \sin^2(\Gamma/2) \qquad (9.15)$$

where I_0 is the intensity of light incident onto a crystal, ψ is the angle made by the oscillations in the polarizer and the major axis of the cross-section ellipse of an optical indicatrix in the plane normal to light propagation, β is the angle between the direction of oscillations in the analyser and the same axis of the ellipse, and Γ is the phase difference of two waves in the crystal studied. Since $2\beta = 180° - 2\psi$, we have $\sin 2\beta = \sin 2\psi$. Optical transmission of the system for crossed ($I_\perp$) and parallel ($I_\parallel$) polaroids is given by the expressions

$$I_\perp = I_0 \sin^2 2\psi \sin^2(\Gamma/2) \qquad (9.16)$$

$$I_\parallel = I_0 [1 - \sin^2 2\psi \sin^2 (\Gamma/2)]. \qquad (9.17)$$

A measure of optical inhomogeneity of lithium niobate crystals is the residual luminous flux equal to the percentage ratio of intensities of light transmitted through a polaroid–crystal–polaroid system with crossed (9.16) and parallel (9.17) polaroids in the absence of an external electric field. The value of residual luminous flux is estimated according to formula (9.1) if the electric field applied to a lithium niobate sample is equal to the wave voltage E_λ and if it produces a phase difference $\Gamma = 2\pi$ for two rays due to the electro-optical effect.

In a perfect crystal for a given experimental geometry $\psi = 45°$ the ratio $I_\perp/I_\parallel$ should be zero. If the applied field is equal to a half-wave voltage $E_{\lambda/2}$, a phase difference $\Gamma = \pi$ arises from the electro-optical effect. The residual luminous flux value is estimated using a ratio which is the reciprocal of (9.1) and which must be zero in a perfect crystal. For actual crystals at $E_0 = 0$ and $E_0 = E_\lambda$, we have $I_\perp/I_\parallel \neq 0$. But if $E_0 = E_{\lambda/2}$ we have $I_\parallel/I_\perp \neq 0$, which just determines the residual luminous flux. At intermediate values of the external electric field a plate $\lambda/4$ is introduced into a system and the residual luminous flux is calculated according to formula (9.1).

The error of RLF measurement in the above optical system does not exceed $\pm 6\%$, while the error of a measured change in RLF due to temperature and external electric field is under $\pm 3\%$.

9.4.2 *Measurement of crystal biaxiality.*

The relationship between crystal biaxiality and its birefringence along the former optic axis is given by the following approximate formula (Indenbom and Tomilovsky 1958):

$$\sin \Omega \approx (\Delta/\Delta_0)^{1/2} \qquad (9.18)$$

where 2Ω is the true angle between the optic axes of a crystal, and $\Delta = n_1 - n_2$ and $\Delta_0 = n_o - n_e$ are the induced and the structure birefringence respectively.

In conoscopic observation the angle 2ε is formed by the rays emerging into the air after they have traversed a crystal along its optic axes. Proceeding to the true angle Ω between the axes, a correction for the refractive index of the crystal is necessary:

$$\sin \Omega = (1/n_0) \sin \varepsilon. \qquad (9.19)$$

Lithium metaniobate single crystals cut along the optic axis were studied. The end faces were polished. Conoscopic patterns were seen with a polarizing microscope. To relate the measured distance d of the optic axes to angle 2 a reference phlogopite plate with $2\varepsilon_{ph} = 10°30'$ and $n_{ph} = 1.55$ was employed. The Mallyar constant thus obtained is

$R = 0.0355$.

It follows from equations (9.18) and (9.19) that

$$\Delta = \Delta_0 R^2 d^2 / n_o^2. \tag{9.20}$$

For lithium metaniobate we have $n_o = 2.2967$, $n_e = 2.2082$ and $\Delta_0 = 0.0885$ for $\lambda = 6000$ Å.

Figure 9.11 plots crystal biaxiality versus boule diameter.

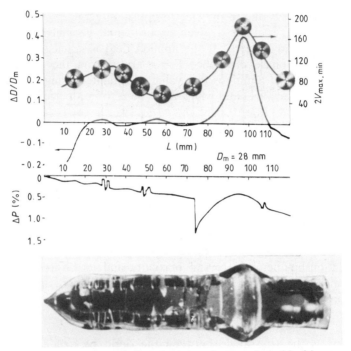

Figure 9.11. Diagrams of distrubutions of anomalous birefringence ($2V_{max}$), crystal diameter deviation ($\Delta D/D_m$), and radiofrequency generator power (ΔP) relative to initial value along boule length L.

9.4.3 Measurement of crystal optical inhomogeneity from the temperature dependence of second harmonic generation.

As is known, random variations in the ability of a crystal to generate second harmonics are related to local variations in the crystal composition that, in turn, change birefringence.

When examining the quality of a crystal for SHG application one should bear in mind that the phase-matching process spreads over a crystal region spanning over light wavelengths (usually 10^4). In this region the relative phase of a wave remains constant. In crystals where the wavefront is distorted markedly in the course of light propagation, no second harmonic is generated.

In lithium niobate crystals, along with random variations in the refractive index arising from temperature fluctuations and thermal stresses, there are observed regular gradients of birefringence along the crystal length. This causes a change in the phase-matching temperature T_m by 1–3 °C per 1 cm crystal length. This effect is assumed to be the result of melt component separation at the melt–solid interface, which contributes to crystal inhomogeneity. The relations needed for experimental data interpretation are presented in chapter 7.

Below we describe a method used to measure refractive indices inhomogeneities from the temperature dependence of second harmonic generation. This method is helpful to study the effect of refractive index inhomogeneities available in a crystal upon the phase-matching temperature T_m and half-width of the curve plotting temperature dependence of the second harmonic yield ΔT_m (Ivleva *et al* 1972).

The effect of transverse inhomogeneity of the refractive indices upon second harmonic generation can be studied with an experimental set-up whose block diagram is presented in figure 9.12.

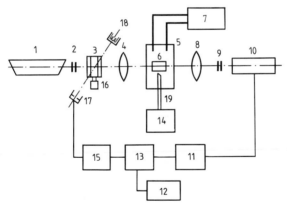

Figure 9.12. Block diagram of the experimental set-up used to study the temperature dependence of second harmonic generation (according to Ivleva *et al* 1972).

The source of fundamental radiation was a trichromatic He–Ne laser (1) oscillating at $\lambda_1 = 1.15\ \mu m$ with an output power of 10 mW. To cut-off the discharge tube radiation, the laser outlet is provided with two IR filters (2). Their total thickness is 4 mm.

Weak-intensity second harmonic generation is recorded synchronously.

To this end, the fundamental laser radiation is modulated by a mechanical chopper (3) (modulation frequency 400 Hz). The chopper is a 20 mm diameter blackened duralumin cylinder having an oval opening.

The modulated radiation is focused by a lens (4) with a focal length $f = 110$ mm near the centre of the lithium niobate crystal (6) placed in a heated chamber (5). The second harmonic radiation ($\lambda_2 = 0.576\ \mu$m) is focused by another lens (8) onto the photocathode of a highly sensitive photomultiplier (10). In front of the photocathode there are blue–green filters (9) cutting off the infrared radiation at $\lambda_1 = 1.15\ \mu$m. To enhance sensitivity, the whole of the light path is enclosed in three interconnected light-proof boxes, which house the modulator, the chamber (5) with a crystal (6), micromotor (16) and a photomultiplier as well as light filter and lenses.

A signal from the photomultiplier is fed to a narrow-band amplifier (11) tuned to the modulation frequency and on to one of the inputs of a sychronous detector (13). The other input receives a reference signal from a photoconductive cell (17) illuminated by an incandescent lamp (18) after it has been amplified by a wide-band amplifier (15). The signal from the synchronous detector is fed to a potentiometer recorder (12). The system is aligned by means of the radiation at $\lambda = 0.63\ \mu$m.

A crystal is placed into a heated chamber (5). The chamber is fitted with an ultrathermostat (7), for a heated silicone liquid to flow through the chamber. The thermostat permits crystal heating up to 250 °C. A crystal together with a brass insert is put into the tube of the chamber around which a heat carrier flows. On the outside the tube is closed by quartz glasses with Teflon layers. The temperature of the crystal is taken by means of a copper–constantan thermocouple (19) whose junction lies in the opening of the brass insert near the crystal. The electromotive force of the thermocouple is measured with a potentiometer (14). The error of temperature measurement is ± 0.1 °C; the rate of heating (cooling) is $1\ \text{K}\,\text{min}^{-1}$.

While recording the temperature dependence of second harmonic generation, temperature marks are applied to the record sheet by a momentary closing of the potentiometer terminal. Measurement is made at isolated points along the x and z axes for each element. To do this, the heated chamber together with the crystal is moved relative to the fundamental radiation along the x and z axes.

The temperature–second harmonic generation plots are used to determine the phase-matching temperature T_m and the half-width ΔT_m at isolated points of each element.

For a congruent lithium niobate crystal the phase-matching temperature T_m of a particular element varied within 0.7 °C, while the half-width ΔT_m varied from 0.7 (for an element from the middle of a boule) to 0.9 °C (for an element from the bottom of the boule).

Along with a relatively high maximum observed at 186 °C, there occurs another maximum at a temperature of 192 °C which is less intense. The appearance of the second maximum is due to the series of adjacent lines in the He–Ne laser radiation spectrum, with wavelengths 1.1118, 1.153, 1.199 and 1.207 μm (Allen and Jones 1967).

Figure 9.13 presents temperature maxima of second harmonics for the three elements of specified refractive indices. The crystal dimensions are $6.5 \times 6.5 \times 10$ mm^3. The laser beam at $\lambda_1 = 1.15$ μm was propagated along the y axis (the long edge of the parallelpiped); the polarization was in the xz plane. The effective lengths calculated according to formula (7.55) are listed in table 9.2. For element 2 the effective length is close to its geometric length; for other crystals it is shorter.

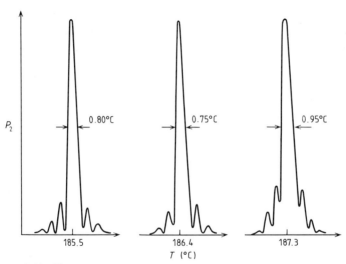

Figure 9.13. Temperature dependence of the second harmonic yield for elements cut from the top, middle and bottom of a lithium niobate crystal boule (according to Ivleva *et al* 1972).

Table 9.2. Characteristics of LiNbO$_3$ crystals grown from the melt with a component ratio Li$_2$O/Nb$_2$O$_5$ = 0.9555 (according to Ivleva *et al* 1972).

Crystal No	Δn (10^{-6})	T_m (°C)	ΔT_m (°C)	L_{ef} (mm)
1	14	185.5	0.8	7.9
2	6	186.4	0.7	9.0
3	34	188.0	0.95	6.3
4	14	185.5	0.85	7.4
5	8	184.4	0.75	8.3

The results obtained indicate that the phase-matching temperature T_m and the half-width ΔT_m are determined by a maximum change in the refractive index and are independent of the location of refractive index inhomogeneity in a bulk crystal. This is evidence that the second harmonic generation process involves a major volume of an irradiated crystal.

9.4.4 Investigation of crystal optical inhomogeneity from the phase-matching angle deviation.

This method consists of measuring the phase-matching angle of second harmonic generation at various points along the growth axis. Since the crystals were grown at an angle of 45° to the optic axis, the growth direction coincides with the channel of wave interaction in the light parametric oscillator. Figure 9.14 shows an optical system for measuring the phase-matching angle for lithium niobate crystals (Ivanova *et al* 1980).

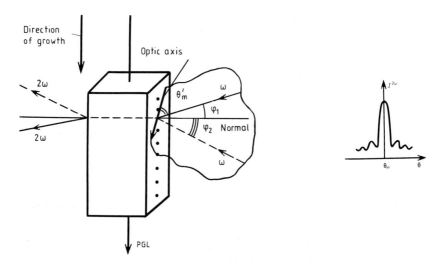

Figure 9.14. Schematic diagram of experiment on the deviation of the phase-matching angle for lithium niobate crystals (according to Ivanova 1979): $\theta'_m + \varphi_1 = 90°$, $\sin\theta_m = (\sin\theta'_m)/n$, θ'_m is the phase-matching angle outside the crystal, n is the refractive index of lithium niobate, PGL is the parametric generation of light.

Variations of crystal birefringence along the channel of interaction of light parametric oscillator waves are related to the variations in its composition, and are responsible for the fact that the phase-matching angle for second harmonic generation is not constant over the crystal length (figure 9.15). Knowing a change in the phase-matching angle, one can calculate a change in birefringence. Since the value of n° for lithium niobate is hardly dependent on the crystal composition (Bergmann *et al* 1968), the change in birefringence $B = n^\circ - n^e$ is primarily attributed to the variations in the refractive index n^e. The relationship between the values of B and phase-matching angle deviation θ_m is found from the expression defining the phase-matching angle:

$$\sin^2\theta_m = [(n_1^\circ)^{-2} - (n_2^\circ)^{-2}]/[(n_2^e)^{-2} - (n_2^\circ)^{-2}] \qquad (9.21)$$

where θ_m is the phase-matching angle inside a crystal, n_2° and n_2^e are the refractive indices for the second harmonic ordinary and extraordinary waves and n_1° is the refractive index for the ordinary pump wave.

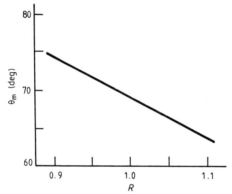

Figure 9.15. Phase-matching angle θ_m inside a lithium niobate crystal versus composition of the melt from which it has been grown: $\lambda = 1.15\ \mu m$ (according to Ivanova 1979).

Differentiating expression (9.21) with respect to θ_m and assuming that only n_2^e is variable, we obtain

$$\Delta n_2^e = \frac{(n_2^e)^3 \cos\theta_m}{\sin\theta_m}\left[\frac{1}{(n_2^e)^2} - \frac{1}{(n_2^\circ)^2}\right]\Delta\theta_m \approx (n^\circ - n^e)\frac{\sin(2\theta_m)}{\sin^2\theta_m}\Delta\theta_m.$$
$$(9.22)$$

To avoid uncertainties due to crystal reorientation, the phase-matching angle is measured on the two sides of the normal to a crystal. Since θ_m is the phase-matching angle inside a crystal, it is in fact the exterior angles φ_1 and φ_2 between the normal to a crystal and two phase-matching directions of the pump beam incident on the crystal that are measured (see figure 9.14). The two angles φ_1 and φ_2 are measured simultaneously only once for a given crystal at its arbitrary point for determining the phase-matching angle θ_m. The relative change in θ_m is defined by measuring the angles φ over the entire crystal length.

Proceeding to the exterior angles φ_1 and φ_2 under the assumption of $n^{\circ} \approx n^{e} \approx n$ and $\sin^2 \theta_m \approx 1$ due to the proximity of θ_m to 90° for lithium niobate, as well as allowing for the small values of φ_1 and φ_2, we obtain

$$\Delta n_2^e \approx (n^{\circ} - n^{e}) \frac{\sin(\varphi_1 + \varphi_2)}{n} \frac{\Delta\varphi}{n} \qquad (9.23)$$

where $\Delta\varphi$ is a maximum change of one of the angles φ over the entire crystal length.

Optical quality of crystals is usually characterized by the parameter of optical inhomogeneity $\partial B/\partial L \approx -\partial n_2^e/\partial L$ (see chapter 7). On the basis of expression (9.23) we derive an expression for the parameter of inhomogeneity

$$\frac{\partial B}{\partial L} \approx -(n^{\circ} - n^{e}) \frac{\sin(\varphi_1 + \varphi_2)}{n^2} \frac{\partial\varphi}{\partial L}. \qquad (9.24)$$

The plot $\varphi_1 = \varphi(L)$ is used to determine the mean value $\partial\varphi/\partial L$ and then an averaged parameter of inhomogeneity of a studied crystal is obtained according to (9.24).

The experimental set-up is shown schematically in figure 9.16. Homogeneity of lithium metaniobate crystals of various compositions has been studied either with a YAG:Nd laser ($\lambda = 1.06\,\mu$m) or with a He–Ne laser ($\lambda = 1.15\,\mu$m). The He–Ne laser is most appropriate for congruently grown lithium niobate, since at room temperature the phase-matching conditions fail with a YAG laser (see figures 9.15 and 9.17). In the work by Ivanova (1979) the pump source was a He–Ne laser providing oscillations at 1.080, 1.152, 1.161, 1.177 and 1.190 μm. In a single transverse mode the laser output radiation was 6 mW, being distributed among lines $1.080\,\mu$m : $1.152\,\mu$m : $1.161\,\mu$m : $1.177\,\mu$m in the proportion 1 : 26 : 6 : 0.4 respectively.

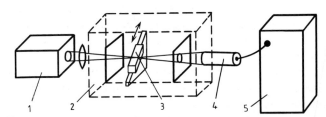

Figure 9.16. Schematic diagram of the experimental arrangement used to measure the phase-matching angle of second harmonic generation: 1, He–Ne laser; 2, opaque box providing angular rotation of a crystal (the input window is formed by IR filters, the output window by blue–green filters); 3, crystal studied; 4, photomultiplier; 5, recording system (according to Ivanova *et al* 1980).

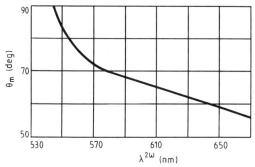

Figure 9.17. Phase-matching angle θ_m of lithium niobate crystals versus second harmonic wavelength $\lambda^{2\omega}$ at room temperature (according to Nelson and Mikulyak 1975).

The laser radiation was focused by a lens ($f = 17$ cm) onto a crystal in a 0.1 mm diameter spot with a confocal parameter $D = 3$ cm. The lithium niobate crystal was put on an alignment table where it was made to turn in the horizontal plane (accuracy 1′) and move across the laser beam in 1.3 mm steps. The alignment table was enclosed in an opaque box, while the controls were brought outwards via rubber seals. The input window of the cell was an IR filter cutting off the external illumination and laser tube emission. The output window was formed by two blue–green filters cutting off the pump radiation. Second harmonic generation and sum frequencies were measured by means of a wide-band amplifier, pulse-amplitude discrimination, pulse counter and photomultiplier. The threshold of the measuring system sensitivity to second harmonic power was $(1-5) \times 10^{-15}$ W. This value ensures reliable measurement of second harmonics and sum frequencies at all generation lines with a signal-to-noise ratio not less than 10^2.

Visualization of the crystal end face via a telescope mounted exactly on the axis of the system provides quick tentative information about the phase-matching angles in a crystal and identification of all the emission lines. Figure 9.18 gives relative intensities and phase-matching angles of all second harmonics and sum frequencies of laser radiation at room temperature, as well as the spectral dependence of θ_m in congruently grown crystals of lithium niobate. The error of phase-matching angles was $\pm 5'$; the error of $\Delta\varphi$ was $\pm 1'$.

Figure 9.19 plots variations of birefringence along the growth axis for some crystals. The characteristics of the crystals and their domain structure are listed in table 9.3. The error of the optical inhomogeneity parameters was 1×10^{-6} for Δn_2^o and 3×10^{-6} for Δn_2^e.

The measured values of $\Delta\theta_m$ indicate two types of inhomogeneity for congruent-composition lithium niobate crystals: continuous variation of

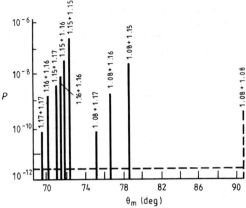

Figure 9.18. Relative intensities and phase-matching angles of the second harmonic and sum frequencies of the He–Ne laser radiation in a congruently grown lithium niobate crystal (the horizontal broken line represents the visual recording threshold) (according to Ivanova *et al* 1980).

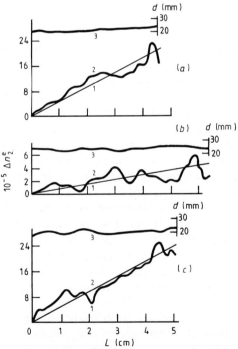

Figure 9.19. Change of birefringence along the crystal axis (1); averaged linear change Δn_2^e (2); change in the crystal diameter (3) of samples No 7(*a*), No 5(*b*) and No 3(*c*) from table 9.3 (*L* is measured from the seed) (according to Ivanova *et al* 1980).

Table 9.3. Domain structure and optical inhomogeneity of lithium niobate crystals (according to Ivanova *et al* 1980).

Crystal No	Domain structure	B (10^{-5} cm^{-1})	θ_m for $\lambda =$ 1.08–1.16 μm
1	Single domain; 9 mm in the bottom multidomain	4.0	72° 42′
2	Single domain; 10 mm in the bottom multidomain	4.0	72° 55′
3	Single domain; 15 mm in the bottom multidomain	4.75	73° 40′
4	Multidomain	6.0	73° 10′
5	Multidomain	0.77	72° 42′
6	Single domain	4.0	75° 40′
7	Single domain	4.5	75° 40′

birefringence along the growth axis and a considerable dispersion in the Δn_2^e values around the mean one.

The inhomogeneity parameter of the samples studied varied between -4×10^{-5} and -6.2×10^{-5} cm^{-1}. The best result was -7×10^{-6} cm^{-1}. The deviation from the mean value $\partial n_2^e/\partial L$ for the crystals studied was 2×10^{-5} to 5×10^{-5} cm^{-1}. The growth conditions, essential as they are in the formation of the domain structure of a crystal, are virtually unimportant for the inhomogeneity parameter. At the same time the value of Δn_2^e and the boule diameter variations are in a clear-cut correlation. Increasing the pulling rate up to 10 mm h^{-1} ensures a stable diameter over the entire crystal length and minimizes the deviation of Δn_2^e from the mean value. A characteristic feature of all the studied crystals is the gradual decrease of Δn_2^e along the growth axis ($\partial n_2^e/\partial L < 0$), which corresponds to an increasing lithium content in a crystal in the course of its pulling. This means that despite the precision preparation of a congruent composition ($R = 0.947 \pm 2 \times 10^{-3}$), the point sought for on the phase diagram has not been obtained.

Ivanova studied the inhomogeneity parameter versus melt composition near the congruent point in the interval $0.900 \leqslant R \leqslant 0.947$.

Despite the great scatter of experimental points, they indicate with sufficient confidence that the smallest inhomogeneity parameter is observed with crystals grown from a melt composition $R = 0.940$ ($\pm 2 \times 10^{-3}$). For a 4 cm long crystal the value of $\partial n_2^e/\partial L$ is 1×10^{-6}, which satisfies the requirements imposed on single-crystal elements of lithium metaniobate applied for light parametric oscillators.

The shift of a congruent composition towards a lower percentage of lithium, observed in this work, may be attributed to the impurities contained in the used reagents.

The lack of correlation between the values of θ_m and $\partial n_2^e/\partial L$ is evidently due to the fact that the phase-matching angle in lithium metaniobate crystals is dependent, not only upon the ratio of major components, but also upon the presence of some impurity ions. The impurity effect on the nonlinear optical properties of lithium meta-niobate has been studied very little, although the effect of changing θ_m upon introduction of certain specific impurity ions is well known (Belasyan *et al* 1977).

The relatively large deviations of Δn_2^e from its mean value indicate the presence of local inhomogeneities in the crystal composition in the region of sharp fluctuations in its diameter. These fluctuations may be the result of unstable operation of the crystallization arrangement, or a non-congruent or contaminated melt composition.

The experimental data obtained do not enable us to estimate crystal composition on the basis of the phase-matching angle. The crystal quality is determined by using the optical inhomogeneity parameter $\partial n_2^e/\partial L$. As the error of the latter is 1×10^{-6} cm^{-1}, the composition is established with an error of $\pm 2 \times 10^{-3}$.

Conclusions

This book is devoted to one crystal—of lithium metaniobate—but its properties and practical applications are so diverse that not all aspects have been taken up here. Preparation of a genuine encyclopedia of lithium niobate would call for collective labour of many physicists, chemists and technologists. Such a book is sure to come out in the future. The first lithium niobate crystals were produced more than twenty years ago but interest in this crystal is not fading; on the contrary, it increases. Fundamental investigations of the effects of various factors on the crystal disclose new phenomena enriching solid state physics and opening up promising new prospects of practical application. There is a specific trend in the development of solid state physics that is related somehow or other to lithium niobate. At present the relevant bibliography contains several thousands of publications. The advent of lithium niobate has been stimulated by research into a wide variety of compounds having a trigonal structure, structures of perovskite, structures of tetragonal tungsten bronze. Common to all of them is the labile crystal structure permiated by many channels and vacant octahedra that can be occupied by various cations, thereby affecting the energy band structure of crystals. Some of the allied compounds of lithium niobate have been widely accepted or are going to enjoy great favour. The authors wish them much success!

References and Further Reading

References

Abrahams S C, Hamilton C W and Reddy J M 1966a *J. Phys. Chem. Solids* **27** 1013

Abrahams S C, Kurts S K and Jamieson P B 1968 *Phys. Rev.* **172** 551–3

Abrahams S C, Levinstein H J and Reddy J M 1966b *J. Phys. Chem. Solids* **27** 1019–26

Abrahams S C, Reddy J M and Bernstein J L 1966c *J. Phys. Chem. Solids* **27** 997–1012

Abramov N A and Voronov V V 1979 *Fiz. Tverd. Tela* **21** 1234

Abramov N A, Voronov V V and Kuz'minov Yu S 1978 *Ferroelectrics* **22** 649

Agheev G V, Bashuk R P, Bebchuk A S, Voicktskayn N S, Gromov D A, Solov'yeva Yu N and Chesnokov A V 1968 *Proc 2nd Symp. on Nonlinear Optics, Novosibirsk* (Nauka, Siberian branch of the USSR Academy of Sciences) pp211–7

Ahrens L 1952 *Geochem. Cosmochem. Acta* 155

Aizy K 1965 *J. Phys. Soc. Japan* **20** 959

Alberts W and Haase C 1969 *Phillips Tech. Rev.* **30** 83, 107, 132

Aleksandrovsky A L 1977 *Domain Structure and Electro-Optical Properties of Barium–Sodium Niobate Single Crystals: PhD Thesis* Moscow University p145

Aleksandrovsky A L, Maskaev Yu A and Naumova I I 1975 *Fiz. Tverd. Tela* **17** 3197–200

Aleksandrovsky A L and Nagaev A I 1983 *Phys. Status Solidi* A**78** 431–8

Allen L and Jones D G C 1967 *Principles of Gas Lasers* (London: Butterworths)

Amodei J J 1971a *RCA Rev.* **32** 185

—— 1971b *Appl. Phys. Lett.* **18** 22

Amodei J J and Staebler D L 1971 *Appl. Phys. Lett.* **18** 540

—— 1972a *RCA Rev.* **33** 71

—— 1972b *Ferroelectrics* **3** 107

Anderson J S 1946 *Proc. R. Soc.* A**185** 62

Anghert N B and Garmash V M 1973 *Elektron. Tekhn. Mater.* No 2 59–63

Anghert N B, Pashkov V A and Solov'yeva N M 1972 *Zh. Eksp. Teor. Fiz.* **26** 1666–72

Antipov V V, Blistanov A A, Sorokin N G and Chizhikov S I 1985 *Kristallogr.* **30** 734–8

Antonov V A, Arsenyev P A, Linda I G and Farshtendiker V L 1975 *Phys. Status Solidi* A**28** 673

Arizmendi L, Cabrera J M and Agullo-Lopez F 1980 *Ferroelectrics* **26** 823–6

—— 1981 *Solid State Commun.* **40** 583–5

—— 1984 *J. Phys. C: Solid State Phys.* **17** 515–29

Armenise M N, Canali C, De Sario M, Carnera A, Mazzoldi P and Celotti G 1983a *J. Appl. Phys.* **54** 62–70

—— 1983b *J. Appl. Phys.* **54** 6223–31

Armstrong J A, Bloembergen N, Ducing J and Pershan R S 1962 *Phys. Rev.* **127** 1918

Ashkin A, Boyd G D, Dziedsic J M, Smith R C, Ballman A A, Levinstein J J and Nassau K 1966 *Appl. Phys. Lett.* **9** 72

Augustov P A and Gotlib V I 1976 *Izv. Latv. SSR* **3** 114

Augustov P A, Gotlib V I, Rubinina N M and Shvarts K K 1977 *Fiz. Tverd. Tela* **19** 1493

Austin I G and Mott N F 1969 *Adv. Phys.* **18**

Avakyan E M, Belobaev K G, Kaminskii A A and Sarkisov K H 1976a *Phys. Status Solidi* A**36** K25–7

Avakyan E M, Belobaev K G and Sarkisov V Kh 1976b *Kristallogr.* **21** 1214–5

Avakyants L R, Kiselev F D and Shchitkov N N 1976 *Sov. Phys.–Solid State* **18** 899–901

Axe J D and O'Kane D F O 1966 *Appl. Phys. Lett.* **9** 58

Azarbayejani G H 1970 *J. Cryst. Growth* **7** 327–8

Ballman A A 1965 *J. Am. Ceram. Soc.* **48** 112–3

Ballman A A, Levinstein H J, Capio C D and Brown H 1967 *J. Am. Ceram. Soc.* **50** 657

Barkan I B, Entin M V and Marennikov S I 1977a *Phys. Status Solidi* A**44** K91–4

Barkan I B, Marennikov S I and Entin M V 1977b *Phys. Status Solidi* A**38** K30

—— 1978 *Phys. Status Solidi* A**45** K17

Barkan I B, Marennikov S I and Pestryakov E V 1977c *Kvant. Elektr.* **4**. 674–6

Batsanov S S 1962 *Electronegativity of Elements and Chemical Bond* (Novosibirsk: Siberian Division of the USSR Academy of Sciences)

Bebchuk A S, Ershov A G, Solov'yeva Yu N, Fadeev V V and Chunaev O N 1968 *Proc. 2nd Symp. on Nonlinear Optics* (Novosibirsk: Nauka) p178

Belasyan R N, Vartanyan E S and Gabrielyan V G 1977 *Abstracts Rep. 3rd Republican Conf. Crystals for Quantum Electronics, Ashtarak, Armenian SSR* vol 1, p89

Belincher V I, Kanaev I F, Malinovsky V K and Sturman B I 1977 *IAN SSSR Fiz.* **41** 733–9

Belincher V I, Malinovsky V K and Sturman B I 1977 *Zh. Eksp. Teor. Fiz.* **73** 692

Belincher V I and Sturman B I 1980 *Usp. Fiz. Nauk* **130** 415

Belobaev K G 1976 *Experimental Investigation of the Nature of Induced Optical*

Inhomogeneities in Lithium Niobate: PhD Thesis (Moscow: IKAN) p160
Belobaev K G, Gabrielyan V G and Sarkisov V Kh 1973 *Kristallogr.* **18** 198–9
Belobaev K G, Markov V B and Odulov S G 1976 *Ukr. Fiz. Zh.* **21** 1820
—— 1978 *Fiz. Tverd. Tela* **20** 2520–2
Belogurov V N, Bylinkin V A, Gotlib I V, Rubinina N M and Sin'kov P E 1976 *Fiz. Tverd. Tela* **18** 142–5
Bergmann G 1968 *Solid State Commun.* **6** 77–9
Bergman J G, Ashkin A, Ballman A A, Driedzic J M, Levinstein H J and Smith R S 1968 *Appl. Phys. Lett.* **12** 92–4
Bernal E, Chen G D and Lee T C 1966 *Phys. Lett.* **21** 259–60
Bernhardt Hj 1979 *Phys. Status Solidi* A**54** 597–603
Bityurin N M, Bredikhin V I and Ghenkin V N 1978 *Kvant. Elektr.* **5** 2453–7
Bjokkholm J E 1968 *Appl. Phys. Lett.* **13** 36–7
Blistanov A A, Danilov A A, Rodionov D A, Sorokin N G, Turkov Yu G and Chizhikov S I 1986 *Kvant. Elektr.* **13** 2536–8
Blistanov A A, Gheras'kin V V and Kudasova S V 1980 *IVUZ Fiz.* 115–7
—— 1981 *Kristallogr.* **26** 356–61
Blistanov A A, Makarevskaya E V, Gheras'kin V V, Kamalov O B and Kablova M M 1978 *Fiz. Tverd. Tela* **20** 2575–80
Blistanov A A, Nosova I V and Taghieva M M 1975 *Kristallogr.* **20** 666–9
—— 1976 *Kristallogr.* **21** 217–20
Bloembergen N 1966 *Nonlinear Optics* (New York: Benjamin) p424
Bloembergen N and Sievers A 1972 *J. Appl. Phys. Lett.* **17** 483
Bocharova N G 1986 *Investigation of Phase Formation at the Surface of Lithium Niobate Crystals: PhD Thesis* (Moscow: Institue of Crystallography of the USSR Academy of Sciences)
Boikova E I and Rozenman G I 1978 *Fiz. Tverd. Tela* **20** 3425
Bokiy G B 1960 *Crystallochemistry* (Moscow: University Press) p357
Bollman W 1983 *Cryst. Res. Technol.* **18** 1147–9
Bollman W and Gernand M 1972 *Phys. Status Solidi* A**9** 301–8
Bollman W and Stöhr M J 1977 *Phys. Status Solidi* A**39** 477–84
Bol'shakov S A, Klyuev V P, Lyapushkin N N, Lyubimov A I and Fedulov S A 1969 *IAN SSSR Neorg. Mater.* **5** 969–71
Bondarenko E I, Zagoruyko V A, Kuz'minov Yu S, Pavolv A N, Panchenko E M and Prokopalo O I 1985 *Fiz. Tverd. Tela* **27** 1037–9
Born M and Huang K 1954 *Dynamical Theory of Crystal Lattices* (London, New York: Academic)
Boyd G D, Ashkin A, Dziedzic J M and Kleinman D A 1965 *Phys. Rev.* **137** A1305
Boyd G D, Miller R C, Nassau K, Bond W L and Savage A 1964 *Appl. Phys. Lett.* **5** 234
Brantov S K and Tatarchenko V A 1983 *Cryst. Res. Technol.* **18** K59
Brice J C 1977 *J. Cryst. Growth* **42** 413–26
Bridenbaugh P M 1973 *J. Cryst. Growth* **19** 45–52
Brown H, Ballman A A and Chin G Y 1975 *J. Mater. Sci.* **10** 1157–60
Burachas S F, Timan B L, Bondar V G and Dubovik M F 1978a *Production and Investigation of Single Crystals* (Kharkov: VNII Monokristallov) vol 1 pp1–6
Burachas S F, Timan B L and Stadnik P E 1978b *Single Crystals and*

Scintillation Materials (Kharkov: VNII Monokristallov) vol 2 pp76–82

Burfoot J C 1967 *Ferroelectrics: An Introduction to the Physical Principles* (Princeton, NJ, Toronto:)

Butyaghin O F 1973 *Investigation of Second Harmonic Generation and Optical Rectification in Lithium Niobate Crystals: PhD Thesis* (Moscow University) p180

Byer R L, Herbst R L, Fiegelson R S and Kway W K. 1974 *Opt. Commun.* **12** 427–9

Byer R L, Park Y K, Feigelson R S and Kway W K 1981 *Appl. Phys. Lett.* **39** 17

Byer R L, Young J F and Feigelson R S 1970 *J. Appl. Phys.* **41** 2320–5

Callaerts R, Denayer M and Nagels P 1972 *Ann. Sci. Rep. BLG* p481

Carpenter R O B 1950 *J. Opt. Soc. Am.* **40** 225

Carruthers J R, Kaminov I P and Stulz L W 1974 *Appl. Opt.* **13** 2333–42

Carruthers J R, Peterson G E, Grasso M and Bridenbaugh P M 1971 *J. Appl. Phys.* **42** 1846–51

Chang I C 1976 *IEEE Trans.* **SU-23** 2–22

Chanussot G, Fridkin V M, Godefroy G and Jannot B 1977 *Appl. Phys. Lett.* **31** 3

Chen F S 1967 *J. Appl. Phys.* **8** 3428

—— 1969 *J. Appl. Phys.* **40** 3389

—— 1970 *IEEE Proc.* **58** 1440–57

Chen F S, Geusic J E, Kurtz S K, Skinner J D and Wemple S H 1966 *J. Appl. Phys.* **37** 388

Chen F S, Macchia Y T and Fraser D V 1968 *Appl. Phys. Lett.* **13** 223–5

Chensky E V 1970 *Fiz. Tverd. Tela* **12** 586

Choy M M and Byer R L 1976 *Phys. Rev.* B**14** 1693

Clark M G, Di Salvo F J, Glass A M and Peterson G E 1973 *J. Chem. Phys.* **59** 6209–19

Cochran W G 1934 *Proc. Camb. Phil. Soc.* **30** 365–75

Cockayne B 1977 *J. Cryst. Growth* **42** 413–27

Courths R, Steiner P, Höchst H and Hüfner S 1980 *Appl. Phys.* A**21** 345–52

Dan'kov I A, Tokarev E F, Kudryashov G S and Belobaev K G 1983 *IAN SSSR Neorg. Mater.* **19** 1165–71

Deighen I F, Odulov S G, Soskin M S and Shanina B D 1975 *Fiz. Tverd. Tela* **16** 1895

De Sario M, Armenise M N, Canali C, Carnera A, Mazzoldi P and Celotti G 1985 *J. Appl. Phys.* **57** 1482–8

Deshmukh K G and Singh K 1972 *J. Phys. D: Appl. Phys.* **5** 1680–5

—— 1973 *J. Phys. D: Appl. Phys.* **6** 1321–3

Di Domenico M Jr and Wemple S H 1968 *Appl. Phys. Lett* **12** 352

—— 1969 *J. Appl. Phys.* **40** 720

Dmitriev V G, Konovalov B E and Shalaev E A 1979 *Kvant. Elektr.* **6** 506

Dubovik M F, Dragaitsev E A and Teplitskaya T S 1973 *IAN SSSR Neorg. Mater.* **9** 334–5

D'yakov V A 1982 *Synthesis and Physico-Chemical Properties of Alkali Metal Metaniobate Single Crystals: PhD Thesis* (Moscow University) p150

D'yakov V A, Luchinsky G V, Rubinina N M and Kholodnykh A I 1981 *Zh.*

Tekh. Fiz. **7** 1557–9

D'yakov V A, Shumov D P, Rashkovich L N and Aleksandrovsky A L 1985 *IAN SSSR Fiz.* **49** 2418–20

Esdaile R J 1985 *J. Appl. Phys.* **58** 1070–1

Evlanova N F 1978 *Domain Structure of Lithium Metaniobate Single Crystals Grown by the Czochralski Technique: PhD Thesis* (Moscow University) p160

Evlanova N F and Rashkovich L N 1974 *Fiz. Tverd. Tela* **16** 555–7

Fay H, Alford W J and Dess H M 1968 *Appl. Phys. Lett.* **12** 89–92

Fedulov S A, Shapiro Z I and Ladyzhinsky P B 1965 *Kristallogr.* **10** 268–9

Feisst A and Kaide P 1985 *Appl. Phys. Lett.* **47** 1125–7

Feisst A and Räuber A 1983 *J. Cryst. Growth* **63** 337

Feng D, Ming N B, Hong J F, Yang Y S, Zhu J S, Yang Z and Wang Y N 1980 *Appl. Phys. Lett.* **37** 607–9

Feng Xi-gi, He Xue-mei and Sche Wei-long 1985 *Chinese Lett.* **2** 417–20

Földvari I, Polgar K and Mecseki A 1984 *Acta Phys. Hungarica* **55** 321–7

Foster N F 1969 *J. Appl. Phys.* **40** 420

Franke H 1984 *Phys. Status Solidi* A**83** K73–6

Franken P A, Hills A E, Peters C W and Weinreich G 1961 *Phys. Rev. Lett.* **7** 118

Fridkin V M 1966 *Pisma Zh. Eksp. Teor. Fiz.* **3** 252

—— 1977 *Appl. Phys.* **13** 357

—— 1979 *Photoferroelectrics* (Moscow: Nauka) p264

Fridkin V M, Boikova E I, Popov B N and Rozenman G I 1978 *Phys. Status Solidi* A**50** K255

Fridkin V B, Grekov A A, Ionov P V, Rodin A I, Savchenko E A and Mikhailina K A 1974 *Ferroelectrics* **8** 433

Fridkin V M and Magomadov P M 1979 *Pisma Zh. Eksp. Teor. Fiz.* **30** 723–6

Fridkin V M and Popov B N 1978 *Ukr. Fiz. Zh.* **126** 657

Fridkin V M, Verchovskaya K A and Popov B N 1977 *Fiz. Tek. Poluprov.* **11** 135

Fukuda T and Hirano H 1975 *Mater. Res. Bull.* **10** 801–6

—— 1976 *J. Cryst. Growth* **35** 127–32

Fushini S and Sugii K 1974 *Japan. J. Appl. Phys.* **13** 1895

Gabrielyan V T 1978 *Investigation of the Growth Conditions and of Some Physical Properties of Electro- and Acousto-Optical Single Crystals—Lithium Niobate, Lead Molybdate and Germanate: PhD Thesis* (Moscow: Crystallography Institute) p196

Gallagher P K and O'Bryan H M 1985 *J. Am. Ceram. Soc.* **68** 147–50

Gan'shin V A, Ivanov V Sh, Korkishko Yu N and Petrova V Z 1986 *Zh. Tekh. Fiz.* **56** 1354–62

Gan'shin V A, Korkishko Yu N and Petrova V Z 1985 *Zh. Tekh. Fiz.* **55** 2224–6

Garmasch V M, Zinkina T P and Lazareva V V 1972 *Cryst. Growth* (Moscow: Nauka) **9** 149–58

Gervais F 1976 *Solid State Commun.* **18** 191–8

Giordmaine J A and Miller R C 1965 *Phys. Rev. Lett.* **14** 973

Glass A M and Von der Linde D 1976 *Ferroelectrics* **10** 163

Glass A M, Von Der Linde D, Auston D H and Negran T J 1975 *J. Elec.*

Mater. **4** 915

Glass A M, Von der Linde D and Negran T J 1974 *Appl. Phys. Lett.* **25** 233

Goldschmidt V M 1926 *Matem. Naturvid Klass* **2** 97

Goodenaugh J B and Kafalas J A 1973 *J. Solid State Chem.* **6** 493–501

Goroshchenko Ya G 1965 *Chemistry of Niobium and Tantalum* (Kiev: Naukova Dumka) p483

Grachev V G and Malovichko G I 1985 *Fiz. Tverd. Tela* **27** 686–9

Graham R A 1976 *Ferroelectrics* **10** 65–9

—— 1977 *J. Appl. Phys.* **48** 2153–63

Granovsky B G 1962 *Kristallogr.* **7** 604

Green G W, Hurle D T J and Joyce G C 1977 *British Patent* 1465191 BO1J 17/18

Gridnev S A, Postnikov V S, Prasolov B N and Turkov S K 1978 *Fiz. Tverd. Tela* **20** 1299–303

Gubkin A N 1978 *Electrets* (Moscow: Nauka) p193

Guenais B, Baudet M, Minier M and Le Cun M 1981 *Mater. Res. Bull.* **16** 643–53

Guinzberg A V, Kochev K D, Kuz'minov Yu S and Volk T R 1975 *Phys. Status Solidi* A**29** 309

Gulyaev Yu V, Proklov V V and Shkerdin G N 1978 *Usp. Fiz. Nauk* **124** 61

Gunter P and Micheron F 1977 *Ferroelectrics* **17** 1

Gurevich G L, Sandler M S and Chertkov Yu S 1973 *Radiotek. Elec.* **18** 2609

Guseva L M, Klyuev V P, Rez I S, Fedulov S A, Lyubimov A P and Tatarov Z I 1967 *IAN SSSR Fiz.* **31** 1161–3

Harris S E 1969 *IEEE Proc.* **56** 5

Hausonne J M 1974 *Bull. Soc. Franc. Ceram.* **103** 33–54

Herrington J R, Dischler B, Räuber A and Schneider J 1973 *Solid State Commun.* **12** 351–4

Hill V G, Gerald K and Limmerman J 1968 *J. Electrochem. Soc.* **115** 978

Ho F D 1981 *Phys. Status Solidi* A**66** 793–806

Hobden M V and Warner J 1966 *Phys. Lett.* **22** 243

Holman R 1978 *US Patent* 4 071 323 23/273 R (BOJD 9/00)

Holman R L, Cressman P J and Revelli J F 1978 *Top. Meet. Integrated and Guided Wave Optics, Salt Lake City, Utah, 1977* (Washington, DC: US Government Printing Office) pp wa 3/1–4

Honigmann B 1958 *Gleichgewichts- und Wachstusformen von Kristallen* (Darmstadt: Steinkopff)

Hordvik A and Schlossberg H 1972 *Appl. Phys. Lett.* **20** 197

Huber W, Granlcher H and Hoffmann R 1970 *Conf. on Single Crystal Growth Oxides, Turnow 1968* (Institute of Monocrystals)

Hulme K H 1971 *IEEE J. Quantum Electron.* **QE-7** 236–9

Hurle D T J 1977 *J. Cryst. Growth* **42** 473–82

Indenbom V L and Tomilovsly G E 1958 *Kristallogr.* **3** 593

Ionov P V 1973 *Fiz. Tverd. Tela* **15** 2827–8

Ishida A, Mikami O, Miyazawa S and Sumi M 1972 *Appl. Phys. Lett.* **21** 192

Ismailzade I G 1965 *Kristallogr.* **10** 287

Ismailzade I G, Nesterenko V I and Mirshly F A 1968 *Kristallogr.* **13** 13

Iu Lin Peng and Bursill L A 1982 *Phil. Mag.* B**45** 911

Ivanova Z 1979 *Investigation of Parametric Oscillation and Amplification in High Optical Quality Lithium Niobate Crystals: PhD Thesis* (Moscow University) p170

Ivanova Z I, Kovrighin A I, Luchinsky G V, Rashkovich L N, Rubinina N M and Kholodnykh A I 1980 *Kvant. Elek.* **7** 1013–8

Ivleva L I and Kuz'minov Yu S 1971a *K. Soobshch. Fiz.* (Moscow: AN SSSR) **8** 3–8

—— 1971b *Kr. Soobshch. Fiz.* (Moscow: AN SSSR) **5** 56–61

Ivleva L I and Kuz'minov Yu S and Osiko V V 1971 *IAN SSSR Neorg. Mater.* **8** 1377–81

Ivleva L I, Kuz'minov Yu S, Osiko V V, Polozkov N M and Prokhorov A M 1983 *DAN SSSR* **268** 69–72

Ivleva L I and Kuz'minov Yu S and Shumskaya L S 1972 *Fiz. Tverd. Tela* **14** 3137–42

Iwasaki H, Toyoda H, Niizeki N and Kubota H 1967 *Japan. J. Appl. Phys.* **6** 1419–22

Jarzebski Z M 1974 *Mater. Res. Bull.* **9** 233–40

Jetschke S and Hehl K 1985 *Phys. Status Solidi* A**88** 193–205

Johnston W D 1970 *J. Appl. Phys.* **41** 3279

Jorgensen P J and Bartlett R W 1969 *J. Phys. Chem. Solids* **30** 2639–48

Jösch W, Munser R, Ruppel W and Würfel P 1978 *Ferroelectrics* **21** 623–5

Kamentsev V P, Nekrasov A V, Ped'ko B B and Rudyak V M 1983 *IAN SSSR Fiz.* **47** 791–3

Kaminov I P 1965 *Appl. Phys. Lett.* **8** 305

Kaminov I P and Carruthers J P 1973 *Appl. Phys. Lett.* **22** 326–8

Kaminov I P and Johnston W D 1967 *Phys. Rev.* **160** 519–22

Kaminov I P and Sharpless W M 1967 *Appl. Opt.* **6** 225–7

Kaminov I P and Turner E H 1966 *Proc. IEEE* **54** 1371–90

Kanaev I F and Malinovsky V K 1974 *Fiz. Tverd. Tela* **16** 3694

—— 1982 *Fiz. Tverd. Tela* **24** 2149–58

Karaseva L G, Bondarenko G P and Gromov V V 1977 *Radiat. Phys. Chem.* **10** 241–5

Katrich M D, Berdinov V F, Bondarenko V S and Shaskol'skaya M P 1975 *IAN SSSR Fiz.* **39** 1044–8

Kaverin L D 1971 *Some Relaxation Processes in Lithum Metaniobate Single Crystals: PhD Thesis* (Voronezh University) p30

Kelly R L 1966 *Phys. Rev.* **151** 721

Ketchum J L, Sweeney K L, Halliburton L E and Armington A F 1983 *Phys. Lett.* A**94** 450–3

Khromova N N 1975 *Effects of Point Defects and Domain Structure on the Properties of Lithium Niobate and Tantalate Crystals: PhD Thesis* (Leningrad) p36

Kim Y S and Smith R T 1969 *J. Appl. Phys.* **40** 4637

King S R, Hartmick T S and Chase A B 1972 *Appl. Phys. Lett.* **21** 312

Kireev P S 1969 *Physics of Semiconductors* (Moscow: Vysshaya Shkola) p590

Klassen-Neklyudova M V 1960 *Mechanical Twinning of Crystals* (Moscow: AN SSSR)

Kleinman D A 1962 *Phys. Rev.* **126** 1977

Klyuev V P, Tolchinskaya R M, Forshtendiker V L and Fedulov S A 1968 *Kristallogr.* **13** 531–3

Kobayashi T, Muto K, Kai J and Kawamori A 1979 *J. Magn. Res.* **34** 459–66

Koch W T H, Munser R and Ruppel W 1976 *Ferroelectrics* **13** 305

Kogelnik H W 1969 *Bell. Syst. Tech. J.* **48** 2909

Kanokov P K, Veryovochkin G E, Goryainov L I, Zaruvinskya L A, Konakov Yu P, Kudryavtzev B B and Tret'yakov G A 1971 *Heat and Mass Transfer in the Production of Single Crystals* (Moscow: Mettalurghiya) p239

Kortov V S, Shvarts K K, Zatsepin A F, Gaprindashvili A I, Gotlib A V and Grant Z A 1979 *Fiz. Tverd. Tela* **21** 1897–9

Kotelyansky I M, Krikunov A I, Medved A V, Mityagin A Yu and Panteleev V V 1977 *Mikroelektr.* **6** 88–9

Kovalevich V I, Shuvalov L A and Volk T R 1978 *Phys. Status Solidi* **A45** 249

Kovrighin A I, Turkin V G, Kholodnykh A I and Chirkin A S 1972 *Opt. Spektrosk.* **33** 752

Kozyreva M S, Arzheukhova N B, Lapshin V I and Rumyantzev A P 1971 *Elektr. Tekhn. Ser. 8 Radiodet.* **4** 75–80

Krätzig E and Kurz H 1977 *J. Electrochem. Soc.* **124** 131–4

Krausslich I and Mohrig H 1974 *Krystall. und Technik.* **9** 811

Krivoglaz M A 1967 *Theory of Scattering of X-Rays and Thermal Neutrons by Actual Crystals* (Moscow: Nauka) p336

Kröger F A 1964 *The Chemistry of Imperfect Crystals* (Amsterdam: North-Holland) p654

Krol D M, Blasse G and Powell R C 1980 *J. Chem. Phys.* **73** 163–6

Kudasova S V 1980 *Investigation of Optical Inhomogeneity in Lithium Niobate Crystals: PhD Thesis* (Moscow: MISiS) p170

Kudasova S V, Bilstanov A A and Geras'kin V V 1979 *Elektr. Tekh. Mater.* **9** 93–6

Kurtz S K and Robinson F N H 1967 *Appl. Phys. Lett.* **10** 62–5

Kurz H, Krätzig E, Keune W, Ergelmann H, Gonser U, Dishler B and Räuber A 1977 *Appl. Phys.* **12** 335–68

Kuz'minov Yu S 1975 *Lithium Niobate and Tantalate—Materials for Nonlinear Optics* (Moscow: Nauka) p224

—— 1982 *Ferroelectric Crystals for Laser Radiation Control* (Moscow: Nauka) p400

Kuz'minov Yu S, Prokopalo O I, Panchenko E M, Zagoruiko B A and Polozkov N M 1983 *Fiz. Tverd. Tela* **25** 758–62

Kuz'minov Yu S, Tikhonov A P and Sakaev R A 1966 *Proc. Institute of Chemical Reagents and Hyperpure Substances* **29** 251–6

Lapitsky A V 1952 *Zh. Obs. Kh.* **22** 36

Lapshin V I and Rumyantsev A P 1971 *Physics of Dielectrics and Prospects for its Development* (Leningrad) vol 1 pp145–7

Laudise R A and Parker R L 1970 *Crystal Growth Mechanisms: Energetics, Kinetics and Transport*, (Englewood Cliffs, NJ: Prentice Hall) p540

Lawless W N 1978 *Phys. Rev.* **17** 1458–9

Lee T C and Zook J D 1968 *IEEE J. Quantum Electron.* **QE-4** 442–54

Lemanov V V 1981 *Ferroelectrics* **35** 123

Lenzo P V, Spencer E G and Nassau K 1966a *J. Opt. Soc. Am.* **56** 633–5

Lenzo P V, Turner E H, Spencer E G and Bullmann A A 1966b *Appl. Phys. Lett.* **8** 81

Lerner P, Legras C and Dumas J P 1968 *J. Cryst. Growth* **3/4** 231–5

Levanyuk A P and Osipov V V 1975 *Fiz. Tverd. Tela* **17** 3595

—— 1977 *IAN SSSR Fiz.* **41** 752

Levanyuk A P, Uyukin E M, Pashkov V L and Solov'yeva N N 1980 *Fiz. Tverd. Tela* **22** 1161–8

Levinstein H J, Bollman A A, Denton R T, Ashkin A and Dziedzic J M 1967 *Appl. Phys.* **38** No 8 3101–2

Linde D and Glass A M 1975 *Appl. Phys.* **8** 85

Lundberg M 1971 *Acta Chem. Scand.* A**25** 3337–46

Maker P D, Terhune R W, Nisenoff M and Savage C M 1962 *Phys. Rev* **8** 21

Maksimov S M, Prokopalo O I and Rayevsky I P 1984 *Proc. 2nd Nat. Conf. on Actual Problems of Production and Application of Ferro- and Piezoelectric Materials, October 1984* (Moscow: NIITEKHIM) p411

Malovichko G I, Karmazin V P, Bykov I P, Laguta V V and Yarunichev V P 1983 *Fiz. Tverd. Tela* **25** 3543–7

Mamedov A M, Osman M A and Hajieva L S 1984 *Appl. Phys* A**34** 189–92

Markov V B, Odulov S G and Soskin M S 1977 *IAN SSSR Fiz.* **41** 821

Marlescu L and Hauret G 1973 *C. R. Acad. Sci., Paris* B**276** 555–8

Matsumura S and Fukuda T 1976 *J. Cryst. Growth* **34** 350–2

Matthias B T and Remeika J P 1949 *Phys. Rev.* **76** 1886

Megaw H D 1954 *Acta Crystallogr.* **7** 187–94

—— 1956 *Ann. Rep. Progr. Chem.* **53** 399

Meisner L B and Rez I S 1969 *Fiz. Tverd. Tela* **11** 2931–8

Michel-Calendini F M, Chermette M and Weber J 1980 *J. Phys. C: Solid State Phys.* **13** 1427–41

Midwinter J E 1967 *Appl. Phys. Lett.* **11** 128–30

—— 1968 *J. Appl. Phys.* **39** 3033–5

Midwinter J E and Warner K 1967 *J. Appl. Phys.* **38** 519

Mikaelyan A L, Koblova M M and Zasovin E A 1971 *Kvant. Elek.* **1** 120–4

Mikaelyan A L, Koblova M M, Zasovin E A and Klyev V P 1970 *Radiotek. Elec.* 937–9

Miller R C 1963 *Phys. Rev.* **131** 95

—— 1964 *Appl. Phys. Lett.* **5** 17

Miller R C, Boyd G D and Savage A 1965 *Appl. Phys. Lett.* **6** 77–9

Miller R C and Nordland W A 1970 *Appl. Phys. Lett.* **16** 174

Miller R C, Nordland W A and Bridenbaugh P M 1971 *J. Appl. Phys.* **42** 4145–7

Miller R C and Savage A 1966 *Appl. Phys. Lett.* **9** 169–71

Ming Nai-ben, Hong Iin-Fen and Feng Duan 1982 *J. Mater. Sci.* **17** 1663

Mirzakhanayan A A 1981 *Fiz. Tverd. Tela* **23** 2452–3

Mustel E P and Paryghin V N 1970 *Methods for Light Modulation and Scanning* (Moscow: Nauka) p295

Mustel E P, Paryghin V N and Solomatin V S 1968 *Vestn. MGU Fiz. Astr.* **1** 109–12

Nagels P 1980 *The Hall Effect and its Application* (New York: Plenum) pp253–80

Nash F R, Boyd G D, Sargent M and Bridenbaugh P M 1970 *J. Appl. Phys.* **41** 2564–8

Nassau K 1967 *Proc. Symp. Ferroelectricity* p259

Nassau K and Levinstein H J 1965 *Appl. Phys. Lett.* **7** 69–70

Nassau K, Levinstein H J and Loiacomo G M 1965 *Appl. Phys. Lett.* **6** 228–9

—— 1966 *J. Phys. Chem. Solids* **27** 989–96

Nassau K and Lines M E 1970 *J. Appl. Phys.* **41** 533–7

Nechaev A S, Pashkov V A and Solov'yeva N M 1972 *Proc. 4th Nat Conf. on Nonlinear Optics, Minsk* pp 53–4

Nelson D F and Lax M 1970 *Phys. Rev.* **24** 379–80

—— 1971 *Phys. Rev.* **33** 2778–94

Nelson D F and Mikulyak R M 1974 *J. Appl. Phys.* **45** 3688

—— 1975 *Appl. Phys. Lett.* **27** 546

Niizeki N, Yamada T and Toyoda H 1967 *Japan. J. Appl. Phys.* **6** 318–27

Ninomiya Y and Motoki T 1972 *Rev. Sci. Instrum.* **43** 519–24

Noda J and Ida I 1972 *Rev. Elec. Commun.* **20** 152–7

Nosova I V 1977 *Investigation of the Mechanical Properties of Crystals Having the Structure of Pseudo-ilmenite: PhD Thesis* (Moscow: Institute of Steel and Alloys) p184

Nowick A S and Berry B S 1972 *Inelastic Relaxation in Crystalline Solids* (New York: Academic Press) p472

Nye J F 1964 *Physical Properties of Crystals* (Oxford: Clarendon) p385

O'Bryan H M, Gallagher P K and Brandle C D 1985 *J. Am. Ceram. Soc.* **68** 493–6

Odulov S G 1982 *Pis'ma Zh. Eksp. Teor. Fiz.* **35** 10–15

Ohmori Y, Yamaguchi M, Yoshino K and Inuishi Y 1976 *Japan. J. Appl. Phys.* **15** 2263–4

—— 1979 *Japan. J. Appl. Phys.* **18** 79–84

Ohmori Y, Yasojoma Y and Inuishi Y 1974 *Appl. Phys. Lett.* **25** 716–7

—— 1975 *Japan. J. Appl. Phys.* **14** 1291–300

Ohnishi N 1977 *Japan. J. Appl. Phys.* **16** 1451

—— 1978 *Proc. 1st Meeting on Ferroelectrics, Kyoto* p145

Ohnishi N and Zizuka T 1974 *Ferroelectrics* **7** 269

—— 1975 *J. Appl. Phys.* **46** 1063

Okamoto E, Ikeo H and Muto K 1975 *Appl. Opt.* **14** 2453

Ormont B F 1969 *Variable-Composition Compounds* (Leningrad: Khimia)

Palatnik L S, Koshkin V M, Belova E K and Rogacheva E I 1969 *Variable-Composition Compounds* (Leningrad: Khimiya)

Pareja R, Gonzales R and Pedrosa M A 1984 *Phys. Status Solidi* A1 179–83

Parfitt H T and Robertson D S 1967 *Br. J. Appl. Phys.* **18** 1709

Pashkov V A and Solov'yeva N M 1978 *Proc. 4th Nat. Meeting on Nonresonance Interaction* (Leningrad) p366

Pashkov V A, Solov'yeva N M and Anghert N B 1979a *Fiz. Tverd. Tela* **21** 92–8

Pashkov V A, Solov'yeva N M and Uyukin E M 1979b *Fiz. Tverd. Tela* **21** 1879–82

Pestryakov E V and Entin M V 1976 *Avtometriya* **4** 13

Peterson G E, Bollman A A, Lenzo P V *et al* 1964 *Appl. Phys. Lett.* **5** 62–3

Peterson G E and Bridenbaugh P M 1968 *J. Chem. Phys.* **48** 3402
Peterson G E and Carnevale A 1972 *J. Chem. Phys.* **56** 4848–51
Peterson G E and Carruthers J R 1969 *J. Solid State Chem.* **1** 98
Petrosyan A K, Khachtryan R M and Sharoyan E G 1984 *Fiz. Tverd. Tela* **26** 22–8
Peurin J C and Tasson M 1976 *Phys. Status Solidi* A**37** 119
Phillips W, Amodei J J and Staebler D L 1972 *RCA Rev.* **33** 94
Pockels F 1906 *Lehrbuch der Kristallooptik* (Leipzig)
Postnikov V S, Kaverin L D, Pavlov V S and Turkov S K 1971 *IAN SSSR Fiz.* **35** 1918–20
Prokopalo O I 1979 *Fiz. Tverd. Tela* **21** 3073
Rakova E V, Bocharova N G, Belughina M V and Semiletov S A 1986 *IAN SSSR Fiz.* **50** 501–4
Räuber Armin 1978 *Chemistry and Physics of Lithium Niobate: Current Topics in Materials Science* (Amsterdam: North-Holland) **1** pp 481–601
Redfield D and Burke W J 1984 *Act. Phys. Hung.* **45** 4566–70
Red'kin B S, Kurlov V N and Tatarchenko V A 1985 *IAN SSSR Fiz.* **49** 2412–4
—— 1987 *J. Cryst. Growth* **82** 106–9
Red'kin B S, Satunkin G A, Tatarchenko V A, Umarov L M and Gubina L I 1983 *IAN SSSR Fiz.* **47** 386–91
Reisman A and Holtzberg F 1958a *J. Am. Chem. Soc.* **80** 35
—— 1958b *J. Am. Chem. Soc.* **80** 6503–7
Roitberg M B, Novik V K and Gavrilova N D 1969 *Kristallogr.* **14** 938–9
Rosa J, Polak K and Kubatova J 1982 *Phys. Status Solidi* B**11** K85–7
Rosenblum B, Braunlich P and Carrico J P 1974 *Appl. Phys. Lett.* **25** 17–9
Rozenman G I and Boikova E I 1979 *Fiz. Tverd. Tela* **21** 1888–90
Rubinina N M 1976 *Investigation of the Mechanism of Iron Incorporation in Nonstoichiometric Lithium Niobate Crystals: PhD Thesis* (Moscow University) p150
Ryvkin S M 1963 *Photoelectrical Phenomena in Semiconductors* (Moscow: Fizmatghuz) p494
Satch B T, Inui J and Iwamoto H 1976 *Fujitsu Sci. Techn. J.* March **93** 133
Satunkin G A, Red'kin B S, Kurlov V N, Rossolenko S N, Tatarchenko B A and Tuflin Yu A 1985 *IAN SSSR Fiz.* **49** 2319–23
Savage A 1966 *J. Appl. Phys.* **37** 3071–2
Schempp E, Peterson G E and Carruthers J R 1970 *J. Chem. Phys.* **53** 306–11
Schirmer O F and Von der Linde D 1978 *Appl. Phys. Lett.* **33** 35–8
Schubert M and Wilhelmi B 1971 *Einführung in die nichtlineare Optik* (Leipzig: B G Teubner) p243
Scott B and Burns G 1972 *J. Am. Ceram. Soc.* **55** 225–9
Selynk B 1973 *Ferroelectrics* **6** 37
Serreze H B and Goldner R B 1975 *Appl. Phys. Lett.* **22** 626
Shapiro Z I, Fedulov S A, Venevtsev Yu N and Rigerman L G 1975 *Kristallogr.* **10** 869–74
Shapiro Z I, Trunov V K and Shupilov V V 1978 *Reagents and Extrapure Substances* (Moscow: NIITEKHIM)
Sheibaidakova T I, Arkhangel'skaya N S, Garmash V M and Anghert N B 1976 *Proc. Moscow Inst. of Steel and Alloys* **88** 133–7

Shi Z J, He Z G and Ying C F 1980 *Ultrasonics* **18** 57–60

Shimura F 1977 *J. Cryst. Growth* **42** 579–82

Shiosaki J and Mitsui T 1963 *J. Phys. Chem. Solids* **24** 1057

Shwartz K K 1977 *IAN SSSR Fiz.* **41** 788

Sirota N N and Yakunishev V P 1974 *IAN BSSR Fiz. Mater.* 119–20

Smakula P H and Claspy P C 1967 *Trans. Metallurg. Soc. AIME* **289** 21–4

Smith R T and Welsh F S 1971 *J. Appl. Phys.* **42** 2219–30

Smolensky G A, Bokov V A, Isupov V A, Krainik N N, Pasynokv R E and Shur M S 1971 *Ferroelectrics and Antiferroelectrics* (Moscow: Nauka) p476

Smolensky G A, Krainik N N, Khucha N P, Zdanova V V and Mylnikova I E 1966 *Phys. Status Solidi* **13** 309

Solov'yeve Yu N 1969 *Optical and Electro-Optical Properties of Lithium Metaniobate: PhD Thesis* (L'vov University) p150

Somekh S and Yariv A 1972 *Opt. Commun.* **6** 301

Sonin A S and Lomonova L G 1967 *Fiz. Tverd. Tela* **9** 3315–7

Sonin A S and Vasilevskaya A S 1971 *Electro-optical Crystals* (Moscow: Atomizdat) p327

Soroka V V, Khromova N N and Klyuev V P 1974 *Zh. Prikl. Spektrosk.* **20** 541–3

Spencer E G, Lenzo P V and Ballman A A 1967 *Proc. IEEE* **55** 5

Stepanova A V 1986 *Effects of Electro- and Mass Transfer on Optical Inhomogeneity of Lithium Niobate Crystals: PhD Thesis* (Moscow: Institute of Steel and Alloys) p185

Sue P 1937 *Ann. Chem. Lett.* **7** 493

Suematsu I, Akiyama K, Ioki H and Sasaki Y 1972 *Trans. Inst. Electron. Commun. Eng.* C**55** 106

Sugii K, Iwasaki, Miyazawa S and Niizeki N 1973 *J. Cryst. Growth* **8** 159–66

Svaasend L O, Eriksrud M, Grande A P and Mo F 1973 *J. Cryst. Growth* **18** 179–84

Svaasend L O, Eriksrud M, Nakken G and Grande A P 1974 *J. Cryst. Growth* **22** 230–2

Sweeney K L and Halliburton L E 1983 *Appl. Phys. Lett.* **43** 336

Sweeney K L, Halliburton L E, Bryan D A, Rice R R, Gerson R and Tomaschke M E 1984 *Appl. Phys. Lett.* **45** 805–7

—— 1985 *J. Appl. Phys.* **57** 1036–44

Tasson M, Legal H and Preusin J C 1975 *Phys. Status Solidi* A**31** 729

Tatarchenko V A and Brener E A 1976 *IAN SSSR Fiz.* **40** 1456–77

Thaniyavaran S, Findakly T, Booher D and Moen J 1985 *Appl. Phys. Lett.* **46** 933–5

Timan B L and Burachas S F 1977 *Physics and Chemistry of Crystals* (Khar'kov: Research Institute of Single Crystals) pp1–4

—— 1978 *Single Crystals and Scintillation Materials* (Khar'kov: Research Institute of Single Crystals) vol 2 pp70–5

—— 1981 *Kristallogr.* **26** 892–4

Townsend R L and La Macchia J T 1970 *J. Appl. Phys.* **41** 5188

Tsinzerling L G 1976 *Proc. Moscow Inst. Steel and Alloys* **88** 108–14

Tsuya H 1975 *J. Appl. Phys.* **46** 4323–33

Turner E H 1966 *Appl. Phys. Lett.* **8** 303–4

Turner E H, Nash F R and Bridenbaugh P M 1970 *J. Appl. Phys.* **41** 5278–82

Ubbelohde A R 1965 *Melting and Crystal Structure* (Oxford: Clarendon) p420

Valyshko E G, Varina T M, Kuz'min R N, Rubinina N M, Smironv V A and Schagdarov VB 1974 *Zh. Prikl. Spektrosk.* **21** 50–4

Vasilevskaya A S, Sonin A S, Rez I S and Plotniskaya T A 1967 *IAN SSSR Fiz.* **31** 1159–60

Vere A K 1968 *J. Mater. Sci.* **3** 617–21

Vinetsky V L, Kukhterev N V, Markov V B, Odulov S G and Soskin M S 1977 *IAN SSSR Fiz.* **41** 811

Vladimirtsev Yu V, Glebova N N, Golenishchev-Kutuzov V A, Migachev S A and Rez I S 1983 *Khim. Fiz.* **3** 358–61

Volk T R, Guinzberg A V, Kovalevich V I and Shuvalov L A 1977 *IAN SSSR Fiz.* **41** 784–7

Volk T R, Kochev K D and Kuz'minov Yu S 1975 *Kristallogr.* **20** 583

Volk T R, Kovalevich V I and Kuz'minov Yu S 1978 *Ferroelectrics* **22** 659–61

Von der Linde D and Glass A M 1975 *Appl. Phys.* **8** 85

Von der Linde D, Glass A M and Rodgers K F 1974 *Appl. Phys. Lett.* **25** 152

Voronov V V 1980 *Photoelectrical and Photorefractive Properties of Ferroelectric Niobate Crystals: PhD Thesis* (Moscow: FIAN) p180

Voronov V V, Ionov P V, Kuz'minov Yu S, Nabatov V V and Osiko V V 1976b *Fiz. Tverd. Tela* **18** 286

Voronov V V, Kuz'minov Yu S and Lukina I N 1976a *Fiz. Tverd. Tela* **18** 1047

Voronov V V, Zharikov E V, Kuz'minov Yu S, Oskio V V, Tobis V I and Schumskaya L S 1974 *Fiz. Tverd. Tela* **16** 162–6

Wainer E and Wentworth C 1952 *J. Am. Ceram. Soc.* **35** 207–14

Wallace C A 1970 *J. Appl. Crystallogr.* **3** 546

Warner A M, Onoe M and Coquim G A 1967 *J. Acoust. Soc. Am.* **42** 1223–31

Wax S J, Chodorow M and Puthoff H E 1970 *Appl. Phys. Lett.* **16** 157

Weber H P, Mathieu E and Meyer K P 1966 *J. Appl. Phys.* A**37** 3584

Weis R S and Gaylord T K 1985 *Appl. Phys.* A**37** 191–203

Wemple S H, Di Domenico M and Camlibel J 1968 *Appl. Phys. Lett.* **12** 209–11

Werner J, Robertson D S and Hulme K F 1966 *Phys. Lett.* **20** 168

Wicks B J and Levis M H 1968 *Phys. Status Solidi* **26** 571

Wiston C D and Smith A J 1965 *Acta Crystallogr.* **19** 169–73

Wood E A 1951 *Acta Crystollogr.* **4** 353–62

Wood E A, Hartman N F and Verber C M 1974 *J. Appl. Phys.* **45** 1449–51

Xu Run Yuan 1983 *J. Chinese Silicon Soc.* **11** 120–2

Yamada T 1981 *Landolt-Börnstein Numerical Data and Functional Relationships in Science and Technology, New Series, Group III* ed K-H Hellwege vol 16a (Berlin: Springer) pp149–56, 489–99

Yamada T, Niizeki N, Iwasaki H and Toyoda H 1968 *Rev. Elec. Commun.* **16** 337

Yamada T, Niikeki N and Toyoda H 1967 *Japan. J. Appl. Phys.* **6** 151–5

Yunovich A Yu 1972 *Radiative Recombination in Semiconductors* ed Ya E Pokrovsky (Moscow: Nauka) p224

Zachariasen W H 1926 *Geochem. Vert. Elem.* **7** 97

—— 1928 *Geochem. Vert. Elem.* **8** 150

Zakharova H Ya and Kuz'minov Yu S 1969 *IAN SSSR Neorg. Mater.* **5** 1086–90

Zernike F and Midwinter J E 1973 *Applied Nonlinear Optics* (New York: Wiley) 261

Zhdanova V V, Klyuev V P, Lemanov V V, Smirnova I A and Tikhonov V V
1968 *Fiz. Tverd. Tela* **10** 1725
Zheludev I S 1966 *Usp. Fiz. Nauk* **88** 253
—— 1968 *Physics of Crystalline Dielectrics* (Moscow: Nauka) p463
Zidnik G 1975 *Mater. Res. Bull.* **10** 9–14
Zook J D, Chen G D and Otto G N 1967 *Appl. Phys. Lett.* **11** 159–60
Zulbersztein A 1976 *Appl. Phys. Lett.* **29** 778–80

Further Reading

Industrial applications of lithium niobate are dealt with in many publications,
including fundamental monographs:

Burfoot J C and Taylor G W 1979 *Polar Dielectrics and their Applications*
(London: Macmillan)
Haus H A 1988 *Waves and Fields in Optoelectronics* (Englewood Cliffs, NJ:
Prentice-Hall)
Lines M E and Glass A M 1977 *Principles and Application of Ferroelectrics and
Related Materials* (Oxford: Clarendon)

Application of lithium niobate crystals to generation of surface acoustical wave
(SAW) is discussed in:

Tominaga H, Masaaki O and Fujiwara Y 1988 *Fujitsu Sci. Tech. J.* **24** 71–99
Whatmore R W 1980 *J. Cryst. Growth* **48** 530–47

As regards growth of crystals without striations, the recommendation concerning
the temperature regime can be found in:

Shigematsu K, Anzai A, Morita S, Yamada M and Yokoyama H 1987 *Japan. J.
Appl. Phys.* **26** 1988–96

Other references include:

Feigelson R S 1986 *J. Cryst. Growth* **79** 669–80
Gallagher P K and O'Bryan H M 1985 *J. Am. Ceram. Soc.* **68** 147–50
Glass A M, Nassan K and Negran T J 1978 *J. Appl. Phys* **49** 4808
Narasimhamurty T S 1981 *Photoelastic and Electro-Optic Properties of Crystals*
(New York and London: Plenum)
O'Bryan H M, Gallagher P K and Brandle C D 1985 *J. Am. Ceram. Soc.* **68–9**
493–6
Weis R S and Gaylord T K 1985 *Appl. Phys.* A**37** 191–203

Index